Financial Econometrics and Empirical Market Microstructure

Anil K. Bera • Sergey Ivliev • Fabrizio Lillo
Editors

Financial Econometrics and Empirical Market Microstructure

Springer

Editors
Anil K. Bera
Department of Economics
University of Illinois
Urbana
Illinois
USA

Sergey Ivliev
Perm State University
Perm
Russia

Fabrizio Lillo
Scuola Normale Superiore
Pisa
Italy

ISBN 978-3-319-09945-3 ISBN 978-3-319-09946-0 (eBook)
DOI 10.1007/978-3-319-09946-0
Springer Cham Heidelberg New York Dordrecht London

Library of Congress Control Number: 2014956190

Printed on acid-free paper

Springer is part of Springer Science+Business Media (www.springer.com)

Acknowledgements

Financial Econometrics and Empirical Market Microstructure edition is a collection of research papers emerged as an outcome of the Perm Winter School conferences.

Perm Winter School takes place in Perm (Russia) and focuses on advanced issues of risk management and financial markets modeling. The school is organized by Prognoz and Perm State University and endorsed by Professional Risk Managers' International Association (PRMIA). Every winter it brings together leading experts from academia and industry professionals in a 3-day program covering latest empirical research, theoretical models, and industry best practice. The four past schools have attracted more than 600 participants from 12 countries. More than 2,000 people were able to join free online broadcasting.

The editors would like to thank Prognoz company and Prof. Dr. Dmitry Andrianov for generous support of this edition.

Contents

Mathematical Models of Price Impact and Optimal Portfolio Management in Illiquid Markets

Nikolay Andreev

Abstract The problem of optimal portfolio liquidation under transaction costs has been widely researched recently, producing several approaches to problem formulation and solving. Obtained results can be used for decision making during portfolio selection or automatic trading on high-frequency electronic markets. This work gives a review of modern studies in this field, comparing models and tracking their evolution. The paper also presents results of applying the most recent findings in this field to real MICEX shares with high-frequency data and gives an interpretation of the results.

Keywords Market liquidity • Optimal portfolio selection • Portfolio liquidation • Price impact

JEL Classification C61, G11

1 Introduction

With the development of electronic trading platforms, the importance of high-frequency trading has become obvious. This requires the need of automatic trading algorithms or decision-making systems to help portfolio managers in choosing the best portfolios in volatile high-frequency markets. Another actual problem in portfolio management field is optimal liquidation of a position under constrained liquidity during a predefined period of time.

Mathematical theory of dynamic portfolio management has received much attention since the pioneering work of Merton (1969), who obtained a closed-form solution for optimal strategy in continuous time for a portfolio of stocks where the market consisted of risk-free bank accounts and a stock with Bachelier–Samuelson dynamics of price. Optimal criterion had the following form:

N. Andreev (✉)
Financial Engineering and Risk Management Laboratory, National Research University Higher School of Economics, Moscow, Russia
e-mail: nandreev@hse.ru

A.K. Bera et al. (eds.), *Financial Econometrics and Empirical Market Microstructure*, DOI 10.1007/978-3-319-09946-0_1

$$(C_t, X_t, Y_t) \in \text{Argmax}\, E\left(\int_0^T e^{-\rho t} U\left(C_t\right) dt + B\left(W_T, T\right)\right),$$

where C_t is consumption rate, X_t, Y_t—portfolio wealth in riskless asset and stocks respectively, $W_t = X_t + Y_t$ is total value of portfolio and $U(C) = \frac{C^\gamma}{\gamma}, \gamma < 1$, or $\log C$—a constant relative risk-aversion (CRRA) utility function, $B(W_t, t)$ is a function, increasing with wealth. This criterion formulates optimality as maximization of consumption and portfolio value at the end of a period. Merton asserted that it is optimal to keep assets in constant proportion for the whole period, that is $\pi_t = \frac{Y_t}{W_t} \equiv const$. This result is known as the Merton line due to the strategy's linear representation in (X_t, Y_t) plane.

2 Contemporary Price Impact Modeling

The ideal frictionless market of Merton (1969) does not adequately simulate the more complex real market. First of all, price dynamics obviously depend on an agent's actions in the market; moreover, there is no single characteristic of an asset's market value (price). Since the 1990s, electronic trading through limit order books (LOB) has been gaining popularity, providing the market with a set of orders with different volumes and prices during any trading period. Inability to close a deal at an estimated price led to the necessity of including transaction costs in portfolio management models and price impact modeling. For the past two decades, research in this field has provided complex models that allow for time varying forms of LOBs, temporary and permanent price impact, resilience etc.

The most sophisticated and yet also fundamental way of estimating transaction costs is estimating the whole structure of LOB. Usually the market is represented as a complex Poisson process where each event is interpreted as the arrival/liquidation/cancellation of orders at specific depth levels. Large (2007) considers the arrival of ten kinds of market events (market bid/ask order limit bid/ask order, cancellation of bid/ask order, etc.) according to a multivariate Hawkes process with intensity depending on the past trajectory. Intensity in Large's model does not depend on order depth (distance from best quote).

Cont and Larrard (2012) introduced a complex Poisson model with time and depth-varying intensity and obtained theoretical results on the subject. Unfortunately, due to the extreme complexity of the general approach, it is extremely difficult to calibrate the parameters. Thus, some simplifying assumptions, based on empirical observations of a particular market, are necessary. On the other hand, the Poisson model must be flexible enough to reflect dynamics of real events, otherwise forecast errors will make the result useless for practice.

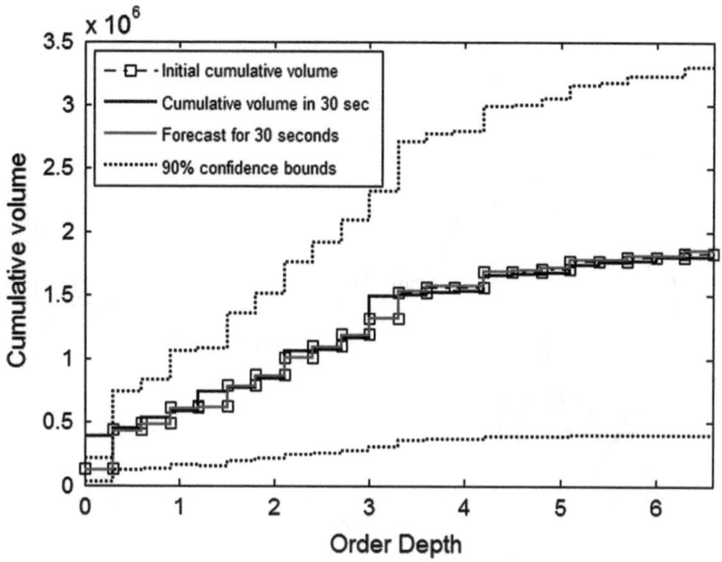

Fig. 1 LOB forecast in terms of cumulative volume as a function of depth for MICEX RTKM shares, 18 January 2006

Consider a simple LOB model with only two types of events: arrival and cancellation of limit order at one side of the book. Intensities are stationary and independent but depend on depth. Volume of each order is a random variable with *a priori* given parametric distribution with unknown parameters depending on depth. Thus, LOB is modelled via compound homogeneous space–time Poisson process. We calibrated the following model to real MICEX data, assuming from empirical observations that

1. event volume distribution is a mixture of discrete and lognormal;
2. intensities as functions of depth are power-law functions.

We estimate parameters θ of the model from order flow history using maximum likelihood and Bayesian methods. Then, using LOB structure L_{t_0} as initial state of the system we model $L_{t_0+T}\,|L_{t_0},\theta$ and take $\widehat{L}_{t_0+T} = E\,(L_{t_0+T}\,|L_{t_0},\theta)$ as a forecast. Results of forecasting structure for 30 s horizon and 90 % confidence bounds are presented in Fig. 1. We see that even for small horizon confidence interval is too wide for any practical use of such forecast. This is partly explained by presence of discrete part in volume mixture distribution, which is usually difficult to estimate from training sample. Atoms of volume distribution stand for volume values preferred by participants (100, 1,000, 5,000 lot etc.), orders with preferred volumes can amount up to 50 % of total number of orders.

Due to technical difficulties and intention to integrate an LOB model into portfolio optimization, a simple *a priori* form of the book is usually considered.

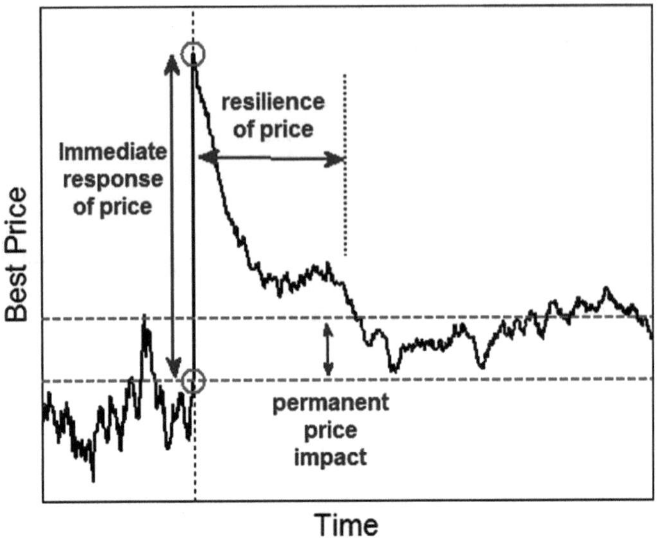

Fig. 2 Price impact aspects

Accent in modeling is made on the price impact function itself. Three main aspects are considered in such an approach:

- Immediate response of best price after a trade, which affects future costs until book replenishes.
- Resilience of LOB, i.e. ability to replenish after a trade; together with immediate response, this is often called temporary price impact. Infinite resiliency means that LOB replenishes instantaneously.
- Permanent price impact, or the effect of replenishment to a level other than pre-trade value; this effect describes the incorporation of information from the trade, which affects market expectations about 'fundamental price' of the asset (Fig. 2).

Permanent price impact is not considered in many classical models of optimal portfolio selection. For a particular case—optimal liquidation—many works assume the simplest dependence, where impact is a linear function of trade volume (i.e., Kyle 1985). Linear approximation can be considered appropriate in most practical cases because of difficulty in calibration of a more complex function in the presence of many agents.

Immediate response function is usually considered linear in volume, which is equivalent to the assumption of the flat structure of LOB (Obizhaeva and Wang 2012), or the assumption that trade volumes *a priori* are less than current market depth. Andreev et al. (2011) consider a polynomial form of immediate response function with stochastic coefficients. Fruth (2011) presents the most general law of immediate response in the form of a diffusion process under several mild conditions.

Resilience has been recently included in impact models and is usually described in exponential form with *a priori* given intensity: Suppose that K_{t_0} is immediate response after a trade at time t_0, then

$$\text{Temporary Impact}_t = K_{t_0} e^{-\int_{t_0}^{t} \rho(u)du}.$$

Almgren and Chriss (1999) considered instantaneous replenishment: $\rho_u = \infty$; Obizhaeva and Wang (2012), Gatheral et al. (2011) and others assumed exponential resilience with constant intensity: $\rho_u \equiv \text{const}$. General law of deterministic resilience rate has been presented in recent papers of Gatheral (2010), Gatheral et al. (2012), Alfonsi et al. (2009), and Fruth et al. (2011).

3 Overview of Contemporary Portfolio Management Models and Their Evolution

Davis and Norman (1990) introduced a consumption–investment problem for a CRRA agent with proportional transaction costs and obtained a closed-form solution for it. Another advantage of the model was allowing for discontinuous strategy. For this purpose, the original Merton framework had to be upgraded to semimartingale dynamics. Portfolio value in each of the assets is described by the following equations:

$$\begin{aligned} dX_t &= (r_t X_t - C_t)\,dt - (1+\lambda)\,dL_t + (1-\mu)\,dM_t, \qquad X_0 = x, \\ dY_t &= \alpha Y_t dt + \sigma Y_t dw_t + dL_t - dM_t, \qquad Y_0 = y, \end{aligned}$$

where coefficients λ, μ define proportional transaction costs, L_t, M_t are cumulative amounts of bought and sold risky asset respectively. Results demonstrated the existence of three behavioral regions for portfolio managers, and are presented in Fig. 3.

Unlike Merton's case, the so-called wait region appears due to transaction costs. That is, it is suboptimal to trade while in the area. Leaving the area leads to immediate buy or sell to get to the wait region's border. Analogous results were also obtained for the infinite horizon problem by Shreve and Soner (1994).

Another extension of the Merton model was presented by Framstad et al. (2001) for jump diffusion price dynamics. It was shown that wait region is absent in this case. That is, this strategy's structure is the same as for Merton's continuous diffusion market.

A number of papers considered a price impact model instead of unrealistic 'fundamental price' dynamics. For example, Vath et al. (2007) presented the following complex price impact function, depending on current price and volume

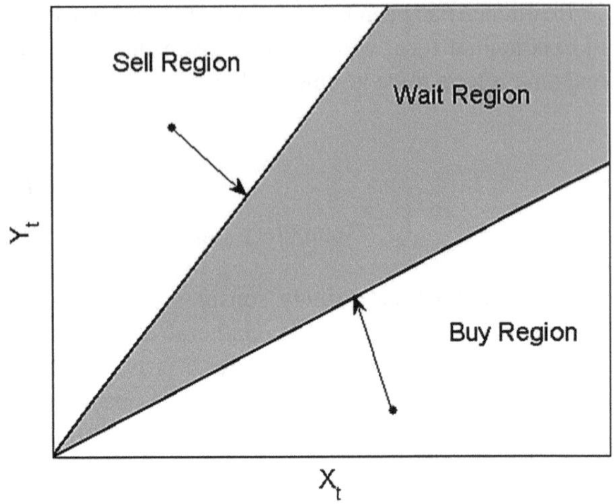

Fig. 3 Buy, Sell and Wait region in a model with proportional transaction costs

of a triggering trade. Around that time, Zakamouline (2002) took another step toward a realistic market model that allowed both proportional and fixed transaction costs. The proportional component described costs due to insufficient liquidity of the market, while the fixed component represented the participation fee for each transaction. Both papers considered discrete trading and produced interesting results. Buy and sell borders were no longer straight lines, as seen in Fig. 3, but still could be obtained beforehand and then used for decision making during trading sessions.

Neither of the abovementioned models considered the form and dynamics of the limit book itself—only the dynamic of an aggregated of a deal, which was considered as price. Microstructure models of electronic limit order markets have become quite popular in literature devoted to the problem of optimal liquidation of a portfolio. This particular case differed from the consumption–investment framework due to the terminal condition—predefined volume of the portfolio to be liquidated. The most notable results in this field are from Almgren and Chriss (1999) and Obizhaeva and Wang (2012). The framework has become quite popular in practice due to the simple models and intuitive results. Both approached considered discrete strategies and defined optimality functional not through utility function, but as a weighted sum of expected value and standard deviation of portfolio value.

The work of Obizhaeva and Wang first appeared as a draft in 2005 and considered a flat static structure of the limit book. Their approach has been adopted by many authors, evolving into several directions. The most realistic models were presented by Predoiu et al. (2011) and Fruth et al. (2011). Predoui et al. consider a general form of order distribution inside a book and non-adaptive strategies of liquidation. Fruth et al. postulate a flat but dynamic form of order distribution while allowing for both discrete and continuous trading in the same framework, linear permanence and general temporary price impact; the described model does not allow several kinds of arbitrage and non-adaptive strategies, which proved to be optimal in the framework. Analytical solutions have been obtained for discrete cases and for continuous trading.

4 Comparison of Portfolio Management Strategies

Despite the great potential of the developed models, most of them have not been applied to real data. To prove the usefulness of portfolio management models for practitioners, we apply some of the contemporary results in this field to real MICEX trading data and give recommendations for their usage. Our database consists of the complete tick-by-tick limit order book for MICEX shares from January 2006 through June 2007. We consider only liquid shares, such as LKOH, RTKM and GAZP, because only during sufficiently intensive trading does it become possible to calibrate models for the real market.

We consider the problem of optimal purchase of a single-asset portfolio over a given period and compare the performance of the following strategies:

1. Immediate strategy—portfolio is obtained via a single trade at the moment of decision-making. This strategy must lead to the largest costs but eliminates market risk completely. It is recommended for high-volatility markets or in case of information about unfavourable future price movements.
2. Fruth et al.'s (2011) strategy—this has the same goal as uniform strategy, i.e. minimization of expected transaction costs but not market risk. The main advantage of the model is its flexibility and consideration of several main microstructure effects, such as time-varying immediate price impact, dynamic model of the order book and time-varying resilience rate. Authors define price impact for buy and sell sides (E_t and D_t) as the difference between best price in the book and unaffected price. Permanent impact is proportional to volume of the order and constant over time while immediate response function $K(t, v) = K_t v$ changes over time. Temporary impact decays exponentially with a fixed time-dependent, deterministic recovery rate ρ_t, so that temporary impact of trade v, occurred at time s, at time t equals $K_s e^{-\int_s^t \rho_u du} v$. General framework considers both continuous and discrete time market models. It generalizes Obizhaeva and Wang's approach and postulates the following strategy: when price impact is low and the agent still has much to buy, she buys until the ratio of impact to remaining position is high enough, otherwise she waits for the impact to lower. After that, the agent can make another deal or wait, etc. So, for each moment of time, the agent has a barrier dividing her "Buy" and "Wait" regions.
3. Andreev et al. (2011) approach—a generalization of the Almgren and Chriss framework. Optimality is considered as minimization of both transaction costs and risk. This model has been obtained specifically for the MICEX market and incorporates a parametric dynamic model of cost function, which provides more accurate results: market model uses fundamental price instead of best bid-ask

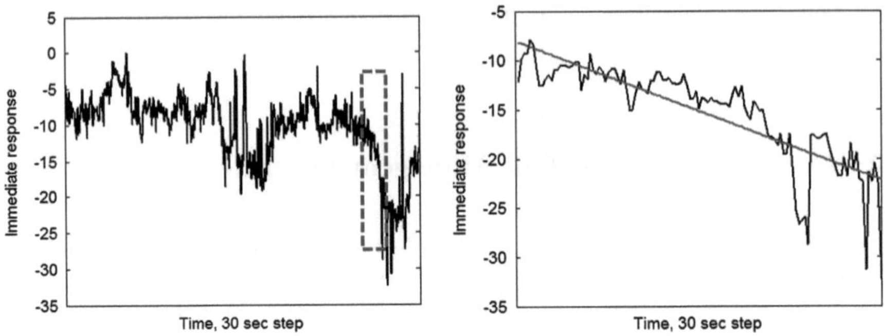

Fig. 4 Immediate response coefficient K_t for the whole trading day (7 February 2006) and dynamics during decline period (LKOH shares)

prices, which follows arithmetic Brownian motion. Transaction costs function has polynomial form (third degree polynom) with stochastic coefficients, which follow simple AR(1) model. No price impact is assumed. The strategy, unlike the previous three, considered agent risk aversion, which is characterized by the weighted sum of two criteria of optimality in minimization of functionality. Thus, problem formulates as minimization of $-E(W_T) + \lambda Var(W_T)$, where W_T is terminal wealth and λ is *a priori* risk aversion parameter.

For example, consider a 100,000 LKOH-share portfolio, liquidated via six consequent trades with 60-s wait periods. Consider also linear immediate response function with coefficient K_t. Rough estimate of K_t is obtained via least-squares method: $K_t = \arg\min \sum_{i=1}^{M} \left(\frac{\partial}{\partial v} C\,(t, v_i) - K_t v_i\right)^2$, where $C(t, v)$ is cost of trade with volume v, reconstructed from order book shape, and $0 < v_1 < \cdots < v_M = \overline{V}$ is *a priori* volume grid, for $\overline{V}$ we take half of available trading volume at the moment. Figure 4 shows dynamics of immediate response coefficient K_t. Liquidation begins when decline in response has been observed for some time (selected region in Fig. 4).

Strategies 2 and 3 are presented in Fig. 5 and have quite different behaviours. The form of the first strategy is obvious from the description. For Strategy 3, we use the simplest calibration assumptions, considering resilience rate a constant and immediate response as linear in time and volume. Assumptions are appropriate for medium periods of time.

We ascertain that the performance of Fruth et al.'s approach is the best of the three, while immediate buy is the worst. This result was expected because Strategy 2 is better adjusted to a specific form of response and can often show better performance if the form was guessed right. The strategy of Almgren and Chriss shows inferior performance and higher aggressiveness (see Fig. 6) due to

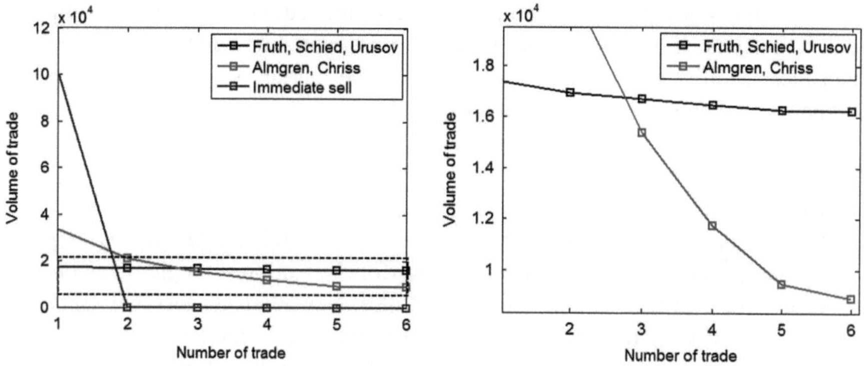

Fig. 5 Trading strategies for immediate strategy, approach by Fruth et al. (2011) and approach by Andreev et al. (2011) with $\lambda = 0.01$ for purchase of portfolio of 100,000 Lukoil shares via six trades with 1-min intervals. Date: February 7, 2006

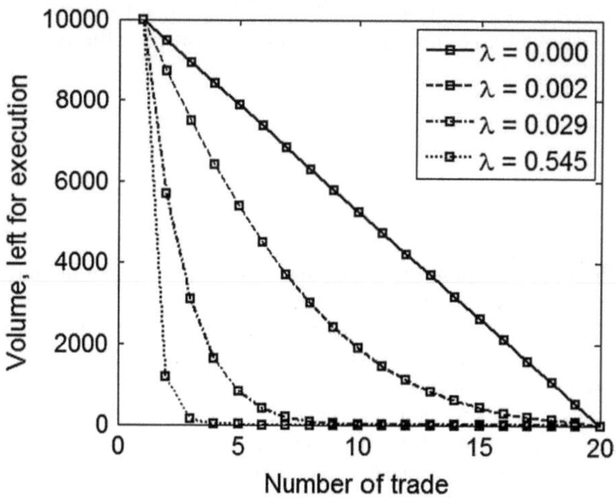

Fig. 6 In Almgren and Chriss framework aggressiveness of the strategy increases with risk-aversion parameter λ. The figure demonstrates how volume left for execution depends on the number of trade for different values of λ. $\lambda = 0$ leads to equal size of trades. Initial volume is 10,000 shares, strategy allows the maximum of 20 trades

minimization of market risk if risk-aversion is sufficiently high.[1] The choice of risk-aversion parameter heavily influences resulting strategy but cannot be chosen automatically. Unfortunately some practitioners interpret this as a misspecification and excessive difficulty of the model and therefore favor simpler strategies. It is also not surprising that the Fruth et al. approach leads to lower costs than immediate

[1]Extreme case of Almgren and Chriss strategy with infinite risk-aversion ($\lambda = \infty$) would be immediate buy.

strategy: the strategies in the model contains immediate buy, and dynamics in the parameters of the market are taken into account. Immediate strategy doesn't consider specifics or the current situation on the market, so it can be frequently outperformed by more elaborate methods.

Conclusion
Due to development of microstructure models and availability of high-frequency historic data, mathematical portfolio selection strategies have been extensively researched since the early 1990s. Nevertheless, very few frameworks were applied by practitioners because underlying models of the market were too unrealistic at the time. The aim of this research is to provide a review of modern accomplishments in the field, including the ongoing work, and demonstrate more realistic market models used in contemporary frameworks. To illustrate the effect of using automatic algorithms of portfolio selection and, in particular, optimal purchase/liquidation, we apply several approaches to real MICEX shares-related trading data and compare the results.

References

Alfonsi, A., Fruth, A., & Schied, A. (2009). Optimal execution strategies in limit order books with general shape functions. *Quantitative Finance, 10*(2), 143–157.

Almgren, R., & Chriss, N. (1999). Value under liquidation. *Risk*. Retrieved 10 March 2013 from http://www.math.nyu.edu/~almgren/papers/optliq_r.pdf.

Andreev, N., Lapshin, V., Naumenko, V., & Smirnov, S. (2011). Opredelenie likvidatsionnoĭ stoimosti portfelya aktsiĭ s uchëtom osobennosteĭ mikrostructury rynka (na primere MMVB) [Equity portfolio liquidation value estimation with market microstructure taken into account (MICEX case)]. *Upravlenie riskom [Risk-Management], 2*(58), 35–53.

Cont, R., & Larrard, A. (2012). *Price dynamics in a Markovian limit order market*. Retrieved 10 March 2013 from http://ssrn.com/abstract=1735338.

Davis, M., & Norman, A. (1990). Portfolio selection with transaction costs. *Mathematics of Operations Research, 15*(4), 676–713.

Framstad, N., Oksendal, B., & Sulem, A. (2001). Optimal consumption and portfolio in a jump diffusion market with proportional transaction costs. *Journal of Mathematical Economics, 35*(2), 233–257.

Fruth, A. (2011). *Optimal order execution with stochastic liquidity* (PhD thesis). Retrieved 10 March 2013 from http://opus.kobv.de/tuberlin/volltexte/2011/3174/pdf/fruth_antje.pdf.

Fruth, A., Schoeneborn, T., & Urusov, M. (2011). *Optimal trade execution and price manipulation in order books with time-varying liquidity*. Retrieved 10 March 2013 from http://www.mathnet.ru/php/seminars.phtml?&presentid=5342&option_lang=rus.

Gatheral, J. (2010). No-dynamic-arbitrage and market impact. *Quantitative Finance, 10*(7), 749–759.

Gatheral, J., Schied, A., & Slynko, A. (2011). Exponential resilience and decay of market impact. In *Proceedings of Econophys-Kolkata V* (pp. 225–236). Milan: Springer.

Gatheral, J., Schied, A., & Slynko, A. (2012). Transient linear price impact and Fredholm integral equations. *Mathematical Finance, 22*(3), 445–474.

Kyle, A. (1985). Continuous auctions and insider trading. *Econometrica, 53*(6), 1315–1336.
Large, J. (2007). Measuring the resiliency of an electronic limit order book. *Journal of Financial Markets, 10*(1), 1–25.
Merton, R. (1969). Lifetime portfolio selection under uncertainty: The continuous-time case. *The Review of Economics and Statistics, 51*(3), 247–257.
Obizhaeva, A., & Wang, J. (2012). Optimal trading strategy and supply/demand dynamics. *Journal of Financial Markets, 16*(1), 1–32.
Predoiu, S., Shaikhet, G., & Shreve, S. (2011). Optimal execution in a general one-sided limit-order book. *SIAM Journal on Financial Mathematics, 2*(1), 183–212.
Shreve, S., & Soner, H. (1994). Optimal investment and consumption with transaction costs. *The Annals of Applied Probability, 4*(3), 609–692.
Vath, V. L., Mnif, M., & Pham, H. (2007). A model of optimal portfolio selection under liquidity risk and price impact. *Finance and Stochastics, 11*(1), 51–90.
Zakamouline, V. (2002). *Optimal portfolio selection with transactions costs for a CARA investor with finite horizon.* Retrieved 10 March 2013 from http://brage.bibsys.no/nhh/bitstream/URN:NBN:no-bibsys_brage_24545/1/zakamoulinevalerii2002.pdf.

[illegible]

Evidence of Microstructure Variables' Nonlinear Dynamics from Noised High-Frequency Data

Nikolay Andreev and Victor Lapshin

Abstract Research of nonlinear dynamics of finance series has been widely discussed in literature since the 1980s with chaos theory as the theoretical background. Chaos methods have been applied to the S&P 500 stock index, stock returns from the UK and American markets, and portfolio returns. This work reviews modern methods as indicators of nonlinear stochastic behavior and also shows some empirical results for MICEX stock market high-frequency microstructure variables such as stock price and return, price change, spread and relative spread. It also implements recently developed recurrence quantification analysis approaches to visualize patterns and dependency in microstructure data.

Keywords Chaos theory • Correlation integral • Microstructure • Price dynamics • Recurrence plot

JEL Classification C65, G17

1 Introduction

Since the nineteenth century, there have been attempts to describe behavior of economic variables via simple linear deterministic systems. Unfortunately, unlike nature phenomena, in many cases financial series could not be reduced to linear dynamic model. Thus, stochastic models proved to be suitable for modeling and prediction. Nevertheless, attempts to find an appropriate deterministic model continued. They had determinism, introduced by Laplace in the early nineteenth century, as a fundamental principle. Poincaré (1912) stated that even if all the underlying laws were known, it would still be impossible to predict the state of the system due to error in estimate of the initial condition. But if the system is not too sensitive to the initial data, we can predict future states up to the error of the same

N. Andreev (✉) • V. Lapshin
Lomonosov Moscow State University, Moscow, Russia

Financial Engineering and Risk Management Laboratory, National Research University Higher School of Economics, Moscow, Russia
e-mail: nandreev@hse.ru

A.K. Bera et al. (eds.), *Financial Econometrics and Empirical Market Microstructure*, DOI 10.1007/978-3-319-09946-0_2

order. This led to the assumption that unpredictable "stochastic" processes could be replaced by fully deterministic but unstable (chaotic) systems. For a detailed review history of nonlinear dynamics research in economics, see Prokhorov (2008).

Particular deterministic processes have been paid a great deal of interest lately due to the quasi-stochastic properties of the generated signals. One of the simple maps producing such effect is the well-known tent map:

$$x_t = \begin{cases} a^{-1}x_{t-1}, 0 \leq x_{t-1} < a, \\ (1-a)^{-1}(1-x_{t-1}), a \leq x_{t-1} \leq 1, \end{cases}$$

which has first and second moment properties that are the same as first-order autoregressive process and thus was called 'white chaos' by Liu et al. (1992). Some values of parameters cause such processes to behave similar to the i.i.d. series (Sakai and Tokumaru 1980). Thus, the natural question arises if stochastic trajectory can be interpreted as completely deterministic and thus perfectly predictable if the underlying map is completely known. It is necessary to note that predictability is not the main goal and cannot be achieved for microstructure variables, as shown below. The main advantage of the chaotic approach is the possibility to describe the data with a more appropriate model that should be more reliable in times of crisis. This intention is justified by the observed similar properties of microstructure data and characteristics of chaotic natural phenomena, such as earthquakes and avalanches. Therefore it is appropriate to assume that underlying laws of dynamics are similar.

2 Smoothing Data for Further Analysis and Preliminary Observations

In this work several microstructure variables were researched, including

- Stock return and price
- Price change and its absolute value
- Spread and relative spread (ratio of spread to price)

Due to systematic noise in microstructure data, it is necessary to smooth the data for further analysis. In this work, one of the modern wavelet methods was used. The basic principle of wavelet smoothing is performing wavelet decomposition and applying a "smoothing" transformation for wavelet coefficients for a certain threshold level. By looking at the smoothed trajectory of a variable, we can already discern whether its behavior is regular or not (Antoniou and Vorlow 2005). One of the results is the visible regular dynamics of stock return, price changes and relative spread, while other variables show randomness. To illustrate the effect, Fig. 1 demonstrates the dependence of Lukoil stock characteristics (intraday data aggregated by 10 s, 13th January 2006) from their delayed values with a rather

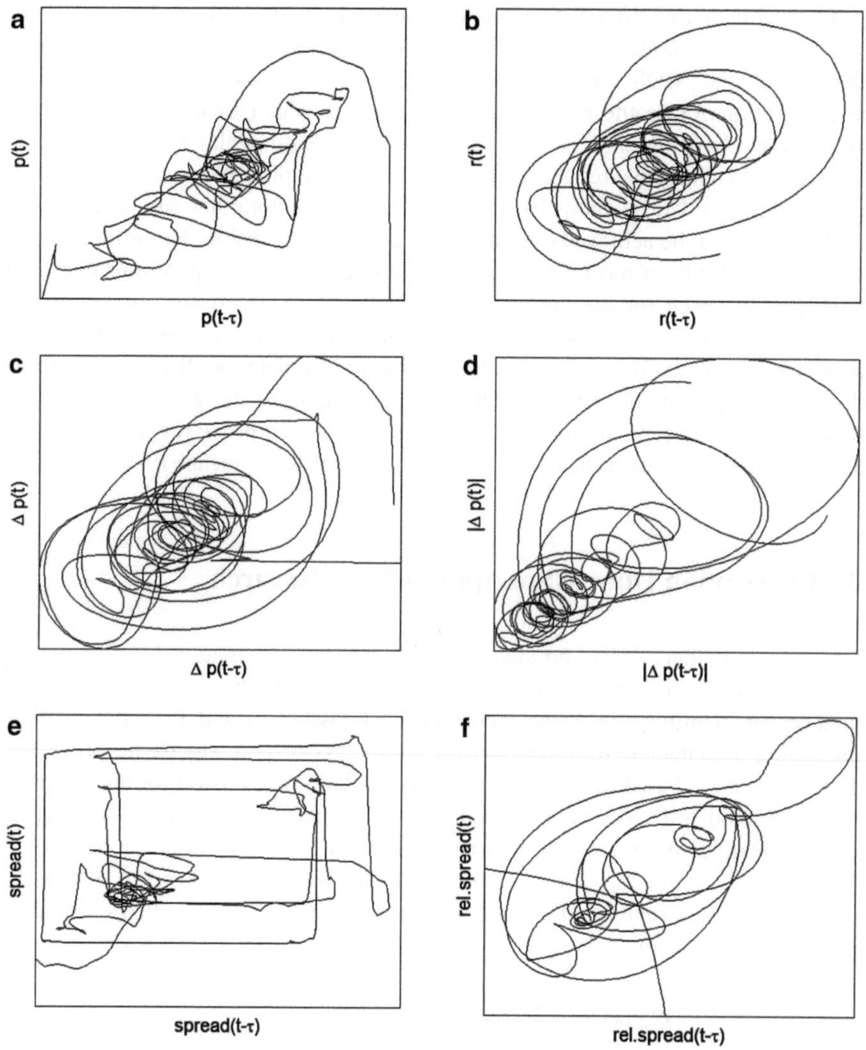

Fig. 1 Phase trajectory for (**a**) price; (**b**) return; (**c**) price change; (**d**) absolute value of price change; (**e**) spread; (**f**) relative spread

large lag of $\tau = 200$ s. Similar results hold for other lag lengths. Obtained results are consistent with the work of Antoniou and Vorlow (2005) for FTSE100 stock returns (daily data). As estimator of price we take arithmetic mean of best bid and ask quotes; return means price change during 10 s divided by price value at the beginning of the period; spread means simple bid-ask spread and relative spread is ratio of spread to price value.

Phase trajectories (b)–(d), (f) produce somewhat regular patterns [unlike (a) and (e)], which can be considered as indirect evidence of nonlinear dynamics.[1] To verify this we calculate BDS statistic for each series which shows at 0.5 % significance level rejection of hypothesis that increments are i.i.d. which means that the data can be generated by a low-dimensional chaotic or nonlinear model. Unfortunately BDS statistic still cannot be a reliable criterion for short samples of data, even for whole trading day (see below about BDS). It is important to note that any practical use of the model can be achieved only in the case of low dimension. High-dimensional systems usually have too many unknown parameters and are quite unstable, thus unpredictable even for the short horizon. In this case, stochastic modeling will be more appropriate. A fine illustration is given by Poincaré (1912), describing atmospheric effects. It is theoretically possible to calculate the distribution of rain drops on the pavement, but due to the complex nature of the generating process, their distribution seems uniform; thus it is much easier to prove this hypothesis by assuming that the generating system is purely stochastic (Poincaré 1912).

3 Correlation Dimension Approach to Research

Unfortunately, there exists no statistical test that has chaos as a hypothesis, nor a characteristic property separating chaos from stochastic process. The basic method for identification is the algorithm by Grassberger and Procaccia (1983), presenting a characteristic property of a wide class of pure stochastic processes. The algorithm is based on the concept of a correlation dimension for the observed m-dimensional trajectory. The main idea of the method is the following: given observable trajectory $x_1, x_2, \ldots x_N$, we reconstruct a series of m-dimensional vectors $y_k = (x_k, x_{k-p}, \ldots x_{k-(m+1)p})^{m \times 1}$. m and p are considered *a priori* given the parameters of the method. Then we find an estimate of the so-called correlation integral of the system:

$$C_m(\varepsilon) = \frac{\textit{number of pairs } (y_i, y_j)\text{: } \|y_i - y_j\| < \varepsilon}{\textit{total number of pairs } (y_i, y_j)}$$

$$= \lim_{m \to \infty} \frac{1}{m(m-1)} \sum_{i,j=1}^{m} \theta\left(\varepsilon - \|y_i - y_j\|\right),$$

where $\theta(x)$ is a Heaviside step function. For small ε correlation integral grows according to power law at the rate of $D(m)$:

$$C_m(\varepsilon) \approx \varepsilon^{D(m)}$$

[1]See, for example, phase trajectories of several well-known simple chaotic systems, such as the Mackey-Glass and Genesio-Tesi systems, and trajectories of purely chaotic system such as Wiener process.

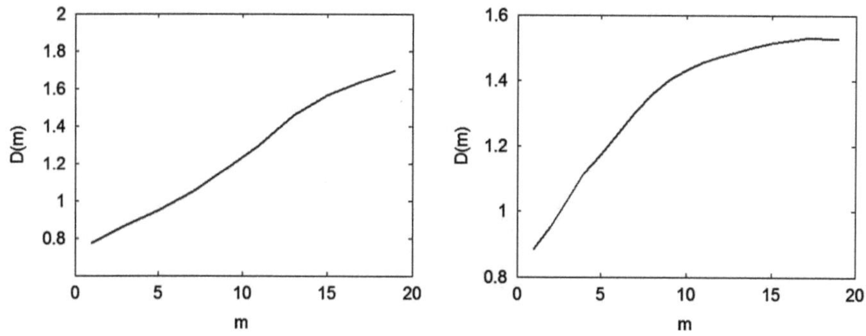

Fig. 2 Correlation dimension for Lukoil stock spread series (*left*) and relative spread series (*right*)

For stochastic white noise $D(m)$ is proportional to m, but for a large class of deterministic systems, correlation exponent $D(m)$ has saturation level D' which can be used as a characteristic of non-stochastic behavior of the variable. Figure 2 demonstrates correlation exponent $D(m)$ for Lukoil stock spread and relative spread. Saturation of correlation exponent D can be seen for return, price changes and relative spread, indicating the existence of complex nonlinear but deterministic behavior. Price and spread show pure stochastic properties. Another advantage of the Grassberger and Procaccia algorithm is that correlation dimension allows us to find upper boundaries for generating system dimensions. Taken's embedding theorem implies that phase dimension of the system cannot be higher than $2D' + 1$, where D' is the saturation level.

Unfortunately, realization of the Grassberger and Procaccia method is quite difficult in practice. One shortcoming is *a priori* value of lag parameter p. The classical solution is to estimate an autocorrelation function of the series and take the first lag value at which autocorrelation turns to zero. The main problem is an insufficient amount of data for correlation integral estimate. While in natural sciences the amount of data used for one test approaches 20,000–30,000, the usual length of a financial series is about several thousand (for example, daily index data or aggregated intraday data). This makes estimates of correlation dimensions unreliable for m higher than 10–15. Moreover, small values of threshold ε lead to insufficient number of summands in estimate and zero value of integral for rather small values. Figure 3 shows a real form of correlation integral for different values of a threshold in a logarithmic scale. For large lengths of input series, the dependency must be close to linear, but in practice the property holds only for a certain range of threshold values that must be chosen very carefully.

Another approach to identifying nonlinear behavior in data was introduced by Brock et al. (1986). The authors presented a statistical test which has i.i.d. of the series as a null hypothesis. Typical use of the method is fitting some *a priori* linear model to given data and testing residuals for i.i.d. property. Necessary statistics uses the correlation integral estimate, which raises all the above mentioned problems, such as large amount of input data.

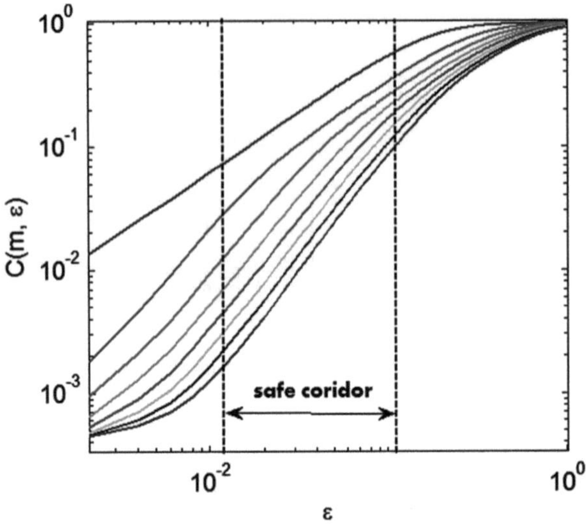

Fig. 3 Correlation integral for Lukoil stock return

Liu et al. (1992) examined the possibilities of a BDS test and found that its power varies for different linear models, e.g., its power, is less for nonlinear moving average models. It is also necessary to emphasize that rejection of the null cannot be interpreted as the presence of chaotic model. It only implies some (probably stochastic) nonlinearity.

4 Scheinkman and LeBaron Test for Predictability

Another interesting use of correlation integral was presented by Scheinkman and Lebaron (1989). As before, $C_M(\varepsilon)$ stands for the correlation integral for M as a phase dimension of reconstructed space, and threshold ε. It is proven that

$$S_{M+1}(\varepsilon) = \frac{C_{M+1}(\varepsilon)}{C_M(\varepsilon)}$$

gives an estimate of conditional probability that

$$\sup_{0 \le i \le M} |y_{1+i} - y_{2+i}| \le \varepsilon,$$

given that

$$\sup_{0 \le i \le M-1} |y_{1+i} - y_{2+i}| \le \varepsilon,$$

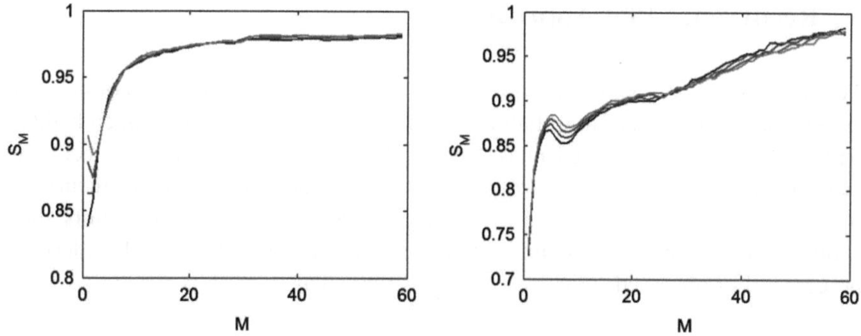

Fig. 4 Scheinkman–LeBaron function for spread (*left*) and return (*right*)

Table 1 Scheinkman–LeBaron function behavior

Variable	$S_M(\varepsilon)$
Return	Increases
Price	Converges to 0.96
Price change	Increases
Absolute value of price change	Increases
Spread	Converges to 0.9
Relative spread	Increases

i.e. the conditional probability that two states of the system are close, given that their past M histories are close.

This result can be implemented to define the measure of predictability and determinism of the data. If $S_M(\varepsilon)$ does not saturate as M grows, then states of the system depend on the information about its history. Otherwise the dynamics are affected by some random factor unrelated to the system itself, which can be interpreted as stochastic behavior of the process. As a result, Scheinkman and Lebaron's function gives the following criteria:

- If states are independent, then $S_M(\varepsilon)$ does not depend on M;
- If past values of the series help predict future values, $S_M(\varepsilon)$ will tend to increase with M.

Figure 4 demonstrates the behavior of $S_M(\varepsilon)$ for spread and return series and four different threshold levels. Growth in case of spread indicates its stochastic nature.

Results for all six microstructure variables are given in Table 1. Predictability is observed for all except spread and price series, which is consistent with previous results.

5 Recurrence Plot Approach

Two main problems with the correlation integral approach are: (1) considerable amount of data necessary for reliable estimates and (2) *a priori* choice of embedding parameters for reconstructed phase space. Recently, a more elaborate technique proved to be useful for analysis of nonlinearity. This new approach uses recurrence plots as a tool for visualization of observed trajectories. Recurrence plots show similarity in dynamics over time without specifying the structure of underlying processes. For observed series x_t it can be expressed as

$$R\left(t_i, t_j\right) = \theta\left(\varepsilon - \left\|x_{t_i} - x_{t_j}\right\|\right),$$

where ε is a specified threshold parameter. Usually the system dimension must be 2 or 3 to allow visualization, otherwise its trajectories can be observed only through projection on two or three dimensional subspaces. A recurrence plot enables us to investigate m-dimensional trajectories through a two-dimensional representation of its recurrences. Figure 5 demonstrates RPs for white noise processes and for predictable periodic sine function (diagonal lines are marked red).

Continuous diagonal lines prevail for sine RP, which is expected for predictable systems. Base structures in the recurrence plot can be easily interpreted: diagonal lines parallel to the main diagonal mean predictability at some periods of time, line length measures period of predictable behavior; horizontal and vertical lines indicate stability of the system state over a period of time. Diagonal lines turn out to be the main characteristic for research of complex deterministic behavior. Unfortunately, real finance data series produce quite complicated RPs that cannot be analyzed visually and need quantitative measures for determinism. Figure 6 shows RPs for Lukoil stock return and price.

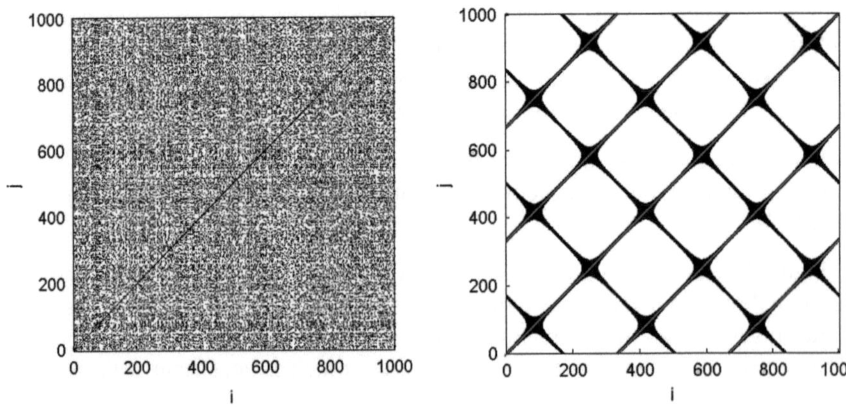

Fig. 5 Recurrence plot for white noise process (*left*) and sine function (*right*)

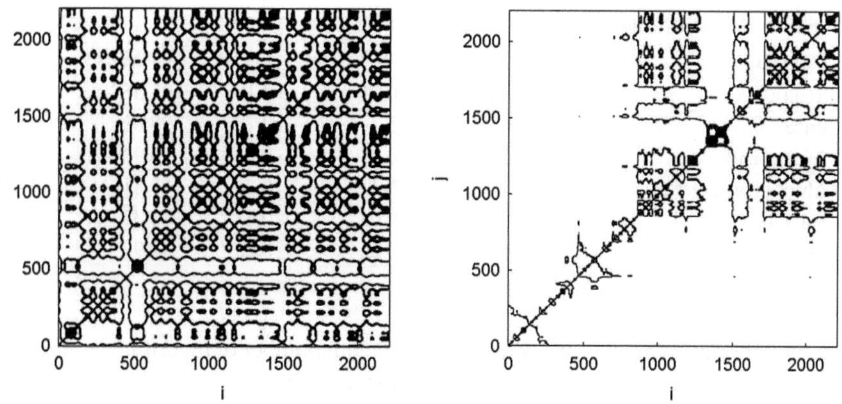

Fig. 6 Recurrence plot for stock return (*left*) and price (*right*)

Table 2 Recurrence quantification analysis measures

Name	Definition	Interpretation
Recurrence rate (*RR*)	Percentage of black points in RP	Correlation integral
Determinism (*DET*)	Percentage of black points which are part of diagonal lines of at least length *L*	Measures predictability
Entropy (*ENTR*)	Shannon entropy of the distribution of diagonal lines *P*(*L*)	Quantifies the complexity of the deterministic structure
Laminarity (*LAM*)	Same as DET for vertical lines	Quantifies the occurrence of laminar states
Trapping Time (*TT*)	Mean length of vertical lines	Measures the mean time that the system sticks to a certain state

White spaces in a price's RP mean abrupt changes in price dynamics, which is the consequence of nonstationarity of the variable. Stationarity of the input signal is one of the implied properties in many nonlinear analysis techniques. Correlation integral methods described the above produced results consistent with our expectations about price. However, as will be shown below for recurrence analysis, applying methods to nonstationary data can lead to counterintuitive results.

A number of measures were introduced with the aim of quantifying structures found in RP to go beyond visual classification. Table 2 shows some of the main characteristics.

Results of quantification analysis are shown in Table 3. A presence of deterministic behavior is shown for stock returns, relative spread and absolute value of price change. The situation is unclear for price change series and no determinism was detected for spread. As we can also see, price series has the best DET value and low entropy, which implies deterministic dynamics. The result is counterintuitive and not consistent with expectations from a simple visual examination of RP, phase trajectories or previous results. The observed effect originates due to the nonstationary

Table 3 Recurrence quantification analysis results for stock dynamics

Variable	RR (%)	DET (%)	ENTR	LAM (%)	TT
Return	6.23	0.91	0.19	56	16
Price	6.95	3.7	0.28	78.02	21
Price change	5.91	0.17	0.32	58.92	16
Absolute value of price change	7.91	0.39	0.26	69.81	17
Spread	8.19	0.03	0.68	78.38	21
Relative spread	6.43	0.51	0.23	73.92	18

nature of price dynamics and, thus, an insufficient number of recurrence points. This leads to unreliable estimates of quantification measures.

Another use of recurrence plot approach was introduced by Thiel et al. (2004). Let $P_\varepsilon(l)$ be the probability to find a diagonal line of at least length l. It can be shown that the following approximate equality holds:

$$P_\varepsilon(l) \approx \varepsilon^\nu e^{-lpK_2},$$

Where p is embedding lag parameter for reconstructed space (introduced in this chapter), ν is the correlation dimension and K_2 is order-2 Rényi entropy of the system. Based on this formula one can estimate Rényi entropy as a slope of $P_\varepsilon(l)$ in log scale and correlation dimension via simple formula:

$$\nu = \ln\left(\frac{P_\varepsilon(l)}{P_{\varepsilon+\Delta\varepsilon}(l)}\right) \cdot \left(\ln\left(\frac{\varepsilon}{\varepsilon+\Delta\varepsilon}\right)\right)^{-1}.$$

Thiel et al. (2004) have shown that both estimates are independent of embedding parameters, which solves one of the correlation integral problems at least to some extent.

Conclusion

A review of modern methods for identifying nonlinear dynamics was given; all algorithms were applied to real microstructure intraday MICEX data while describing difficulties of implementation in practice. It is worth mentioning that all the procedures are applicable without *a priori* knowledge of the underlying model or class of models. Results can be structured as follows:

- According to all identification techniques, return, price changes and relative spread show signs of a complex nonlinear underlying structure. Thus a random walk model isn't appropriate for them (such as Merton model for returns). The Scheinkman–LeBaron procedure shows that the history

(continued)

of these variables helps to predict future values. Unfortunately, obtained results indicate but do not imply deterministic behavior of the variables.

- Price and spread dynamics in the correlation integral approach show purely stochastic behavior, which can also be due to the large amount of noise in initial data. Future values are not fully predicted by information in history.
- Recurrence quantification analysis shows the presence of determinism in returns, relative spread and absolute price change dynamics, but no determinism for spread. Results for price are clearly incorrect due to nonstationarity of the initial series.

Nonstationarity also questions the reliability of obtained price results of other methods. This can explain the contradictory conclusion: return and price change appears to be deterministic in nature, while price is purely stochastic—though it is a deterministic function of return/price change. A simple explanation can be proposed: due to integral transformation of return/price change, the price series loses stationarity and becomes inappropriate for given methods. Dependency on time makes it impossible to recognize similar patterns in data, so the price series is identified as stochastic process.

References

Antoniou, A., & Vorlow, C. E. (2005). Price clustering and discreteness: Is there chaos behind the noise? *Physica A, 348*, 389–403.

Brock, W. A., Dechert, W. D., & Scheinkman, J. (1986). *A test for independence based on the correlation dimension*. Manuscript. Madison/Chicago: University of Wisconsin-Madison/University of Chicago.

Grassberger, P., & Procaccia, I. (1983). Measuring the strangeness of strange attractors. *Physica, 9D*, 189–208.

Liu, T., Granger, C. W. J., & Heller, W. P. (1992). Using the correlation exponent to decide whether an economic series is chaotic. *Journal of Applied Econometrics, 7, Supplement: Special Issue on Nonlinear Dynamics and Econometrics*, S25–S39.

Poincaré, A. (1912). *Calcul des probabilités*. Paris: Gauthier-Villars.

Prokhorov, A. (2008). Nonlinear dynamics and chaos theory in economics: A historical perspective. *Quantile, 4*, 79–92.

Sakai, H., & Tokumaru, H. (1980). Autocorrelations of a certain chaos. *IEEE Transactions on Acoustics, Speech, and Signal Processing, ASSP-28*(5), 588–590.

Scheinkman, J. A., & LeBaron, B. (1989). Nonlinear dynamics and stock returns. *Journal of Business, 62*(3), 311–337.

Thiel, M., Romano, M. C., Read, P., & Kurths, J. (2004). Estimation of dynamical invariants without embedding by recurrence plots. *Chaos, 14*(2), 234–243.

Revisiting of Empirical Zero Intelligence Models

Vyacheslav Arbuzov

Abstract This paper describes a zero-intelligence approach implementation for the modeling of financial markets. We construct a mechanism of order flow and market engine simulation. We analyze stylized facts to estimate the quality of our models. The research is based on a 1 month order and execution history data of the Moscow Exchange (MOEX) for one stock (JSC "Aeroflot").

Keywords Daniels model • Market microstructure • Mike–Farmer model • Order flow • Stylized facts • Tail exponent • Zero-intelligence models

JEL Classification G15, G17

1 Introduction

Agent-based models play an important role in understanding the mechanisms of financial markets driven by the advances in technologies that allow the creation and calibration of complex and very detailed models. An adequate replication of the mechanism of the price formation in those models is of the same or greater importance than the replication of the behavior of the agents. As first shown in Daniels et al. (2003), zero-intelligence (ZI) agent models are able to reproduce statistical regularities of the market with the Continuous Double Auction (CDA). ZI models are based on the hypothesis that the behavior of all agents can be described by random order flow with empirically estimated parameters. We studied the implementation of a ZI model on the Russian market. We reconstructed Daniels and Mike–Farmer versions.

After the description of Farmer and Daniels models, we try to change some details in the model and compare all our models with the real market.

V. Arbuzov (✉)
Department of Economics, Prognoz Risk Lab, Perm State National Research University, Perm, Russia
e-mail: arbuzov@prognoz.ru; arbuzov1989@gmail.com

A.K. Bera et al. (eds.), *Financial Econometrics and Empirical Market Microstructure*, DOI 10.1007/978-3-319-09946-0_3

2 Data

Our study is based on detailed market data, which includes the order history (order log) for Aeroflot stock (AFLT). Aeroflot is the largest Russian airline company and its equity is referred to as blue chip and is included in the MICEX index. During the observed period (21 trading days) there were, 2,765,074 orders which arrived and 31,572 which were executed (15,786 trades). Over this period, 15.3 million stocks were bought and sold yielding a 779.4 million ruble (approximately $26 million) turnover. All the data comes from the Moscow Exchange and is based in Perm State National Research University clusters (Computer cluster for reverse engineering, agent-based modeling and market microstructure researches of the Russian capital market). Most calculations were made using statistical environment R (Core Team R 2013). For calculations of the best bid and best asking prices from the order flow we used an Rcpp package with low-level programming.

3 The Daniels Model (2003)

The first example of this model was presented in Daniels et al. (2003). After that there were a few papers published with an analysis of this model (Farmer et al. 2005, 2006). The main assumption of this model was that orders are come onto the market randomly. There are market orders that are executed immediately and limit orders, which are placed at a fixed price level and executed only when there is a counter-party on the market which wants to trade at this price. All orders have an intensity of incoming, an intensity of canceling, a volume and a price (see Fig. 1). All these parameters can be measured using empirical data, and this is the main advantage of this model. We try to create a model as in the original papers.

For the estimation of parameter α we calculated the difference between the best prices and incoming orders (relative price). The model suggested that most orders come near the best prices (see Fig. 2a). According to the model, we estimated only 58 % of the distribution of the relative price for effective limit order placement.

We calculated $Q_t^{upper} = 12$ *tick size* (the 60 percentile of distribution of the relative price for effective limit order) and $Q_t^{lower} = -11$ *tick size* (the two percentile of distribution of the relative price for effective limit order) and the total number of effective limit orders 1,655,646 in this interval, so $\alpha = 3,427$ orders/per day $\cdot$ per price or $\alpha = 0.108$ orders/per second$\cdot$per price. The total number of effective orders was 869 market orders and 3,390 limit orders with immediate execution, so $\mu = 0.0064$ orders /per second (for more details of parameter estimation see Appendix A.1 in Farmer et al. 2005). There is one important remark that we estimated parameter δ in terms of time (not in terms of events as in the original work). We found that the average time of an order's life is equal to 3.5 s before cancellation.

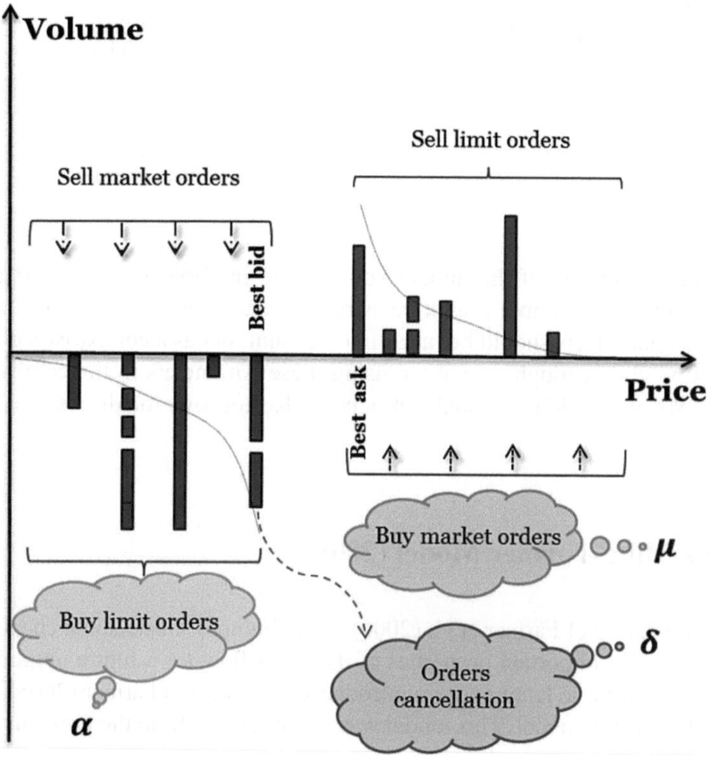

Fig. 1 Scheme of the Daniels model

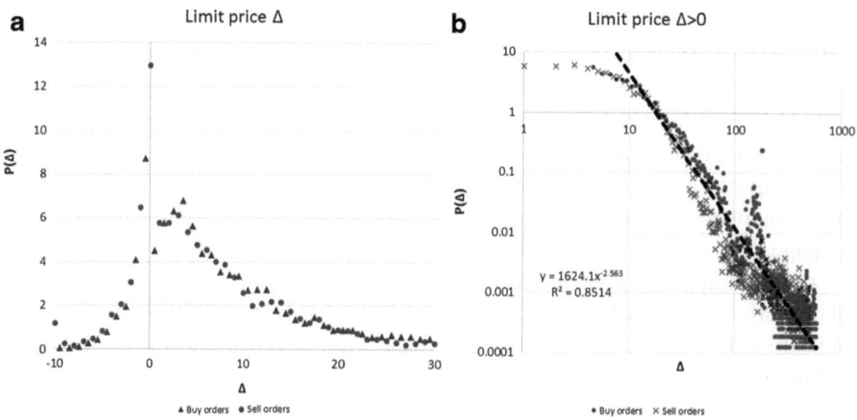

Fig. 2 Histogram of (**a**) entering order price differences from the best price and (**b**) power-law tails of order price

Table 1 Parameters of the Daniels model on the Russian market (AFLT, January 2012)

Parameters	Description	Value
α	Intensity of limit orders	0.108
μ	Intensity of market orders	0.006
δ	Intensity of cancelations	0.287
dp	Tick size	0.01
σ	Volume of orders	1,184

All the parameters of this model you can see in Table 1. We understand that the 1 month of our sample cannot be as representative as the length of 1 year and some seasonal effects should be taken into account, but as a comparison of models it would be good enough, so we are using these parameters in the comparison of models. We would like to thank Oksana A. Kostousova for discussions on this model.

4 The Mike–Farmer Model (2008)

In the publication of Farmer et al. (2006) in the Future Enhancement chapter, there were announced important properties of the order flow for a future upgrade of the model. Parts of these features were introduced in Mike and Farmer (2008). We call this model the MF model. This model was distinguished from the previous one in:

- Trending of order flow
- Power placement of limit prices
- Non-Poisson order cancellation process

Later, this model was upgraded and analyzed in Chakraborti et al. (2011), Gu and Zhou (2009), and He and Wen (2013). The first and most important assumption that signifies order flow is a long memory process (Bouchaud et al. 2004; Lillo and Farmer 2004; Lillo et al. 2005).

The first step for the construction of the model is the estimation of the Hurst exponent using methods in Achard and Coeurjolly (2009) and realized in the package *dvfBm* of the R environment (for the estimated parameters see Table 2).

Another important point in this research is the distribution of the order price. For all the variables we use the same names as in Mike and Farmer (2008). In calculating and fitting the relative distance from the best price (the best bid for buying orders and the best ask for selling orders) we find that Student's t-distribution is not the best theoretical distribution for the description of our data. The positive tail of distribution is quite definitely less than the theoretical tail of distribution. This means that effective market orders will appear more often than in reality (see Fig. 3). The negative tail of distribution does not differ too greatly from the theoretical values, but it does describe the power-law tail of order price (see Fig. 2b).

Table 2 Parameters of the Mike–Farmer model on the Russian market (AFLT, January 2012)

Parameters	Description	Value
H_s	Hurst exponent of the order sign series	0.73
α_x	Degrees of freedom of the order placement distribution	2.08
$\sigma_x \cdot 10^{-3}$	Scale parameter of the order placement distribution	6.76
A	Parameter for the equation of order cancellation	0.0167
B	Parameter for the equation of order cancellation	57.12
D_1	Parameter for the equation of order cancellation	0.283
D_2	Parameter for the equation of order cancellation	27.4
T	Tick size	*0.01*

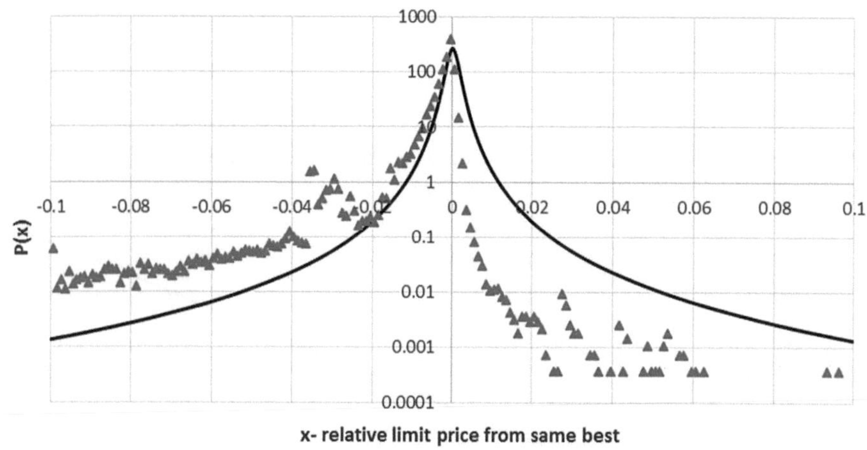

Fig. 3 Fitting of the empirical price distribution using t-distribution

This crude assumption of our data can lead to bigger spreads than in reality, and bigger returns, because the number of effective market orders would be more, and these orders take away liquidity from the market. Later in our research, we try to upgrade a theoretical description of this distribution.

In the MF model there are advanced cancellation processes, which differ from the Poisson process. We calculate probability conditioned on position in the order book as in the original paper (see Fig. 4).

We find that we are not able to have a good fit of this curve without redesigning the functional form as:

$$P\left(C_i \middle| y_i\right) = K_1 \left(1 - D_1 exp^{-y_i}\right)$$

After the estimation parameters, we calculated another important factor, which determined an imbalance between buyers and sellers on the market: order book imbalance (see Fig. 5).

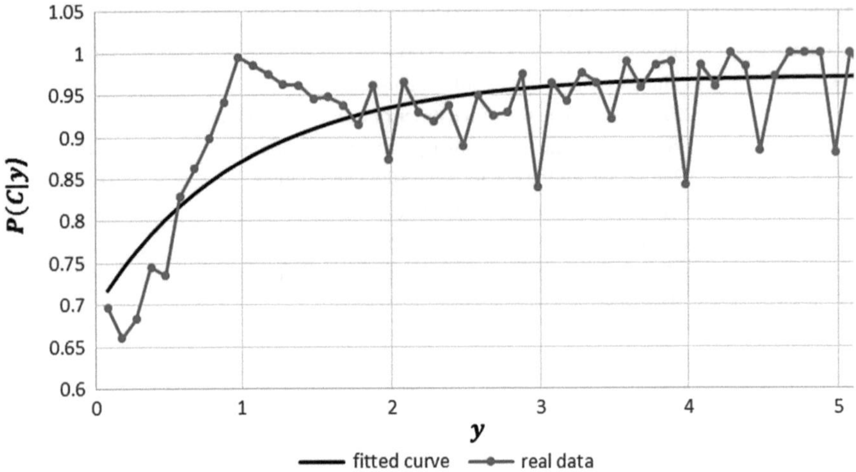

Fig. 4 The probability of cancellation conditioned on the position in the order book

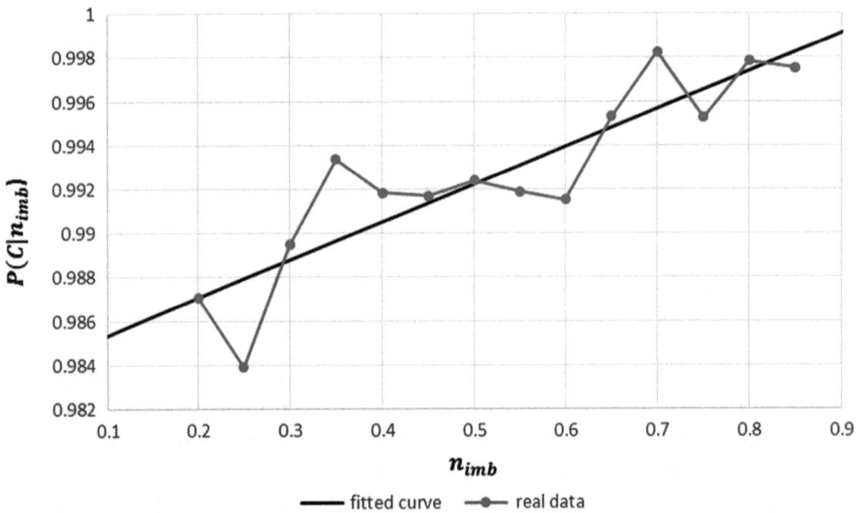

Fig. 5 The probability of cancellation conditioned on order book imbalance

After that we try to estimate probability conditioned on number of orders in the order book and it was very surprising for us, because in Mike–Farmer data there was inverse relationship (see Fig. 6).

In order to fit our data we bring an analytical form for the curve as in the process conditioned on the position in the order book. Total conditional probability was calculated as:

$$P\left(C_i \middle| y_i, n_{imb}, n_{tot}\right) = \mathrm{A}\left(1 - D_1 exp^{-y_i}\right)\left(n_{imb} + \mathrm{B}\right)\left(1 - D_2 exp^{-n_{tot}}\right)$$

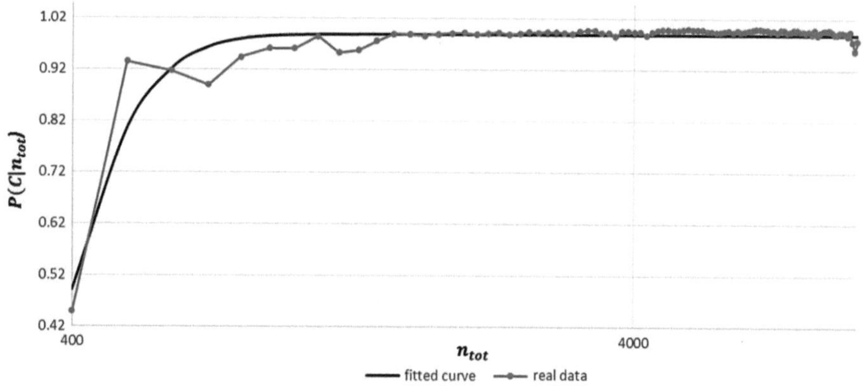

Fig. 6 The probability of cancellation conditioned on number of orders in the order book

At each step (in our case, each second) we generate one order with sign, volume and price. After that, we calculate the conditions of the order cancellations. For details of this realization see He and Wen (2013). It is interesting that during the process of evolution, the structure of the financial market has undergone changes, especially with the emergence of high frequency and algorithmic trading. Now algorithms trade on financial markets at the "speed of light," and many orders are cancelled after the fact of entry onto the market. Most orders in our sample close after their submission and so the probability of cancellation is very high.

5 The Mike–Farmer Model Without the Cancellation Process (MFWC)

It is an interesting question about what there would be on the market if there were no cancellations. Would trading or the market be stable or not? We realize the MF model without cancellations (we call it MFWC).

6 Model Upgrading

The most important thing that we try to improve in the MF model is the distribution of order price. We cut distribution into two parts: one with a positive tail and one with a negative tail. We find that both tails of distribution fit a good by power-law distribution with a tail exponent $= -2.15$ for positive values and a tail exponent $= -2.493$ for negative values (we inversed the negative tails and after that the estimate coefficients). Power-law poorly describes the center of distribution, when orders are put at the best prices. We fit ± 10 ticks from the best prices $(x = 0)$

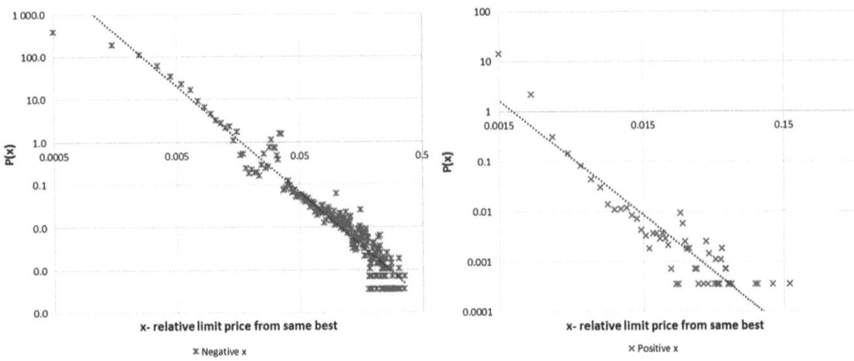

Fig. 7 Fitting of tails of empirical distribution using the power-law

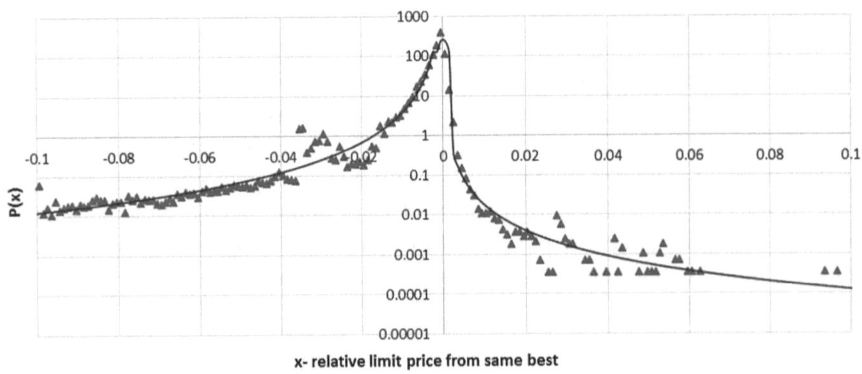

Fig. 8 Fitting of the empirical price distribution using two power-law distributions with Student's t in the center

using Student's t-distribution. On the Russian market traders see only the first ten prices for buy and sell, so this part of the orders should have another distribution (for example t-distribution) (Figs. 7 and 8).

Another additional improvement related to the order cancellation process is trying to take into account another metric of liquidity, for example RTCI:

$$RTCI = \frac{\sum_{i=1}^{k} |p_i - p| \bullet n_i}{\sum_{i=1}^{k} p_i n_i}$$

where

i: order position in the order book, $i = 1 \ldots k$,
k: total number of limit orders in the book,
p_i: price of order i,
n_i: volume of order i, n_i <0 for buy side orders,
p: current market price.

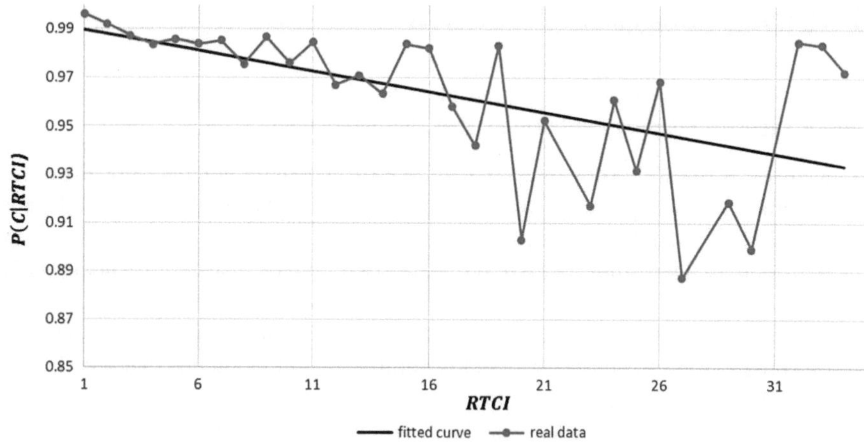

Fig. 9 The probability of cancellation conditional on the RTCI liquidity metric

Table 3 Parameters of the upgrade model on the Russian market (AFLT, January 2012)

Parameters	Description	Value
$\alpha_{positive}$	Exponent of the positive tail	−2.15
$\alpha_{negative}$	Exponent of the negative tail	−2.493
$A \cdot 10^{-5}$	Parameter for equation of the orders cancellation	2.87
D_3	Parameter for the equation of the order cancellations	−583.1
$\beta_{positive} \cdot 10^{-6}$	Scale factor for the positive tail	40
$\beta_{negative} \cdot 10^{-6}$	Scale factor for the negative tail	0.9

This metric allows the measurement of the sparseness of the order book. The order book may contain a large number of orders, but all the orders are far away from each other (in this case book it would be rarefied). For more details of this metric, see Arbuzov and Frolova (2012).

We calculated the probability of cancellation conditional on RTCI and found that it could be approximated by a linear function as in case of order book imbalance. In Fig. 9, we can see a reasonably expected result, that when orders in the order book are located far from each other, traders have no reasons to cancel their orders (Table 3).

We calculated an RTCI metric at each step of our simulation. The total conditional probability was calculated as:

$$P\left(C_i \middle| y_i, n_{imb}, n_{tot}, RTCI\right) = \mathrm{A}\left(1 - D_1 exp^{-y_i}\right)\left(n_{imb} + \mathrm{B}\right) \times \left(1 - D_2 exp^{-n_{tot}}\right)\left(RTCI + D_3\right)$$

7 Quality Analysis of the Models

Stylized facts are a good test for the identification of model quality, but another important aspect is parity of basic market characteristics:

1. Returns. It is a well-known fact that simple Brownian motion does not allow the generation of heavy tails of distribution. The ZI model can generate fat tails, but the MF and Daniels models (in our case) can generate more heavy tails than in reality. It is interesting that MFWC generated returns, but without heavy tails (Fig. 10).
2. Distribution of spread. Farmer et al. (2005, 2006) in their research concentrated on spread. The spread of our model is not like the empirical one, but with heavy tails in their distribution (Fig. 11).
3. Cancellation time. The order cancellation process plays an important role in asset pricing, so it is important that its lifetime has heavy tails. The order cancellation process in the MF model shows complicated behavior, which is conditional on different market characteristics (just this process leads to a fat tail in an order's life) (Fig. 12).

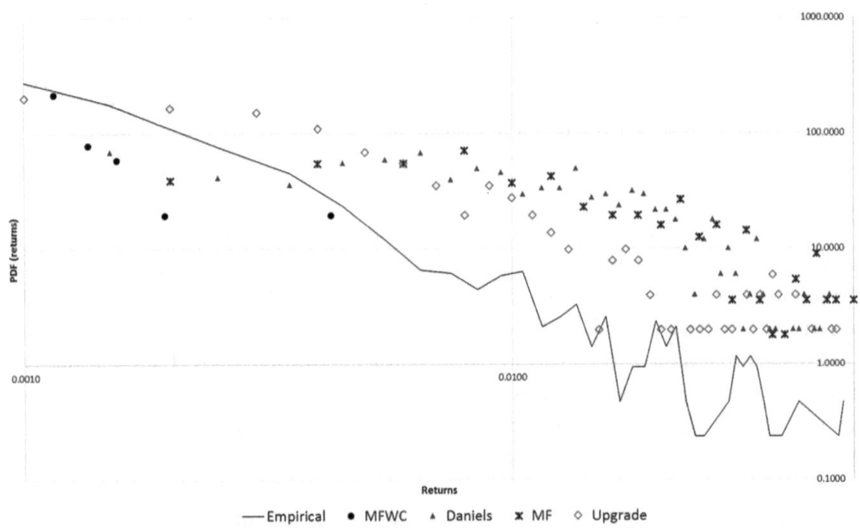

Fig. 10 Distribution of minute returns of analyzing models

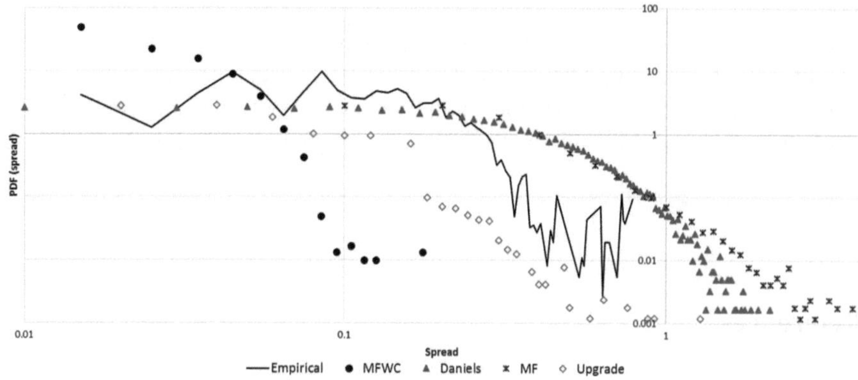

Fig. 11 Spread distribution of analyzing models

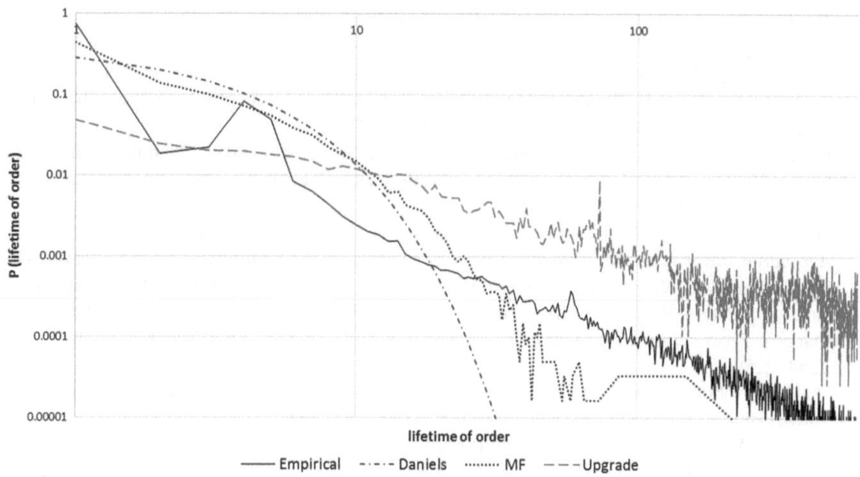

Fig. 12 Order lifetime distribution of analyzing models

Conclusion

We construct and estimate the parameters of two well-known models: Daniels and Mike–Farmer. During the process of the estimation of parameters, we find that distributions of price and probability of cancellation are conditional on the number of orders in the order book being quite different from the MF model. It is important that this model is very sensitive to small details in realization and small bugs in the code. Parameters being not carefully estimated can lead to a significant worsening of model results. We have tried to upgrade the model for our data, including an additional parameter for the

(continued)

order cancellation process and fitting prices using two power-law distributions with t-Student's center. The upgrade model for our sample shows the best results. It is important that the model represents only the microstructure of the market of Aeroflot stocks in January and cannot be spread to other instruments.

References

Achard, S., & Coeurjolly, J.-F. (2010). Discrete variations of the fractional Brownian motion in the presence of outliers and an additive noise. *Statistics Surveys, 4*, 117–147.

Arbuzov, V., & Frolova, M. (2012). *Market liquidity measurement and econometric modeling. Market risk and financial markets modeling*. Heidelberg: Springer.

Bouchaud, J.-P., Gefen, Y., Potters, M., & Wyart, M. (2004). Fluctuations and response in financial markets: the subtle nature of 'random' price changes. *Quantitative Finance, 4*(2), 176–190.

Chakraborti, A., Toke, I., Patriarca, M., & Abergel, F. (2011). Econophysics review: II. Agent-based models. *Quantitative Finance, 11*(7), 1013–1041.

Daniels, M. G., Farmer, J. D., Gillemot, L., Iori, G., & Smith, E. (2003). Quantitative model of price diffusion and market friction based on trading as a mechanistic random process. *Physical Review Letters, 90*(10), 108102.

Farmer, J. D., Gillemot, L., Iori, G., Krishnamurthy, S., Smith, D. E., & Daniels, M. G. (2006). *A random order placement model of price formation in the continuous double auction. The economy as an evolving complex system III* (pp. 133–173). New York: Oxford University Press.

Farmer, J. D., Patelli, P., & Zovko, I. I. (2005). The predictive power of zero intelligence in financial markets. *Proceedings of the National Academy of Sciences of the United States of America, 102*, 2254–2259.

Gu, G.-F., & Zhou, W.-X. (2009). On the probability distribution of stock returns in the Mike-Farmer model. *European Physical Journal B, 67*(4), 585–592.

He, L.-Y., & Wen, X.-C. (2013) Statistical Revisit to the Mike-Farmer Model: can this model capture the stylized facts in real world markets? *Fractals, 21*(2), 1–8. http://www.worldscientific.com/doi/abs/10.1142/S0218348X13500084

Lillo, F., & Farmer, J. D. (2004). The long memory of the efficient market. *Studies in nonlinear dynamics & econometrics, 8*(3), 1–33.

Lillo, F., Mike, S., & Farmer, J. D. (2005). Theory for long-memory of supply and demand. *Physical Review E, 7106*, 287–297.

Mike, S., & Farmer, J. D. (2008). An empirical behavioral model of liquidity and volatility. *Journal of Economic Dynamics and Control, 32*, 200–234.

R Core Team (2013) *R: A language and environment for statistical computing*. Vienna, Austria: R Foundation for Statistical Computing.

Construction and Backtesting of a Multi-Factor Stress-Scenario for the Stock Market

Kirill Boldyrev, Dmitry Andrianov, and Sergey Ivliev

Abstract Nowadays stress-testing is a popular framework for the analysis of the financial stability of different markets' institutes and objects. This work proposes a new approach to trading book stress-testing by building price paths based on generalized autoregressive conditional the heteroskedasticity (GARCH) model with Pareto distribution for the random fluctuation of prices and t-copula for describing the dependency structure between factors.

Keywords Copula theory • Extreme value theory • GARCH • Pareto distribution • Stress-testing • Stylized facts

JEL Classification C49, G17

1 Introduction

Stress-testing is a set of various techniques which allows the gauging of an institute's vulnerability to "severe, but plausible" events (Basel Committee on Banking Supervision 2009). Nowadays most interest in this comes not from financial market participants, but from regulators. The recent crisis shows that risk estimation methods have to be more flexible and versatile if we would like to see the real picture (Sorge 2004). We have to take into account not only of large single events which shock a situation, but also of their aftermath. Therefore we should consider the dynamics of the market's conditions, and stress-testing allows us to do that.

This paper provides an approach to the stress-testing of a trading portfolio. As a test portfolio for consideration we use the MICEX-10 index, which includes major Russian blue chip stocks (HYDR, GMKN, VTBR, ROSN, GAZP, SNGS, URKA, LKOH, SBERP, SBER). The proposed approach is based on two models: the risk factors evolution model and the risk factors interrelation model.

K. Boldyrev (✉) • D. Andrianov • S. Ivliev
Department of Economics, Perm State National Research University, Perm, Russia
e-mail: boldyrev@prognoz.ru

A.K. Bera et al. (eds.), *Financial Econometrics and Empirical Market Microstructure*, DOI 10.1007/978-3-319-09946-0_4

2 The Risk Factors Evolution Model

To describe this evolution, the following AR(1)-GARCH(1,1) model (Posedel 2005) is applied for each risk factor:

$$\begin{aligned} r_t &= \mu + \eta_1 r_{t-1} + \varepsilon_t \\ \varepsilon_t &= \sigma_t \delta_t \\ \sigma_t^2 &= \omega + \beta_1 \varepsilon_{t-1}^2 + \alpha_1 \sigma_{t-1}^2 \end{aligned}$$

where r_t—return at time t, μ—basic value of return, ε—model error, which is decomposed to δ_t—stochastic component and σ_t—conditional standard deviation at time t, ω—basic value of σ_t.

The stochastic component of error δ_t is often considered as a simple random variable with standard normal distribution. However from empirical data one can clearly see that this cannot be true, because it's distribution usually has heavy tails.

In this paper we use Pareto distribution from extreme value theory to simulate this feature in the following way:

- The AR-GARCH model fitted onto historical returns gives historical values for δ_t;
- Historical data on δ_t allows to build its distribution;
- The modeled distribution of δ_t used for the AR-GARCH forecast. Distribution δ_t was constructed in the following way:
- The central part of the density curve obtained with univariate kernel density estimator in the form:

$$\widehat{f}(x,h) = \frac{1}{nh}\sum\nolimits_{i=1}^{n} K\left(\frac{x - X_i}{h}\right),$$

where X_i—sample, K—smoothing kernel (function which satisfies $\int K(x)dx = 1$), h—bandwidth parameter.

Here we used Gaussian kernel:

$$K(x) = \frac{1}{\sqrt{2\pi}} \exp\left(-\frac{x^2}{2}\right)$$

- Tails are fitted separately with a Pareto distribution. It is a base distribution from an extreme value theory in the sense that every distribution of any extreme value can be transformed into a Pareto distribution. It has the form:

$$GP_{\xi,\beta}(x) = 1 - \left(1 + \xi\frac{x}{\beta}\right)^{-\frac{1}{\xi}},$$

where β—scaling, $1/\xi$—tail index.

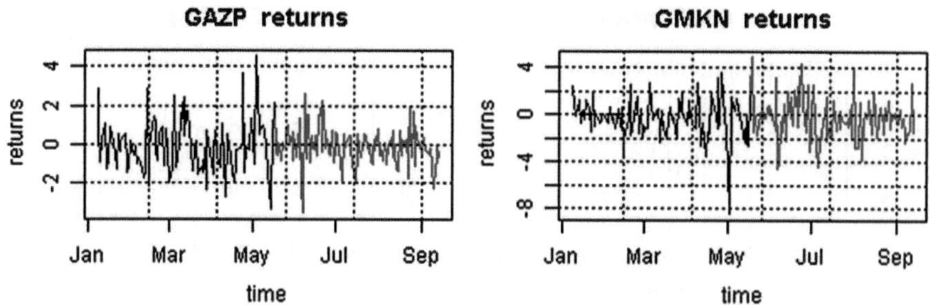

Fig. 1 Returns simulated by the AR(1)-GARCH(1,1) model (fitted onto the second half of 2008): *black line*—historical data, *red line*—simulated data (starts with 15th of May)

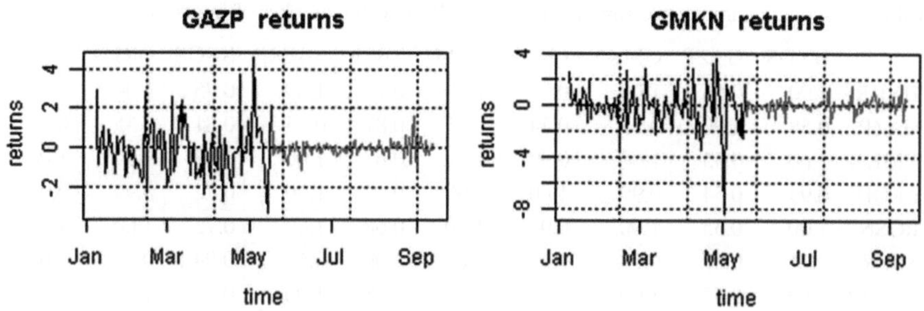

Fig. 2 Returns simulated by the AR(1)-GARCH(1,1) model (fitted onto the first half of 2013): *black line*—historical data, *red line*—simulated data (starts with 15th of May)

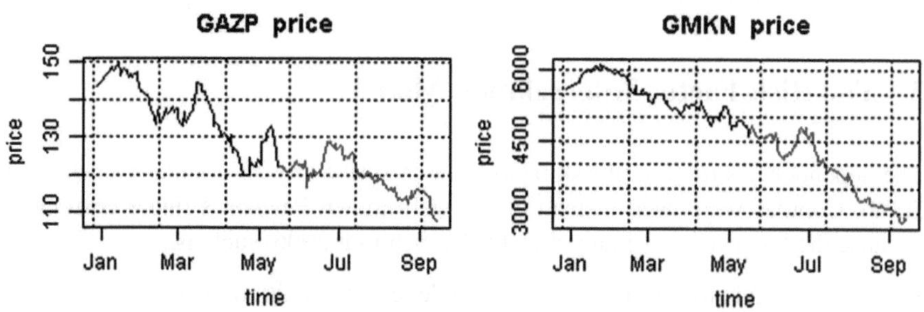

Fig. 3 Prices simulated by the AR(1)-GARCH(1,1) model (fitted onto the second half of 2008): *black line*—historical data, *red line*—simulated data (starts with 15th of May)

The proposed evolution model was applied to two historical periods:

1. Second half of 2008 (crisis conditions);
2. First half of 2013 (stable conditions).

Samples of returns and price dynamics forecast by this model are shown in Figs. 1, 2, 3, and 4. One can see that it catches the volatility clustering effect and correctly transfers initial the historical market conditions to the forecast period.

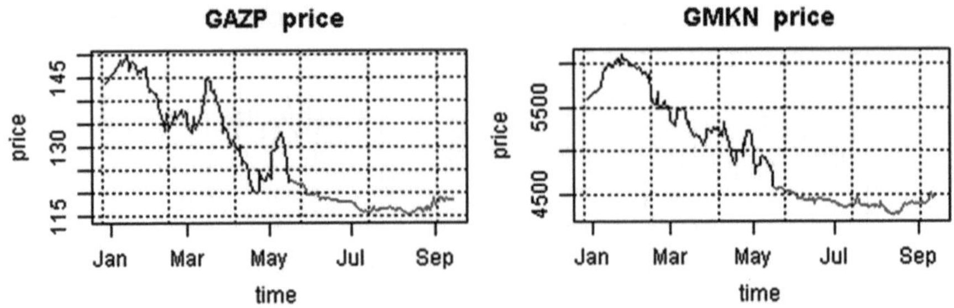

Fig. 4 Prices simulated by the AR(1)-GARCH(1,1) model (fitted onto the first half of 2013): *black line*—historical data, *red line*—simulated data (starts with 15th of May)

Table 1 *t*-copula parameter estimation (second half of 2008), number of freedom degrees = 5

	HYDR	GAZP	GMKN	LKOH	ROSN	SBER	SBERP	SNGS	URKA	VTBR
HYDR	1.00	0.58	1.00	0.99	1.00	0.71	1.00	0.75	1.00	1.00
GAZP	0.58	1.00	0.55	0.54	0.55	0.87	0.55	0.81	0.55	0.55
GMKN	1.00	0.55	1.00	0.99	1.00	0.68	1.00	0.73	1.00	1.00
LKOH	0.99	0.54	0.99	1.00	0.99	0.67	1.00	0.72	0.99	0.99
ROSN	1.00	0.55	1.00	0.99	1.00	0.68	1.00	0.73	1.00	1.00
SBER	0.71	0.87	0.68	0.67	0.68	1.00	0.68	0.94	0.68	0.69
SBERP	1.00	0.55	1.00	1.00	1.00	0.68	1.00	0.72	1.00	1.00
SNGS	0.75	0.81	0.73	0.72	0.73	0.94	0.72	1.00	0.73	0.73
URKA	1.00	0.55	1.00	0.99	1.00	0.68	1.00	0.73	1.00	1.00
VTBR	1.00	0.55	1.00	0.99	1.00	0.69	1.00	0.73	1.00	1.00

3 The Risk Factor Interrelation Model

The dependence structure of risk factors was described by a *t*-copula (Genest et al. 2009). Copulas were used instead of the well-known Pearson's linear correlation because the latter one has many drawbacks (Schimdt 2006) such as:

- It is impossible to capture the full dependency composition of risk factors;
- If the correlation is equal to zero it does not mean that the factors are independent;
- It does not work correctly for distributions with heavy tails because it supposes that risk factor variances are finite (which contradicts the empirical data).

The maximum likelihood method was used to estimate the parameters of the *t*-copula (Charpentier 2006). Historical data for stochastic component δ_t of the AR(1)-GARCH(1,1) model error was used as a sample for this estimation. The results of the estimation are illustrated in Table 1, Table 2 and Fig. 5.

Table 2 *t*-copula parameter estimation (first half of 2013), number of freedom degrees = 5

	HYDR	GAZP	GMKN	LKOH	ROSN	SBER	SBERP	SNGS	URKA	VTBR
HYDR	1.00	0.61	1.00	0.94	0.79	1.00	1.00	0.72	1.00	1.00
GAZP	0.61	1.00	0.61	0.52	0.79	0.61	0.61	0.85	0.61	0.60
GMKN	1.00	0.61	1.00	0.94	0.79	1.00	1.00	0.72	1.00	1.00
LKOH	0.94	0.52	0.94	1.00	0.72	0.94	0.94	0.64	0.95	0.95
ROSN	0.79	0.79	0.79	0.72	1.00	0.79	0.79	0.93	0.79	0.79
SBER	1.00	0.61	1.00	0.94	0.79	1.00	1.00	0.72	1.00	1.00
SBERP	1.00	0.61	1.00	0.94	0.79	1.00	1.00	0.72	1.00	1.00
SNGS	0.72	0.85	0.72	0.64	0.93	0.72	0.72	1.00	0.72	0.72
URKA	1.00	0.61	1.00	0.95	0.79	1.00	1.00	0.72	1.00	1.00
VTBR	1.00	0.60	1.00	0.95	0.79	1.00	1.00	0.72	1.00	1.00

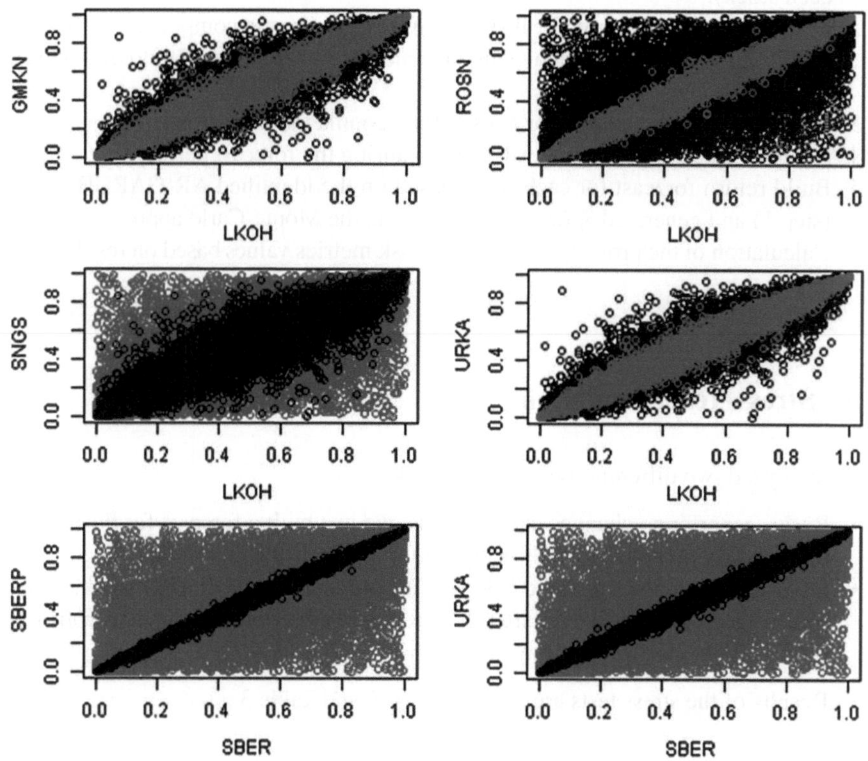

Fig. 5 Simulated correlation between different stocks (as scatter plots) obtained on the basis of *t*-copula estimation and simulation: *red color*—second half of 2008, *black points*—first half of 2013

The dependence structure simulation in Fig. 5 reproduces the well-known empirical fact that in a period of crisis and instability the correlation among separate stocks increase. But also the simulation shows a discorrelation in some cases. Perhaps this can be explained by links between the stocks in the portfolio under consideration.

4 General Scheme of the Model Workflow

The model workflow includes the following steps:

1. The AR(1)-GARCH(1,1) model estimation of each risk-factor (i.e. returns on each stock).
2. Distribution construction for historical δ_t (the stochastic component of the AR-GARCH model error) with a Gaussian smoothing kernel and Pareto distribution for the tails.
3. A constructed distribution for δ_t used for t-copula identification.
4. t-copula used for generating values of δ_t during the forecast period.
5. Build return forecast for each stock based on the identified AR-GARCH model (step 1) and generated δ_t (step 4) N times via the Monte-Carlo approach.
6. Calculation of the profit-loss profile and risk metrics values based on results from step 5.

5 Stress-Test Simulation

We analyzed two different use case scenarios:

- Basic scenario: no changes in conditions; we just built a forecast for the next 30 days and calculated a profit-loss profile for the portfolio on the 30th day.
- Stress-scenario: we simulated a 40 % idiosyncratic drop in GMKN stock, built a forecast for the next 10 days after this drop, and after that calculated a profit-loss profile for the portfolio on the 10th day.

Results of the stress tests are shown in Fig. 6 and Table 3.

6 Analysis and Backtesting

The backtesting of simulated return and price time-series shows that our approach is able to reproduce some stylized facts (Andersen and Davis 2009). There is an auto-correlation in the absolute values of simulated returns (Malmsten and Terasvirta 2004), but it decays very fast (Fig. 7).

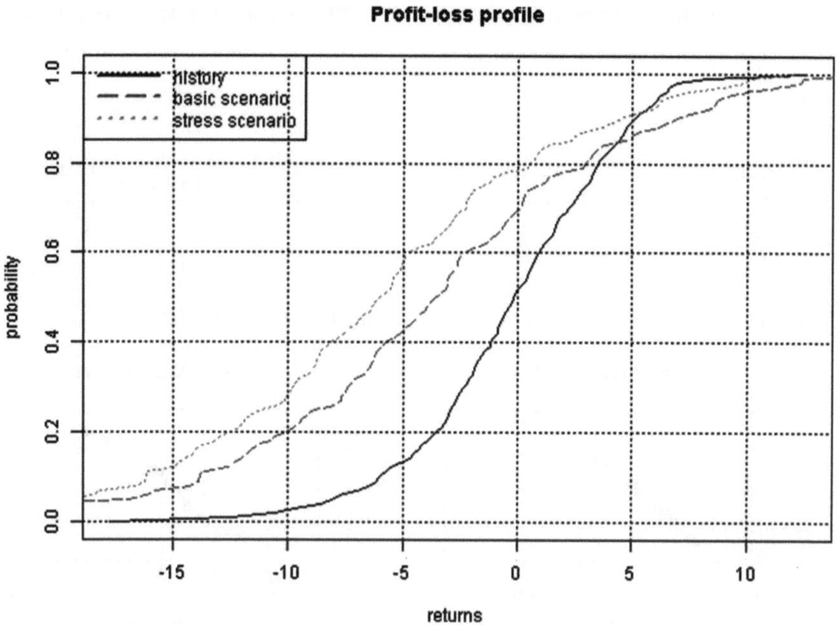

Fig. 6 Profit-loss distribution: historical, basic scenario, stress scenario "GMKN-40 %"

Table 3 Estimated risk-metrics on scenarios

	Basic scenario	Stress scenario "GMKN-40 %"
Maximum loss	39.90 %	42.30 %
Maximum profit	59.40 %	57 %
90 % VaR	−13.90 %	−16.30 %
95 % VaR	−17.80 %	−20.20 %
99 % VaR	−26.80 %	−29.20 %
90 % ES	−21.10 %	−23.50 %
95 % ES	−26.90 %	−29.30 %
99 % ES	−39.90 %	−42.30 %

QQ-charts show that the distribution of simulated returns differs from normal and (Fig. 8) and demonstrates heavy-tails fairly close to the distribution of historical returns (Fig. 9).

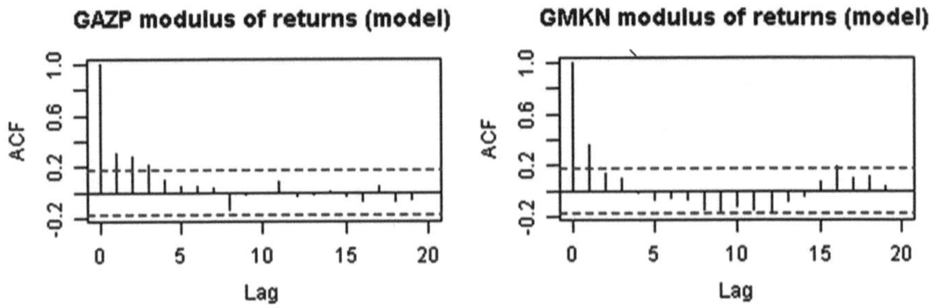

Fig. 7 Auto-correlation function for the modulus of simulated returns for different stocks

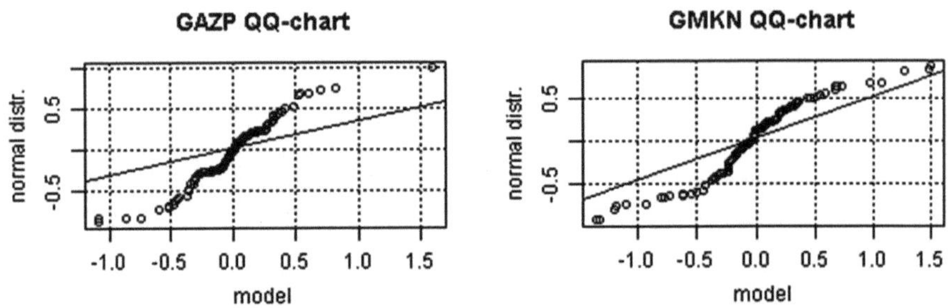

Fig. 8 QQ-charts for the distribution of simulated returns in comparison with normal distribution

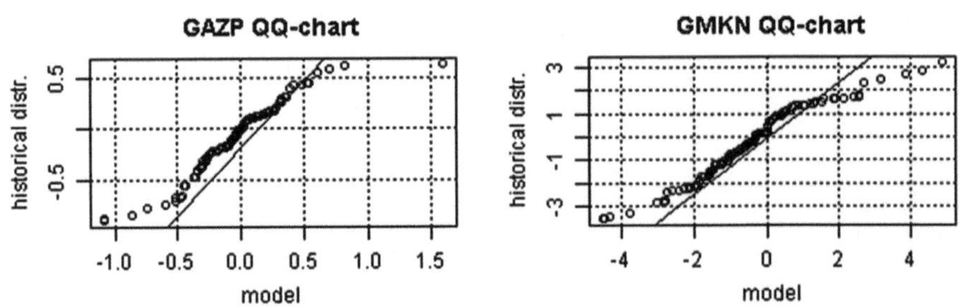

Fig. 9 QQ-charts for the distribution of simulated returns in comparison with the distribution of historical returns

Conclusion

We propose a new approach for stress-testing of a given investment portfolio based on the application of the GARCH model with a particular specification for the model's error together with the copula's description of risk factors dependency structure. This method can be backtested by the reproduction of stylized facts known for returns and price time-series:

(continued)

- There is autocorrelation in simulated return time-series, but it decays fairly fast;
- The distribution of simulated returns differs from normal and has heavy tails pretty close to the distribution of historical returns;
- There is a volatility clustering effect in the simulated returns;
- By using the copula for a dependency structure description it is possible to catch various and complicated changes in dependencies between the risk factors.

The model allows us to simulate the returns of the portfolio according to the variations in risk factors for use for profit-loss distribution estimation, as well as market risk measurement under stress conditions.

References

Andersen, T. G., & Davis, R. A. (2009). *Handbook of financial time series*. Berlin: Springer.

Basel Committee on Banking Supervision. (2009). *Principles for sound stress testing practices and supervision*. Basel consultative documents. www.bis.org/publ/bcbs155.pdf.

Charpentier, A. (2006). *The estimation of copulas: theory and practice*. http://perso.univ-rennes1.fr/arthur.charpentier/chapter-book-copula-density-estimation.pdf.

Genest, C., Gendron, M., & Bourdeau-Brien, M. (2009). The advent of copulas in finance. *The European Journal of Finance, 15*, 609–618.

Malmsten, H., & Terasvirta, T. (2004). *Stylized facts of financial time series and three popular models of volatility*. http://ljsavage.wharton.upenn.edu/~steele/Resources/FTSResources/StylizedFacts/MalmstenTerasvirta04.pdf.

Posedel, P. (2005). *Properties and estimation of GARCH(1,1) model*. http://www.ressources-actuarielles.net/EXT/ISFA/1226.nsf/0/73d982a644ea2f6dc1257609006edb99/$FILE/posedel.pdf.

Sorge, M. (2004). *Stress-testing financial systems: an overview of current methodologies*. http://www.bis.org/publ/work165.pdf.

Schimdt, T. (2006). *Coping with copulas*. http://www.math.uni-leipzig.de/~tschmidt/TSchmidt_Copulas.pdf.

Modeling Financial Market Using Percolation Theory

Anastasiya Byachkova and Artem Simonov

Abstract Econophysics is a relatively new discipline. It is one of the most interesting and promising trends in modeling complex economic systems such as financial markets. In this paper we use the approach of econophysics to explain various mechanisms of price formation in the stock market. We study a model, which was proposed by Jean-Philippe Bouchaud and Dietrich Stauffer (Bouchaud 2002; Chang et al. 2002; Stauffer 2001; Stauffer and Sornette 1990), and used to describe the agents' cooperation in the market. The most important point of this research is the calibration of the model, using real market conditions to proof the model's possibility of setting out a real market pricing process.

Keywords Agent modeling • Econophysics • Financial markets modeling • Percolation theory • Quantitative finance

1 Elements of the Percolation Theory

Physics and finance are both based on the theory of random walks and on the collective behavior of large numbers of correlated variables (Sornette et al. 1999).

The considered model is based on percolation theory, which describes phase transition in physical systems. It regards the square lattice from $L * L$ sites. Every site can be "occupied" or "free"; the site can be occupied with probability p randomly. The groups of neighboring occupied sites are formed in clusters.

The main task of the percolation theory is to search for an infinite cluster—cluster, which extends from one side of the lattice to another. In this situation, most parts of cells belong to one cluster. In this case p_c is the percolation threshold, the critical probability of infinite cluster appearance and the offensive of phase

A. Byachkova (✉)
JSC Prognoz,, Perm, Russia
e-mail: abyachkova@gmail.com

A. Simonov
EY Advisory, Moscow, Russia
e-mail: art.simonov@gmail.com

A.K. Bera et al. (eds.), *Financial Econometrics and Empirical Market Microstructure*, DOI 10.1007/978-3-319-09946-0_5

transition (Gould and Tobochnik 1990). Next, we turn to the application of the model in finance.

2 Percolation Model of Stock Market Prices

It is well known that we often observe “clustering” (or herding) phenomena in the financial markets—the situation when agents in the market prefer to make the same decisions. This behavior is clear in terms of psychology, because people are used to behaving dependently with each other; we might be easily influenced by others in many aspects of our lives. This correlation comes from random clustering. In our model, there are clusters of agents, i.e. groups of traders in the market that prefer to act together, i.e. to buy or to sell securities simultaneously.

For market simulation, each occupied site is regarded as an agent, and clusters are groups of traders who randomly decide to buy or to sell together (Stauffer 2001). Argument a is a measure of a one-time step. Small value of a corresponds to a small interval, and a value near the maximum of 1/2 corresponds to a large-time interval. Argument has influences on an agent’s decisions: each cluster decided randomly to sleep with probability $1-2a$ or to be active with some probability. Argument p_{buy} is the probability that an active agent prefers to buy. Argument $p_{sell} = 1 - p_{buy}$ is the probability that an active agent prefers to sell. This parameter helps us to consider influence of past and present trends on the market. It’s important to note that in this model, we have an assumption that agents have only two possible activities—to buy or to sell—and its sum is a full group of events.

Thus, for every time step, we analyze the existing clusters and find the number n_s of clusters containing s investors each. The distribution of n_s closely follows the percolation threshold of the scaling law:

$$n_s \sim s^{-\tau} f\left[(p-p_c)\, s^{\sigma}\right] \tag{1}$$

with two critical σ, τ exponents, and a function f decaying exponentially in its tails. Then each cluster randomly decides to buy, sell or sleep with some defined probabilities $p_{buy},\ \ p_{sell}$, and $(1-2a)$ probability of sleeping. The price change in the market in a one-time step, which is labeled as $\Delta(t)$, is proportional to the difference of demand and supply in this market:

$$\Delta(t) = \sum_{buy} n_s s - \sum_{sell} n_s s, \tag{2}$$

where the total demand is sum of all agents in all clusters that decide to buy, and total supply is the sum of all agents in all clusters that decide to sell (Stauffer 2001).

The important part of this research is the analysis of model behavior and price change at the critical moment of percolation threshold occurrence. It’s possible to explain this market mechanism: when $p < p_c$, price rises and more people enter the

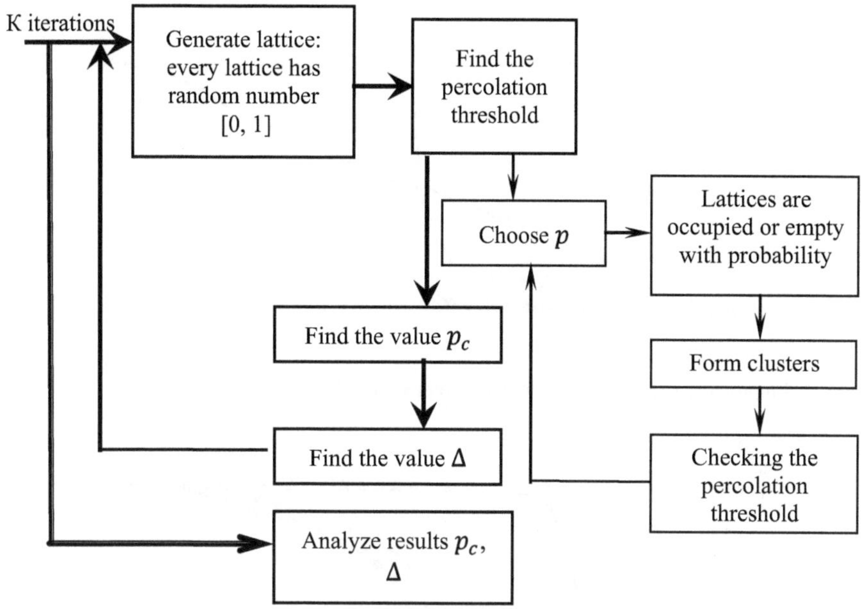

Fig. 1 Algorithm of the single iteration of Monte Carlo simulation

market. Therefore p rises until a big crash occurs at $p = p_c$. In this moment the price falls sharply, agents suffer losses and leave the market. As a result p falls and the cycle starts again at low p. The market crash during the moment of percolation threshold occurrence means that the most agents have the same opinion about their strategy. It leads to mass selling or buying; such a situation causes a market crash or a market boom (Chang et al. 2002).

Thus the basic purpose of the percolation model is to analyze the percolation threshold, which characterizes the threshold probability of a market crash. The model studies Δ empirical distribution as a distribution of price change in the market.

For the modeling of percolation theory, we use the Monte-Carlo method, which was realized in statistical environment R.

Results of our modeling were processed in MS Excel. The steps for the single iteration of Monte Carlo simulation are presented in Fig. 1.

We study the Δ empirical distribution with different values of model parameters. We have discovered a strong interrelationship of statistical characteristics of the received distribution of size Δ from parameters p_{buy}, a (Figs. 2 and 3). It is possible to note various curves shapes of the received functions. Sharper excess of function is marked at the maximum difference between probability of purchase p_{buy} and probability of sale $1 - p_{buy}$ and measure of time interval a.

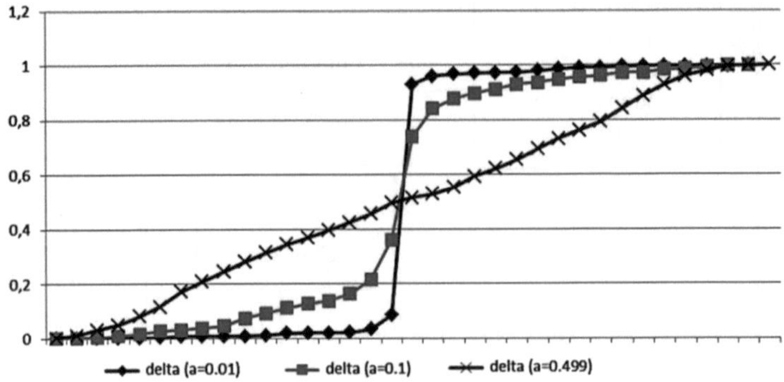

Fig. 2 Empirical distributions of Δ with different value of a

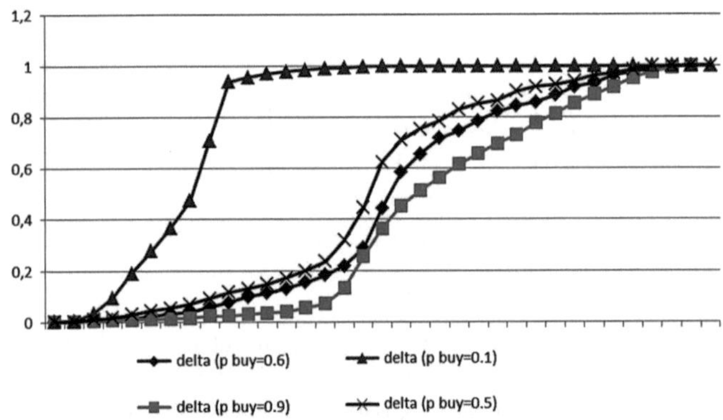

Fig. 3 Empirical distributions of Δ with different value of pbuy

If it's necessary to receive authentic distribution of market price change, it is important to pick up values of the parameters defining current market trends and preferences of agents in this market.

3 Model Calibration

In order to understand this model's properties and its advantages, it is necessary to analyze how the model can reflect real data conditions. Thus, there is an issue of calibration of the model and applicability of the model for the description of a real market situation.

The percolation model allows us to simulate price change distribution in a one-time step as a hypothetical situation of interaction of agents for a certain time interval. The task of calibration is to select values of parameters and receive the

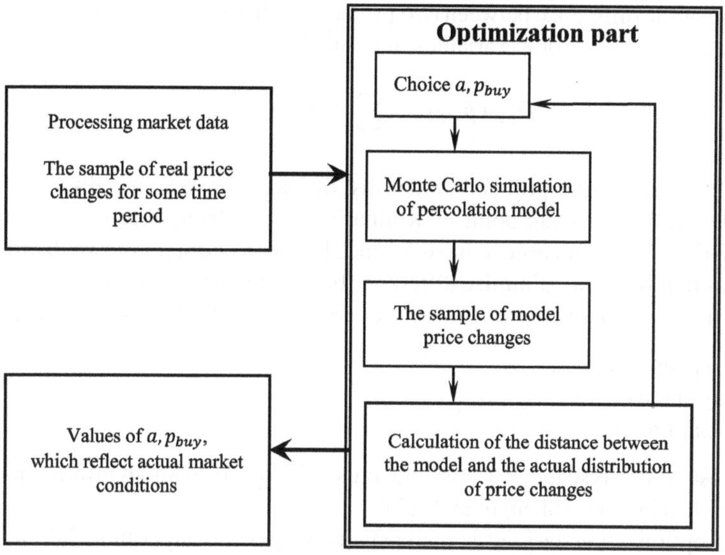

Fig. 4 Algorithm of reverse engineering calibration

model's empirical distribution, which is similar to the real-world market distribution in terms of some pre-selected measures Wiesinger et al. (2010).

This is the method of reverse engineering in the context of financial time-series. With its parameters and strategies, it optimizes the similarity between the actual data and simulated data.

An algorithm of the reverse engineering calibration of our model is presented in Fig. 4. We have already noticed that values of a, p_{buy} have very strong influence on the empirical distribution of Δ. Because of this, we will find the values of a, p_{buy} which will give the required similarity.

At the first stage, we do the processing of real market data. We consider hourly log returns of RTS index (leading Russian stock index) during the period of January 1st, 2008 to December 31st, 2009. This period could be characterized as an instable stage in the financial market. Thus, at calibration we are expecting a condition of infinite cluster occurrence, which most precisely characterizes a crisis situation in the market.

The next stage is optimization. There is the minimization of distance between a real sample of price changes and the model sample of Δ, as a result of the Monte Carlo simulation. The algorithm changes values of required parameters a, p_{buy} and generates a new percolation model as a result. We have a new sample of model price changes as a result of this iteration step.

The part of calculation the distance between modeling and the fact sheet assumes using various measures of distance between two probability distributions. In this research we decide to use Kullback–Leibler divergence. This is non-symmetric measure of the difference between two probability distributions and used for discrete and for continuous random variables. It defined to be:

$$D_{KL}(p,q) = \sum_{x\epsilon} p(x) \ln \frac{p(x)}{q(x)}, \quad (3)$$

where $p(x)$, $q(x)$—the probability density of the corresponding discrete random variables X, Y. The main point of Kullback–Leibler divergence is that it base on information theory and reflect the difference between entropies of two distributions. It means that we try to minimize difference between indeterminacies of two samples of information. There are some other properties of Kullback–Leibler divergence:

- non-symmetric
- always nonnegative
- non-parametric.

In case of this research it's possible to use divergence without information about form of distributions (Shengqiao 2012).

The optimization task was realized, using genetic algorithm. We minimize of Kullback–Leibler divergence with DEoptim R package, which is a global optimization algorithm from class of genetic algorithms, which uses biology-inspired principles. The main argument for this choice is the possibility to work with discontinuous and nondifferentiable functions, because we haven't got enough information about function we have to minimize (Ardia et al. 2012).

Results of calibration are empirical distribution of modeling price change with parameters $a = 0,02$ and $p_{buy} = 0,31$. The results of empirical function are presented in Fig. 5.

The small value of parameter $a = 0,02$ is interpreted as a short time interval when market was observed. That's why we can conclude, that high frequency traders are presented in this market. The value of probability to buy $p_{buy} = 0,31$, which can

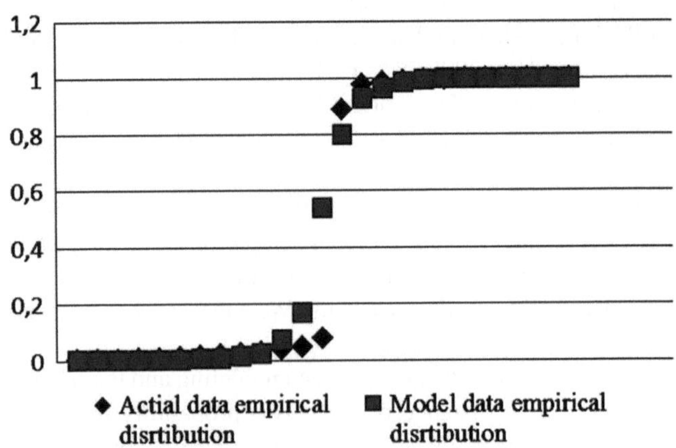

Fig. 5 Empirical distribution function of model and actual price changes

be interpreted as a small asymmetry between demand and supply and the most of agents prefer to sell in the market because of the critical crisis situation.

Conclusion
In this paper percolation model was used to describe agent's cooperation in the financial market. The results of Monte Carlo simulation allow analyzing model price changes distribution and concluding about price change distribution and model parameters dependence. This dependence suggests the possibility of model calibration. Using optimization procedures help to find model parameters values which describe real market pricing process. The result shows that presented model generally comply with real-market data.

References

Ardia, D., Muller, K., Peterson, B., & Ulrich, J. (2012). *Global optimization by differential revolution*. Available via Internet: http://cran.r-project.org/web/packages/DEoptim.

Bouchaud, J.-P. (2002). An introduction to statistical finance. *Physica A: Statistical Mechanics and Its Applications, 313*, 238–251.

Chang, I., Stauffer, D., & Pandey, R. B. (2002). Asymmetries, correlations and fat tails in percolation market model. *International Journal of Theoretical and Applied Finance, 5*(6), 585–597.

Gould, H., Tobochnik, J. & Christian W. (1990). An introduction to computer simulation methods: Applications to Physical Systems Third Edition, Addison-Wesley, 2006, p. 796

Li, S. (2012). *Fast nearest neighbor search algorithms and applications*. Available via Internet: http://cran.r-project.org/web/packages/FNN.

Sornette, D., Stauffer, D., & Takayasu, H. (1999). *Market fluctuation II: Multiplicative and percolation models, size effects and prediction* (Vo1, p. 30). Available via Internet: arXiv:cond-math/9909439.

Stauffer, D. (2001). Percolation models of financial market dynamics. *Advances in Complex Systems, 4*(1), 19–27.

Stauffer, D., & Sornette, D. (1990). Self-organized percolation model for stock market fluctuation. *Physica A: Statistical Mechanics and Its Applications, 271*, 496–506.

Wiesinger, J., Sornette, D., & Satinover, J. (2010). *Reverse engineering financial market with majority and minority games using genetic algorithm*. Available via Internet: arXiv:1002.2171v1.

How Tick Size Affects the High Frequency Scaling of Stock Return Distributions

Gianbiagio Curato and Fabrizio Lillo

Abstract We study the high frequency scaling of the distributions of returns for stocks traded at NASDAQ market as a function of the tick-to-price ratio. The tick-to-price ratio is a measure of an effective tick size. We find dramatic differences between distributions for assets with large and small tick-to-price ratio. The presence of returns clustering is evident for large tick size assets. The statistical differences between large and small tick size assets appear to reduce at higher time scales of observation. A possible way to explain returns dynamics for large tick size assets is the coupling of returns with bid-ask spread dynamics. A simple Markov-switching model is able to reproduce the properties of the distribution of returns for large tick size assets.

Keywords Bid-ask spread • Markov-switching models • Returns clustering • Returns distribution • Scaling • Tick size

1 Introduction

In financial markets, the price of an order cannot assume arbitrary values but it can be placed on a grid of values fixed by the exchange. The tick size is the smallest interval between two prices, i.e. the grid step, and it is measured in the currency of the asset (Ascioglu et al. 2010). It is institutionally mandated and sets a limit on how finely prices may be specified. All price information is discretized by the tick size. Historically, the tick size of most securities has been consecutively reduced, resulting in tick sizes of 1/100th or smaller. This process is often referred to as decimalization (Gibson et al. 2003; He and Wu 2004; Chung et al. 2004;

G. Curato
Scuola Normale Superiore, Pisa, Italy
e-mail: gianbiagio.curato@sns.it

F. Lillo (✉)
Dipartimento di Fisica e Chimica, University of Palermo, Palermo, Italy

Scuola Normale Superiore, Pisa, Italy
e-mail: fabrizio.lillo@unipa.it; fabrizio.lillo@sns.it

A.K. Bera et al. (eds.), *Financial Econometrics and Empirical Market Microstructure*, DOI 10.1007/978-3-319-09946-0_6

Loistl et al. 2004; U.S. Securities and Exchange Commission 2012). The current tick size for stocks traded in US stock exchanges, such as the New York Stock Exchange (NYSE) or the National Association of Securities Dealers Automated Quotations (NASDAQ), is typically \$0.01. An argument for maintaining the tick size is that it serves to maintain a minimum level of profits for market makers and thus guarantees the provision of liquidity (MacKinnon and Nemiroff 2004; Huang and Stoll 2001; Bollen and Busse 2006), but a too large tick size increases the transaction cost to investors by increasing the bid-ask spread. It is controversial whether a smaller tick size generally improves market quality.

Tick size can affect prices in a direct way on different time scales, starting from the microstructural scale to the daily scale. In this study we analyze the midprice process, i.e. the dynamics of midpoint between bid and ask quotes, in transaction time and in continuous time. We want to study the scaling of the distributional properties of price fluctuations at different time scales, starting from the smallest time scale, e.g. price changes and log-returns caused by 1 transaction. In this way we can see the connection between high frequency dynamics of prices, i.e. 1 s or 1 min dynamics, and low frequency dynamics, i.e. 1 h dynamics. The basic observation is that at the smallest time scale the distributions of returns are very far from Gaussian or Levy stable distributions, that are instead used to model price fluctuations at higher time scales (Bouchaud and Potters 2009; Hautsch 2012; Dacorogna et al. 2001). The return distribution at the smallest time scales strongly depends on the value of the tick-to-price ratio. We have large or small effective tick size assets if this ratio is high or small. As it is known in the literature, the value of the tick size is not the best indicator for understanding and describing the high frequency dynamics of prices. The tick-to-price ratio is one of the definitions of the notion of an effective tick size, introduced in order to account and quantify the different behavior of price fluctuations. Another useful definition is based on the bid-ask spread. In this case the measure is given by the frequency the spread is equal to one tick and we have a large tick size if the spread is almost always equal to one tick. Usually these measures of the effective tick produce the same ranking between different securities.

The key observation is that for large tick assets the price changes are clustered on the grid of the possible integer values that they could assume. Specifically, we find that even price changes are more populated than odd values. This property is found to hold from small to high time scales. Instead for small tick size assets the clustering of price changes is not present. The high frequency dynamics of price for a large tick asset is characterized by the presence of clustering. A similar property has been reported in literature (Harris 1991; Onnela et al. 2009) for daily closing price series. The presence of clustering affects also the distribution of returns for large effective tick size assets, instead this effect is negligible for small tick asset.

We want to quantify empirically the distortion of the shape of distributions of price changes and returns as a function of the effective tick size, measured by the tick-to-price ratio or by the frequency of bid-ask is equal to 1 tick. We expect that, after a certain time scale of aggregation, the shape of distributions becomes independent from the effective tick size of the asset. On one hand the distortion can be characterized by measuring how far the distributions are from the Gaussian, and on the other hand by fitting a microstructural model, developed for large tick assets, on our data in order to reproduce the statistical properties of price changes and returns at different time scales.

We start in Sect. 2 by reviewing the effect of tick size on the market microstructure and the statistical properties of price fluctuations. In Sect. 3 we study the influence of the effective tick size on the return distributions for four assets traded on the NASDAQ market. In Sect. 4 we fit a recently introduced microstructural model (Curato and Lillo 2013) on data of a large tick asset in order to reproduce the statistical properties that we have measured. We summarize the results in section "Conclusions".

2 Literature Review

Most of the studies about tick size present in the literature are case studies of the impact of a reduction of tick size on market quality, i.e. on microstructural quantities like the narrowing of the bid-ask spread (Loistl et al. 2004) or liquidity provision (Goldstein et al. 2000; Ahn et al. 2007). The part of literature more related with our work is composed by papers that have revealed how the investors actually use the price resolution allowed by the tick size. We focus also on statistical properties of price fluctuations (Onnela et al. 2009; Münnix and Schäfer 2010; La Spada et al. 2011; Gopikrishnan 1999; Plerou et al. 1999) and on the connection between bid-ask spread and midprice dynamics (Dayri and Rosenbaum 2013; Wyart et al. 2008; Robert and Rosenbaum 2011).

The concept of price clustering is known in the literature for daily price time series. It appears that instead of making full use of the available price spectrum, investors stick to a subset of it and use coarser prices instead. There are at least two alternative explanations for this: natural clustering (Harris 1991; Osborne 1962) or collusion (Christie and Schultz 1994; Christie et al. 1994). Harris (1991) studied the frequency distribution of the integer portion of CSRP daily closing price stocks for the years 1963 to 1987, including NYSE, AMEX and NASDAQ stocks. In this case the minimum ticks size ranged from \$1/8 to \$1/16, and the tick size was smaller for stocks with lower prices. He argued that stock price clustering is pervasive and that clustering distributions from the mid-nineteenth century appear very similar to those observed in the late twentieth century. Clustering increases with price level and volatility and occurs if traders use discrete price sets to simplify their negotiations. He claimed also that clustering must affect price changes distributions and bid/ask quote distributions. Collusion instead refers to the idea that market makers quote prices only in certain fractions in order to increase bid-ask spreads. Christie and Schultz (1994), and Christie et al. (1994) show that many NASDAQ stocks exhibit a paucity of odd-eights quotes and quote prices mainly in even-eights. Bessembinder (2000, 1997, 1999, 2003) provides empirical evidence on relations between trade execution costs and price rounding practices on the NYSE and NASDAQ. His results indicate that higher execution costs are associated with the rounding of quotations and trade prices, and finds that the effect of clustering on trading costs decreases as the tick size decreases.

Onnela et al. (2009) study the effect of changes in tick size, enabled by the decimalization process, on asset log-returns. They analyze a set of NYSE and TSE (Toronto Stock Exchange) cross-listed stocks that were traded under different tick sizes. The data were daily closing prices from Jan-1-1990 to Jun-30-2003. They show that investors do not use all price fractions uniformly as allowed for by the tick size, leading to a clustering of prices on certain fractions, a phenomenon that could potentially affect the way returns are distributed. This phenomenon persists after decimalization. They observed that approximately 57 % of cases exhibit a price clustering such that the effective tick size deviates from the nominal tick size. In this study the tick-to-price ratio, i.e. a measure of the effective tick size, appears to be indicative of the zero returns frequency. They conjectured that large effective ticks lead to a distortion of the shape of return distribution, and this effect should be particularly strong when the price of stock is low, i.e. when tick-to-price ratio is high.

Münnix and Schäfer (2010) demonstrate that the tick size has a large impact on the structure of financial return distributions. They analyze a basket of stocks from the S&P 500 index ranging from 1 min to 1 day frequency during the first half of 2007. They find returns clustering at 1 min frequency but do not connect their statistical properties to an effective tick size. They observe that the discrete distribution of price changes could lead to think that the transition from integer price changes to relative price changes, i.e. returns, remove the discretization from the distribution. A closer analysis instead reveals that the discretization effect are still visible when considering returns. They argue that the discretization affects returns on any time scale. They perform an approximate analysis that reveals a sort of mapping between the discrete distribution of price changes and the distribution of returns. They decompose the set of returns according to the absolute price changes, i.e. one value of price change corresponds to a specific set of returns. Their computations lead to the conclusion that the width of this sets are proportional to the absolute value of price changes, while the distance between their centers remains almost constant. In this way the sets of values of returns are increasingly overlapping for larger values of absolute price changes. From their viewpoint the discretization is only visible for small absolute price changes, i.e. one could see an unusual distortion of return distribution near its center. Moreover they find that the shape of the distribution of normalized returns compared to the underlying normalized price changes are quite similar for time scales ranging from 5 min to 1 day. According to Münnix and Schäfer (2010) the meaning of clustering is that the distribution of returns is defined on specific sets of the real line, i.e. we do not have a smooth distribution like the Gaussian or Lévy distributions. This effect is less and less visible if we have a large number of possible different values for price changes, because we have the overlap of the different sets.

Wyart et al. (2008) use a theoretical framework to obtain a linear relation between the bid-ask spread and the instantaneous impact of market orders and then use this relation to justify a strong empirical correlation between the spread and the volatility per trade. They test this on empirical data and find good agreement with the predicted bound for small tick electronic markets. The case of large tick stocks is different since in this case the spread is nearly always one tick, with very large volumes at both the bid and the ask, leading to a spread that is substantially larger than that predicted for small tick stocks.

Curato and Lillo (2013) develop a statistical model in order to reproduce the statistical properties of the discrete process of price changes for a large tick asset. Large tick assets display a dynamics in which price changes and spread are strongly coupled. They introduce a Markov-switching modeling approach that describes this coupling and the dynamics of spread and return in transaction time. The latent Markov process is the transition process between spreads. Montecarlo simulations of this model reproduce remarkably well the statistical properties of time series representing stocks on NASDAQ market.

3 Empirical Analysis

In this section we study the role of the effective tick size on the distributional properties of price changes and log-returns at different time scales. We make use of two simple definitions of the effective tick size, the first one is the tick-to-price ratio and the second one is the unconditional frequency to have the bid-ask spread equal to one tick. They are usually used in equivalent way in order to classify assets in large and small tick assets (Eisler et al. 2012; Dayri et al. 2011). We use the first definition mainly in Sect. 3, instead the second one is used to define a statistical model in Sect. 4.

3.1 NASDAQ Data

In this paper we study high frequency data of highly liquid stocks traded at NASDAQ market in the period from 01/07/2009 to 31/08/2009 (42 trading days). We analyze the stocks: Apple Inc. (AAPL), Amazon (AMZN), Microsoft Corporation (MSFT), Cisco Systems (CSCO). Our data contain time stamps corresponding to order executions, trade prices, bid-ask quotes, size of trading volume and direction of trading. The time resolution is millisecond. The trading activity at NASDAQ starts at $9:30$ and ends at $16:00$. We decide to discard all transaction data corresponding to first and last 6 min of the day. During these minutes we observe bursts of trading activity and an abnormal high price fluctuations that could affect the statistical analysis of returns distributions.

It is important to point out that, since when a market order hits several limit orders, it results in several trades being reported, we choose to aggregate together all such transactions and consider them as one trade if the millisecond time stamps of trades in our database are the same. We are going to use these transactions as our "events", meaning that all relevant values are calculated at the time just before each transaction. Hereafter we define the transaction or trade time as an integer counter of events defined by the execution of a market order.

For each asset, we define the following time series in trade time:

- t_i is the time of i-th trade, $i \in \mathbb{N}$ is the transaction or trade time.
- $b(i) = b(t_i)$ and $a(i) = a(t_i)$ are respectively the best bid and ask prices just before the i-th trade.
- $p(i) = p(t_i) = (b(t_i) + a(t_i))/2$ is the midpoint price just before the i-th trade, $p \in \mathbb{N}$ is measured in units of half tick.
- $s(i) = s(t_i) = a(t_i) - b(t_i)$ is the spread just before the i-th trade, $s \in \mathbb{N}$ is measured in units of tick.
- $\Delta p(i) = \Delta p(t_i) = p(t_{i+1}) - p(t_i)$ is the price change caused by the i-th trade, $\Delta p \in \mathbb{Z}$ is measured in units of half tick.
- $r(i) = r(t_i) = \log(p(t_{i+1})) - \log(p(t_i))$ is the log-return, $r \in \mathbb{R}$.
- $\Delta p(i, n) = p(t_{i+n}) - p(t_i)$ is the price change caused by n consecutive trades, n is the trade time scale at which we observe the price change process.
- $r(i, n) = \log(p(t_{i+n})) - \log(p(t_i))$ is the log-return caused by n consecutive trades.

We want to study the price process also in continuous time. To this end we define the midprice process $p_c(t)$ assuming that the price between two transactions is given by the midprice just before the second transaction. This defines a piecewise constant function like that shown in Fig. 1.

- $p_c(t) = p(t_{i^*})$, where $t_{i-1^*} \leq t < t_{i^*}$ is the time between the two subsequent transactions $i - 1^*$ and i^*.
- $\Delta p_c(t, \Delta t) = p_c(t_{i^*+n^*}) - p_c(t_{i^*})$, is the price change observed sampling the time series $p_c(t)$ at a time scale Δt. This change is caused by n^* consecutive trades. The number of trades n^* is a stochastic variable for each fixed value of the time scale Δt.
- $r_c(t, \Delta t) = \log(p_c(t_{i^*+n^*})) - \log(p_c(t_{i^*}))$, is the log-return caused by n^* consecutive trades.

We develop an algorithm that samples the time series of trade time t_i in order to determine the index i^* and n^* of trades that we need to observe the series in continuous time $p_c(t)$, Δp_c and r_c. We report in Table 1 some sample statistic about log-returns corresponding to the smallest time scales studied in this work. When we observe prices in continuous time, the empirical returns distribution is more fat-tailed than that defined in trade time. The increase of kurtosis can be explained if we think to price process as a subordinated random process. We give same details on the subordination hypothesis in Sect. 3.2.

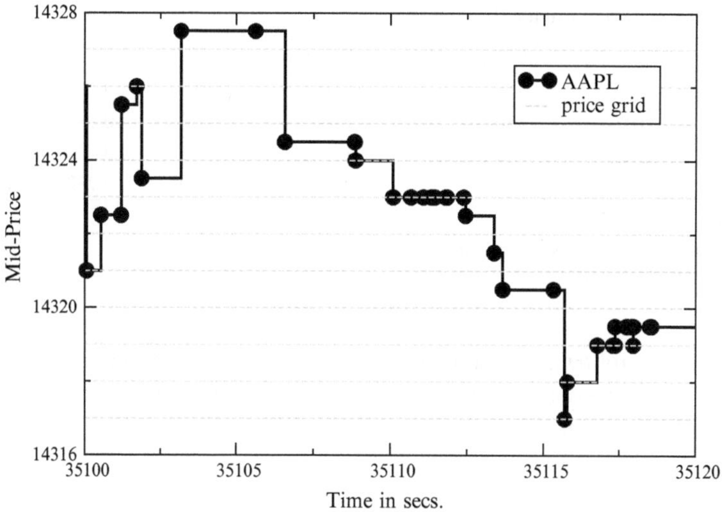

Fig. 1 Midprice process for the stock AAPL is the black piecewise constant curve. The price grid is measured in units of one tick. The time is the number of seconds from the beginning of the trading day. *Circles* indicate executions of market orders. We can observe trades that do not cause price changes

Table 1 Sample statistic for log-returns

Stock	Time scale	Mean	Std.deviation	Skewness	Ex.kurtosis
AAPL	1 trade	6.602e−08	7.553e−05	0.01434	5.268
	1 s	6.641e−08	8.043e−05	−0.00105	22.16
AMZN	1 trade	6.497e−08	1.484e−04	0.0442	8.177
	1 s	4.590e−08	1.097e−04	0.3164	41.25
MSFT	1 trade	1.038e−07	1.190e−04	0.0112	7.773
	1 s	5.949e−08	8.406e−05	−0.0146	50.50
CSCO	1 trade	1.173e−07	1.427e−04	0.00505	7.008
	1 s	5.587e−08	9.791e−05	0.1285	50.34

We make use of two definitions of the effective tick size:

- $T_r = 1/\langle p_t \rangle$ is the tick-to-price ratio, where the mean trade price $\langle p_t \rangle$ is measured in 0.01\$, i.e. the value of one tick.
- $T_s = \#[s(i) = 1]/N_t$ is the fraction of times the spread is equal to 1 tick, N_t is the total number of trades in 42 days.

The symbol $\langle \cdots \rangle$ denotes a temporal average over the entire length of the time series. We can observe in Table 2 that these two measures divide the stocks in the same manner in two groups: AAPL and AMZN are small tick size stocks, instead MSFT and CSCO are large tick size stocks. It is important to observe that the two measures lead to the same classification between large and small tick assets. As we

Table 2 Effective tick size for NASDAQ stocks

Stock	Tick size	# Trades N_t [a]	Duration [b]	T_r [c]	T_s	Class
AAPL	0.01$	918294	1.037	0.64	0.256	SMALL
AMZN	0.01$	530076	1.797	1.2	0.243	SMALL
MSFT	0.01$	532795	1.788	4.1	0.932	LARGE
CSCO	0.01$	420963	2.263	4.9	0.932	LARGE

[a] 42 days of transactions
[b] mean time value in sec between 2 trades
[c] measured in basis points

will see in the following these two classes are different from the point of view of the statistical properties of their price changes and returns distributions from small to large time scales of observation.

3.2 Distributions

We start with the study of the shape of distributions of price changes $\Delta p\,(i,n)$ and log-returns $r\,(i,n)$ as a function of the value of the tick-to-price ratio. In this qualitative discussion we refer to price changes and returns computed in trade time because we want to describe only the differences between discrete distributions and distributions defined on a continuous support. Our findings are the same for the continuous time case. We want to show that the effect of a discrete tick size is more substantial for a high tick-to-price ratio. We choose to show the results for AAPL and MSFT stocks because they exemplify the two types of qualitatively different behavior.

The first important observation is the presence of price changes clustering when we observe the process in trade time or in continuous time. The price change clustering is the phenomenon for which we have an uneven use of price fractions of the price grid. In Fig. 2 we show the histogram of price changes at an aggregation scale $n = 128$ for a large and a small tick asset. A large tick asset has a distribution of price changes in which odd values are less populated then even values. Our empirical observations indicate that the process $\Delta p\,(i,n)$ shows clustering for each value of n in the case of large tick assets. For example if we observe the process $\Delta p\,(i,n = 8192)$ the clustering is still present and 8192 transactions are a significant part of the total transactions that we could have in one day of trade, e.g. in the case of MSFT they correspond to an average execution time of four hours. So this effect is not only visible at high frequency time scales, and we want to stress that its origin comes from the price dynamics that we observe at the scale of single transactions.

The effect of price changes clustering is not present at all in the case of a small tick-to-price ratio. From the smallest time scale to the largest, i.e. from 1 to 8192 transactions, we observe a usual occupation of even and odd levels of price changes.

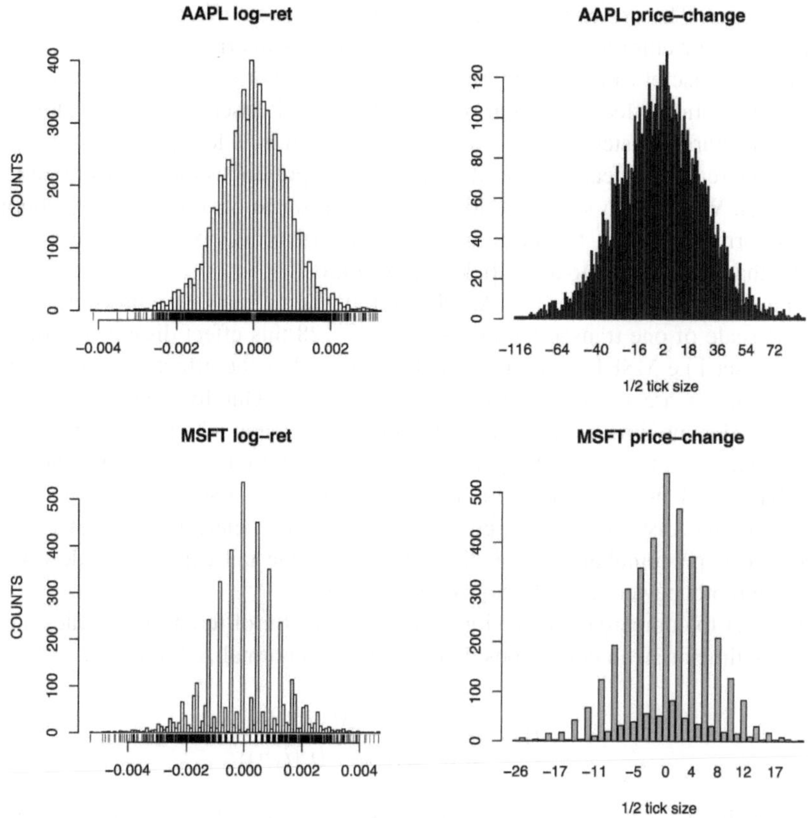

Fig. 2 Histograms of price changes $\Delta p\,(i, n = 128)$ and log-returns $r\,(i, n = 128)$ for AAPL and MSFT stocks. Price changes for MSFT are clustered on even values. The effect of clustering for log-returns is clearly visible

When we refer to a usual occupation we mean an absence of systematic differences in populations of price changes and a presence of a smooth discrete distribution like a binomial or a Poisson distribution. The differences in the shape of price changes distributions between small and large tick size assets are clear in Fig. 2 on the right column of the panel. The observation of clustering at a daily time scale is already known in literature (Onnela et al. 2009; Münnix and Schäfer 2010; Harris 1991) but it is not clearly connected to a measure of the effective tick size. Our observations, instead, connect this property of prices directly to the effective tick.

In this way we conclude that we do not have a universal shape of distributions of price changes, but we have a dependence from an effective tick size, measured by T_r or T_s. There is a growing consensus that distributional properties of returns are quite universal, i.e. the shape of the distribution is the same for all the assets, especially for relatively large time scales (Cont 2001). How can we reconcile the empirical observation of price change clustering with a universal shape of returns distributions? The presence of a discrete tick size has an effect on the distribution of returns. Our empirical analysis shows that returns distributions are affected by an effect of discretization and by returns clustering.[1] For small tick size asset only the discretization effect is present, while for large tick asset there is the additional effect of returns clustering. The effect of discretization is less present as the time scale n increases, instead returns clustering is less present as the tick-to-price ratio decreases. We want to stress the idea that the presence of discretization, coupled with returns clustering, for an high tick-to-price ratio disappears at time scales n higher than that relative to a low tick-to-price ratio.

In a small tick size asset like AAPL discretization effects on returns are visible at time scale of one transaction but when $n \approx 128$ this effect disappears. Instead for an asset like MSFT at the time scale of $n \approx 128$ the effect of discretization and clustering are present as we can see in Fig. 2. Our hypothesis is that we should find some time scale in trade or continuous time at which the discretization and clustering of returns disappear and we could find a universal shape for distributions of price returns. This means that we should study the properties of scaling of the distributions of returns. We made this analysis by means of the empirical hypercumulants Λ_q of distributions and the tail exponent α describing the asymptotic power-law behavior of distributions.

In order to compare the behavior of distributions for different time scales, i.e. n for trade time or Δt in continuous time, we define a normalized return g:

$$g(i,n) = \frac{r(i,n) - \langle r(i,n) \rangle}{\sqrt{\langle r^2(i,n) \rangle - \langle r(i,n) \rangle^2}}. \tag{1}$$

The definition for the continuous case is similar. We analyze the scaling by the moments defined by a fractional index q, i.e. the hypercumulants (Bouchaud and Potters 2009; Gopikrishnan 1999; Plerou and Stanley 2007), of the distributions of normalized returns $g(i,n)$:

$$\Lambda_q(n) = \langle |g(i,n)|^q \rangle, \tag{2}$$

in this way this quantity is defined as a function of the time scale n.

We estimate also the tail exponent α for the normalized returns g. This exponent describes the asymptotic power-law behavior of probability density functions in the following way:

$$P(x) \sim x^{-(1+\alpha)}, \tag{3}$$

[1] Notice that for returns the discretization effect is different from clustering: discretization is a consequence of the fact that price is defined on a grid, while clustering denotes the preference for some price variations over others.

where $\alpha > 0$. This estimate gives information on the existence of the moments of a given distribution. A necessary condition for the qth moments to exist is that the probability density $P(x)$ should decay faster than $1/|x|^{q+1}$ for $|x|$ going towards infinity, then all the moments such that $q > \alpha$ are infinite. The asymptotic behavior of the density $P(x)$ is also connected to properties of random variables under summation. Consider the sum $S_m = \sum_{i=1}^{m} x_i$ of independent identically distributed (i.i.d) random variables x_i. If the x_i's have finite second moments, the central limit theorems holds and S_m is distributed as a Gaussian in the limit $m \to \infty$. If the random variables x_i are characterized by a distribution having asymptotic power-law behavior like that in Eq. (3) where $0 < \alpha < 2$, then S_m converges to a Lévy stable stochastic process of exponent $0 < \alpha < 2$ in the limit $m \to \infty$.

A common problem when studying a distribution that decays as a power-law is how to obtain an accurate estimate of the exponent characterizing its asymptotic behavior. We use a method developed by Clauset et al. (2009) in order to estimate the exponent α and the value of x ,i.e. x_{min}, beyond which we have the power-law behavior. Their approach combines maximum-likelihood fitting methods with goodness-of-fit tests based on the Kolmogorov-Smirnov statistics and likelihood ratios. They studied the following probability density function defined on $x > x_{min}$:

$$P(x) = \frac{\alpha - 1}{x_{min}} \left(\frac{x}{x_{min}} \right)^{-\alpha}, \tag{4}$$

where x_{min} is for us the lower bound of power-law behavior and $\alpha > 1$. Here we should make attention about the range of values of α because the power-law exponent of Eq. (4) correspond to $\alpha + 1$ of Eq. (3). They use the well-known Hill maximum likelihood estimator:

$$\hat{\alpha} = 1 + m \left[\sum_{i=1}^{m} \ln \frac{x_i}{x_{min}} \right]^{-1}, \tag{5}$$

where $x_i,\ i = 1, \cdots, m$, are the observed values of x such that $x_i > x_{min}$.

The Hill estimator is known to be asymptotically normal and consistent, i.e. $\hat{\alpha} \to \alpha$ in the limit of large m.

We compute the hypercumulants for our NASDAQ stocks at different time scales. In trade time we use for $\Lambda_q(n)$ the following set of values for n: 1, 4, 8, 16, 32, 64, 128, 256, 512, 1024, 2048, 4096, 8192. In continuous time we use for $\Lambda_q(\Delta t)$ the following set of values for Δt measured in seconds: 1, 2, 5, 15, 30, 60, 120, 240, 480, 960, 1800, 3600 and 7200. Since the mean number of transactions in one day ranges from 12000 to 21000, the highest values of n represent a significant fraction of the entire daily transactions. In this way we can study the hypercumulants of returns process from the scale of one transaction to a significant fraction of the daily scale. In continuous time we investigate from the scale of 1 s to 2 h.

We can observe from Figs. 3 and 4 that the behavior of the hypercumulants for the two classes of assets, i.e. large and small effective tick size, is different at small

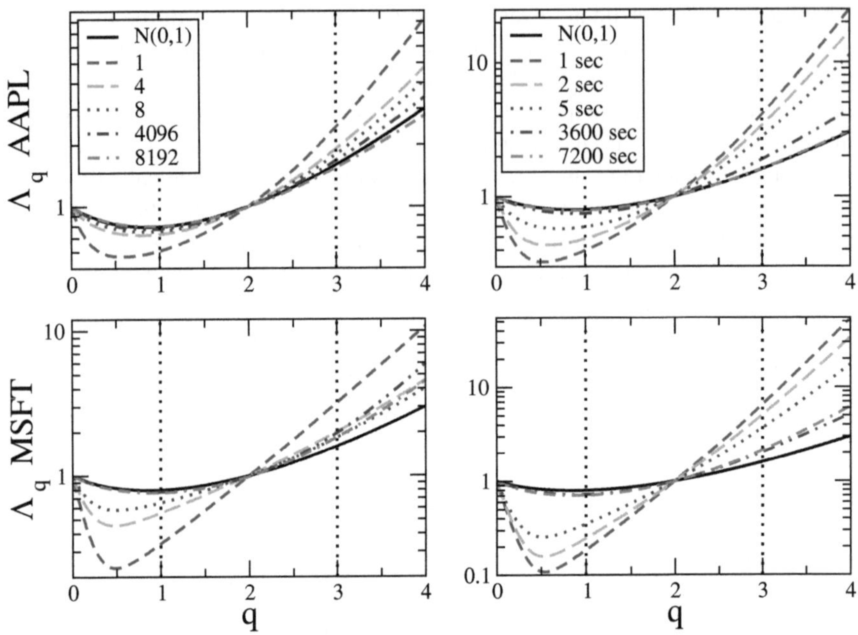

Fig. 3 Linear-log plot of the scaling of hypercumulants Λ_q of normalized returns g for stocks AAPL and MSFT, i.e. respectively a small and a large tick size asset. On the left we have the case of normalized returns $g(i, n)$ in trade time and on the right the case of normalized returns $g(t, \Delta t)$ in continuous time. The different style of lines indicates the different time scales Δt. The *vertical lines* indicate the values of q, i.e. $q = 1, 3$, for which we illustrate the time scale dependence $\Lambda_q(n)$ or $\Lambda_q(\Delta t)$ in Fig. 4

time scales but that they seem to converge at higher time scales. In these figures we compute the corresponding values of Λ_q for the standard Gaussian distribution in order to control the convergence of distributions of price returns as a function of the time scale of aggregation n or Δt. We show the dependence of the hypercumulants from the power index q in Fig. 3 for some fixed values of time scales. When we have a large effective tick asset, the convergence of normalized returns g to a Gaussian behavior is slower with respect to returns computed in presence of a small effective tick asset. A possible motivation could be the presence of clustering of returns for large tick size assets, i.e. we have a distortion of distribution that is almost absent for the case of small tick size assets. We make this hypothesis on the basis of the computed histograms of distributions of returns. Here we can observe that the discretization effects starts to disappear after $n = 32$ for a small tick size stock, instead this threshold is higher for large tick size stocks (where we have also the presence of clustering), i.e. we find a value of $n = 512 \div 1024$. If we observe the returns process in continuous time the discretization effect disappears after a time scale around $\Delta t \simeq 30$ s for small tick stocks, instead for a large tick stock it starts to disappear after around 16 min.

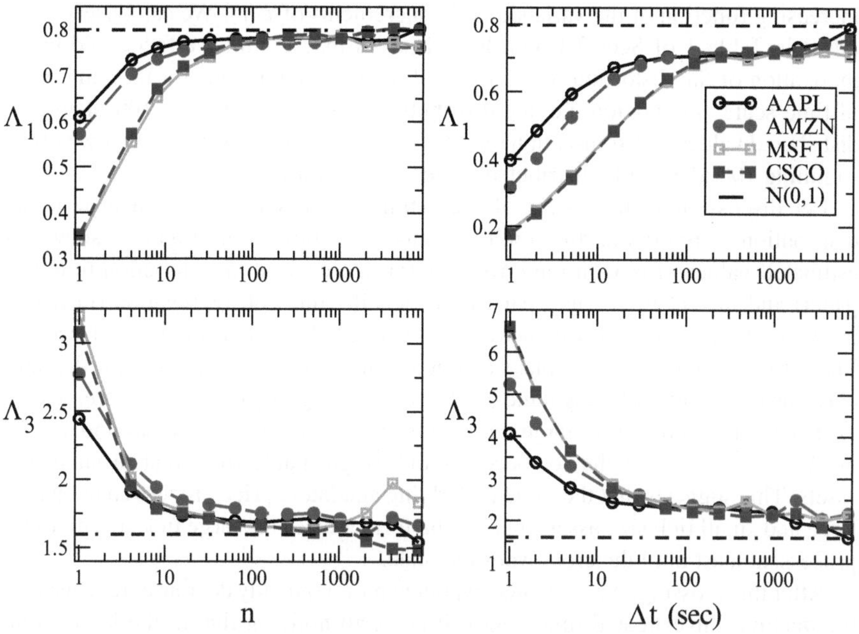

Fig. 4 Log-linear plot of the scaling of hypercumulants $\Lambda_{q=1,3}(n)$ and $\Lambda_{q=1,3}(\Delta t)$ for stocks: AAPL, AMZN, MSFT and CSCO. We observe a different speed of convergence to a Gaussian behavior between small and large tick size stocks. There is also a different behavior if we observe the price process in trade time (*left panels*) or continuous time (*right panels*)

The behavior of Λ_q as a function of trade time or of continuous time is showed in Fig. 4 for two values of q. We observe that, independently of the effective tick size, in continuous time the convergence toward a Gaussian behavior is slower with respect to trade time. This observation may be explained by the subordination hypothesis. The original idea dates back to a paper by Mandelbrot and Taylor (1967) that was later developed by Clark (1973a,b). Mandelbrot and Taylor proposed that prices could be modeled as a subordinated random process $Y(t) = X(\tau(t))$, where Y is the random process generating returns, X is a Brownian motion and $\tau(t)$ is a stochastic time clock whose increments are i.i.d. and uncorrelated with the process X. Clark hypothesized that the time clock $\tau(t)$ is the cumulative trading volume in time t, but more recent works indicated that the number of transactions, i.e. trade time, is more important than their size (Ane and Geman 2000). Gillemot et al. (2006) and La Spada et al. (2011) showed that the role of the subordination hypothesis in fat tails of returns is strongly dependent on the tick size. From our point of view it is important to stress that the stochastic clock $\tau(t)$ could modify the moments of distributions for the increments ΔX and ΔY. For example Clark (1973a) showed that if X is a Gaussian stochastic process with stationary independent increments, and $\tau(t)$ has stationary independent positive increments with finite second moment which are independent from X, then the kurtosis of the increments of $X(\tau(t))$ is an increasing function of the variance of the increments of $\tau(t)$. We can observe such effect in Table 1 of Sect. 3.1. Our hypothesis is that a similar effect could be the motivation of the distortion of the value of the hypercumulants for the same value of time scale aggregation, i.e. the distribution of the sum of $n = 100$ identically distributed values of 1 transaction returns is not the same distribution that we obtain summing 100 identically distributed values of 1 s returns.

Let discuss now the results of the estimation of the tail exponent α of the distribution of returns at different time scales of aggregation. Figure 5 shows the estimated values of α with the error bars for aggregation in transaction time (left panel) and in real time (right panel). For small values of aggregation (in real or transaction time) a clear difference appears between large and small tick size assets. The former type of assets displays a large estimated value of α, while for the latter class the exponent is already quite small. When the aggregation scale increases, the estimated tail exponent for large tick size assets rapidly decays and around $n \simeq 30$ or $\Delta t \simeq 30$ s their behavior becomes indistinguishable from that of small tick assets. This suggests that the same underlying and latent price process characterizes large and small tick size assets, but for the former class the large tick size hides the process, at least until the crossover time scale.

After this crossing, the estimated exponent monotonically decreases (at least until the maximal investigated time scale). It is worth noticing that at the largest time

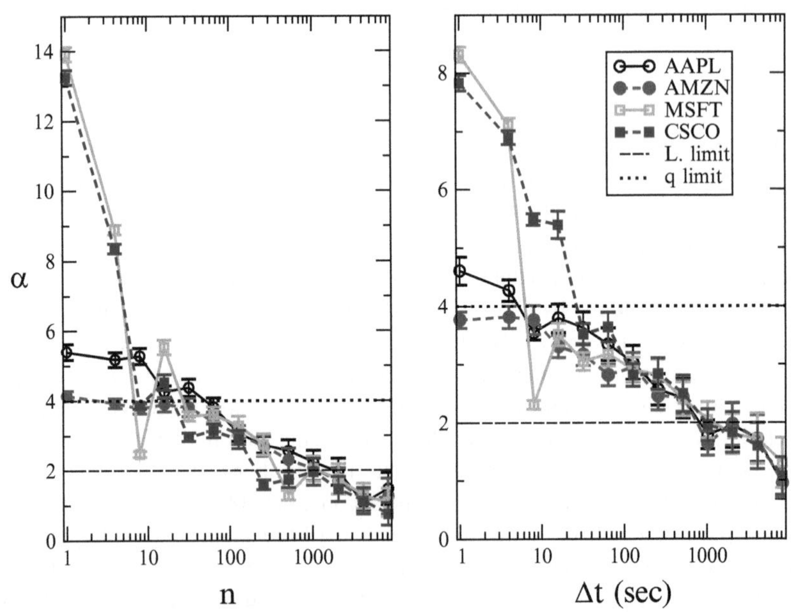

Fig. 5 Log-linear plot of the scaling of asymptotic tail exponent α, defined in Eq. (3), for distributions of normalized returns g as a function of trade time n and continuous time Δt for stocks: AAPL, AMZN, MSFT and CSCO. The *straight dashed line* $\alpha = 2$ is the upper bound of Lévy behavior, i.e. $0 < \alpha < 2$. The *horizontal dotted line* is the upper bound of the index for which we computed Λ_q, $q \in [1, 4]$

scales of aggregation the estimated exponent, both in real and in transaction time, is smaller than 2, indicating a Levy-like regime. Before commenting this result it is important to stress that the values shown in Fig. 5 are *estimated* exponents, i.e. there is no guarantee that the distributions have a true power law tail. Clauset et al. (2009) algorithm gives the best estimate of α and x_{min} *assuming* that the tail is power law. Moreover as noticed in Clauset et al. (2009) when the sample is small, the method can give incorrect estimations. This is the case of the last two points of Fig. 5, where the sample size is around 100 data points. These two considerations highlight the problems that could arise when estimating the tail exponent of a distribution which has a discrete (and finite) support. In this case the tail is obviously not power law, but the method gives in any case an estimated value. In fact numerical simulations of iid models with finite support and thus finite variance (the model i.i.d. discussed in the next section) display a similar behavior of the estimated tail exponent, including a value smaller than 2 for large aggregation. This is clearly a misestimation result.

In conclusion, the analysis of the tail exponent shows two regimes, one in which large and small tick size assets show a markedly different behavior and one where tick-to-price ratio does not play any role. Moreover numerical simulations and empirical analyses suggest to be very cautious when estimating tail exponent of a distribution that is either defined on a discrete support (as for price changes) or has an hidden discretization (as for log-returns). Arbitrarily small values of the exponent could be (mis)estimated as a result of an improper use of statistical methods.

4 Statistical Models for Large Tick Assets

In this section we present briefly the statistical models recently introduced by Curato and Lillo (2013) describing the high frequency dynamics of price changes for a large tick size asset in trade time. We want to show that these models are able to reproduce the phenomenon of clustering for log-returns and the scaling of hypercumulants $\Lambda_q(n)$ in trade time.

The building blocks of these models are simple: the distribution of price changes caused by 1 transaction, i.e. $\Delta p\,(i, n = 1)$, and the statistical properties of the dynamics of the bid-ask spread $s\,(i)$. In our model we impose a coupling between the process of the price changes and of the spread in order to reproduce the price-change clustering.

We consider first a benchmark model, hereafter called i.i.d. model, in which this coupling is absent and where we use only the information contained in the distribution of $\Delta p\,(i, n = 1)$.[2] Our empirical analysis indicates that for a large tick asset the distribution of Δp is mainly concentrated on $\Delta p = 0$. This observation allows us to limit the discrete set on which we define the distribution at the scale of 1 transaction, i.e. $\Delta p \in \{-2, -1, 0, 1, 2\}$ in units of half tick size. In the i.i.d.

[2] Hereafter we use $\Delta p\,(i)$ or Δp instead of $\Delta p\,(i, n = 1)$.

model the $\Delta p\,(i)$ process is simply an i.i.d. process in which each observation has the distribution estimated from data. Numerical simulations and analytical considerations show that this model is unable to reproduce price change clustering at any scale, i.e. when we aggregate n values we recover a bell shaped distribution for $\Delta p\,(i,n) = \sum_{k=1}^{n} \Delta p\,(i)$.

Our solution to recover price changes clustering is to use the process of spread $s\,(i)$. The key intuition behind our modeling approach is that for large tick assets the dynamics of mid-price and of spread are intimately related and that the process of price changes is conditioned to the spread process. For large tick assets the spread can assume only few values. For example, for MSFT and CSCO spread size is only 1 or 2 ticks. The discreteness of mid-price dynamics can be connected to the spread dynamics if we observe that when the spread is constant in time, price changes can assume only even values in units of half tick size. Instead when the spread changes, price changes can display only odd values. This effect is visible in Fig. 2 for MSFT stock where even values of price change are more populated than odd values, because spread changes are relatively rare. The dynamics of price changes is thus linked to dynamics of spread transitions. It is well known that spread process is autocorrelated in time (Ponzi et al. 2009; Plerou et al. 2005; Dayri and Rosenbaum 2013). In our models the spread process $s\,(i)$ is represented by a stationary Markov(1) process:

$$P\,(s\,(i) = k|s\,(i-1) = j, s\,(i-2) = l, \cdots) = P\,(s\,(i) = k|s\,(i-1) = j) = p_{jk}, \tag{6}$$

where $j, k, l \in \{1, 2\}$ are spread values and $i \in \mathbb{N}$ is the trade time. The spread process is described by the transition matrix $S \in M\,(2,2)$:

$$S = \begin{pmatrix} p_{11} & p_{12} \\ p_{21} & p_{22} \end{pmatrix}$$

where the normalization is given by $\sum_{k=1}^{2} p_{jk} = 1$. For example for CSCO we estimate $\hat{p}_{11} = 0.97$ and $\hat{p}_{21} = 0.58$, i.e. the transitions in which the spread changes are not frequent. In this model the spread could assume 2 values so we could have 4 possible transitions $t\,(i)$ between two subsequent transactions, that we could identify with an integer number from 1 to 4. For example, the transition $s\,(i) = 1 \rightarrow s\,(i+1) = 1$ is described by the state $t\,(i) = 1$, etc. In this way we can derive a new Markov(1) process that describe the process $t\,(i)$. At this point the mechanistic constraint imposed by a price grid, defined by the value of the tick size, allows us to couple the price changes $\Delta p\,(i)$ with the process of transitions $t\,(i)$. In this way we are able to define a Markov-switching model (Hamilton 2008) for price changes $\Delta p\,(i)$ conditioned to the Markov process $t\,(i)$ by the conditional probabilities:

$$P\,(\Delta p\,(i)\,|t\,(i) = m)\,, \tag{7}$$

where $m \in \{1, 2, 3, 4\}$. The estimation of such conditional probabilities enable us to simulate the process for price changes. In order to compute log-returns we follow a simple procedure. We generate the simulated series of price changes from the Markov-switching model calibrated from data, then we integrate it choosing as starting point the first mid-price recorded on the measured series. At this point we have a synthetic discrete series of mid-price on which we can compute log-returns $r(i, 1)$ correspondent to 1 transaction. Then we aggregate individual transaction returns on non-overlapping windows of width n to recover the process at a generic time scale n. This model is able to reproduce clustering for price changes and for log-returns. In fact as we can observe in Fig. 6 this model reproduces the returns clustering at different time scales. The clustering starts to disappear beyond the time scale of aggregation $n = 512$.

The Markov-switching model is not able to explain the empirically observed correlation of squared price changes, that is related to the presence of volatility clustering. Usually in financial econometrics an autoregressive conditional heteroskedasticity model (ARCH) (Bera and Higgins 1993; Engle et al. 2008) can account for volatility clustering and non-Gaussianity of returns. We do not make use of this class of models because they are defined by continuous stochastic variables. Instead, we have seen in Sect. 3.2 that the high frequency return distribution is characterized by the presence of discretization and clustering. For this reason we have chosen to define our model directly on discrete variables as price changes, but this choice prevents us from using models like ARCH. Therefore in Curato and Lillo (2013) we develop a second model based on an autoregressive switching model for price changes, which preserves the ideas of ARCH type models that past squared returns affect current return distribution. This means that the conditional

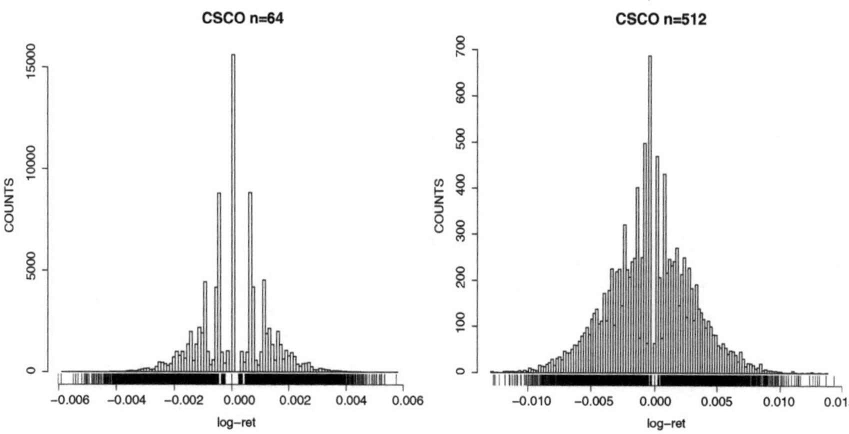

Fig. 6 Histograms of log-returns computed from the data generated by the Markov-switching model defined by Eqs. (6), and (7). On the left we have a time scale of aggregation $n = 64$, on the right we have $n = 512$. We could observe that the effect of clustering is less and less present for higher values of n

probabilities of Eq. (7) can depend not only on the last spread transition, but also on the recent past values of price changes. In the case in which regressors are defined only by past squared price changes, our model can be viewed as a higher-order double chain Markov model of order p (Berchtold 1999). For our purposes we remember only that we have fitted this model for an order $p = 50$ on our data. Here we use it to generate a series in order to study the scaling of hypercumulants $\Lambda_q(n)$. We refer to our original article (Curato and Lillo 2013) for the details and definitions for this autoregressive model.

In order to fit our three models we split our daily time series in two series, the first displays low volatility instead the second displays high volatility. Here we focus on the low volatility series that starts at 10:30 and ends at 15:45, but our findings are the same for the series with high volatility. We compute the log-returns from the Montecarlo simulations of our models and then compute the normalized log-return $g(i,n)$ in trade time. The Montecarlo simulations generate 5961600 data points that correspond to 1 year of transactions for a mean duration time, i.e. the interval of time between two transaction, of 1 s and 6 h of trading activity each day. The sample for the stock CSCO instead covers 2 months of trading for a total of 275879 transactions. We can observe from the Figs. 7 and 8 that the proposed models converge to a Gaussian behavior. The Markov-switching model and the double chain

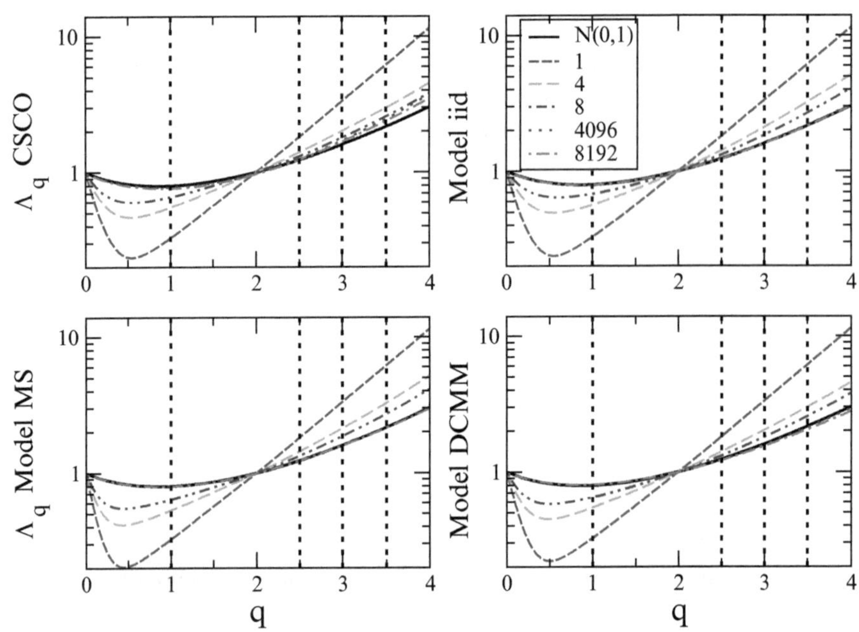

Fig. 7 Linear-log plot of the scaling of hypercumulants Λ_q of normalized returns $g(i,n)$ for the stock CSCO and the correspondent simulated returns processes. The different style of lines indicates the different time scales n. The *vertical dotted lines* indicate the values of q, i.e. $q = 1, 2.5, 3, 3.5$, for which we illustrate the time scale dependence $\Lambda_q(n)$ in Fig. 8

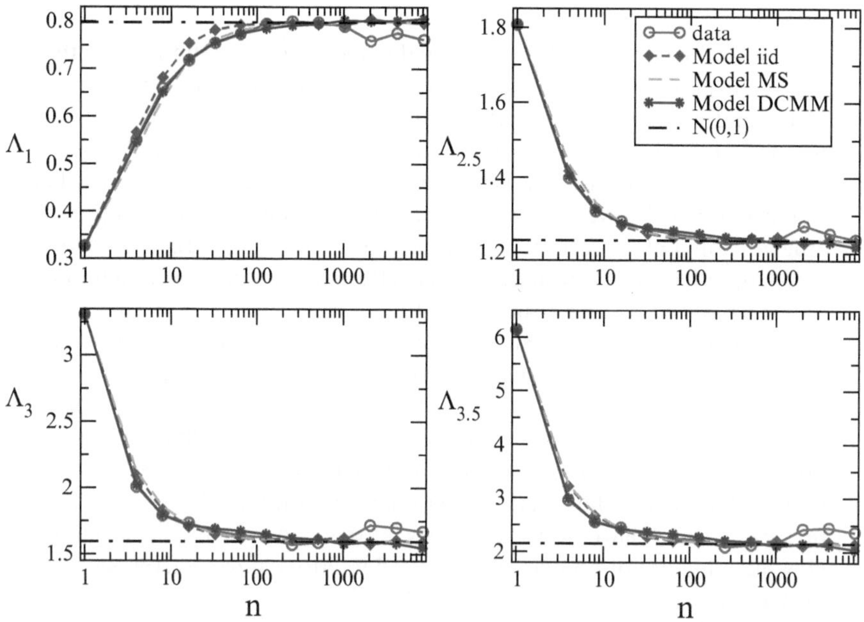

Fig. 8 Log-linear plot of the scaling of hypercumulants $\Lambda_{q=1,2.5,3,3.5}(n)$ for the stock CSCO and the related simulated returns processes

Markov model, i.e. DCMM in the figures, reproduce slightly better the scaling of the hypercumulants respect to the simple i.i.d. model, although all models are very close to the empirical moments. We could observe a little difference in Fig. 8 for high values of n between simulation and real data. We think that this distortion rises from the reduced number of the data sample used to compute Λ_q for the stock CSCO with respect to the number of data points available for simulated data. The Markov-switching model results to be the simplest model able to reproduce at the same time the clustering of price changes and log-returns together with the correct scaling of hypercumulants toward a Gaussian behavior.

Conclusions

In this work we show empirically that the effective tick size affects strongly the statistical properties of distributions of mid-price changes and log-returns of stocks traded on a decimalized market like NASDAQ. The effect of clustering of price changes on even values of the price grid is particularly strong for stocks with a large effective tick size. The clustering of price changes persists at each time scale in presence of a large effective tick. This effect is absent in presence of a small effective tick. On the other hand, the

(continued)

discrete nature of prices affects also the distribution of log-returns. The effect of discretization for log-returns disappears when the time scale of observation of the mid-price process is high enough, both for large effective tick than for small effective tick. The difference between the two kinds of stock is that for a large tick asset we have also the effect of returns clustering. For example, if we observe the price process in continuous time for a time scale around 15 min, the effect of discretization and clustering starts to disappear for a large effective tick, instead for small effective tick sizes the discretization disappears at shorter time scales, i.e. 30 s.

The analysis of the hypercumulants of distributions of log-returns indicates a converge toward a Gaussian behavior for stocks with large and small effective tick sizes. The presence of a large effective tick seems to slow down the convergence toward a Gaussian behavior, coherently with a progressive disappearance of clustering of returns. The analysis of convergence by means of the Hill estimator shows a crossover time scale after which large and small tick size assets behave in the same way. However our analysis characterize how the blind use of estimators could lead to errors in the determination of the tail exponents.

We develop statistical models in trade time for large tick size asset that are able to reproduce the presence of clustering for price changes and log-returns. We find that in order to reproduce the clustering effect we need a model in which the dynamics of mid-price changes is coupled with the dynamics of the bid-ask spread. A Markov-switching model, where the switching process is defined by the possible transitions between subsequent values of the spread, is able to reproduce the effect of clustering and the scaling of hypercumulants computed from empirical data. This simple high frequency statistical microstructural model, defined by quantities like bid and ask prices on a discrete price grid determined by the value of the tick size, is able to recover a Gaussian behavior for returns at the macroscopic time scale of the hours.

Acknowledgements Authors acknowledge partial support by the grant SNS11LILLB Price formation, agents heterogeneity, and market efficiency.

References

Ahn, H. J., Cai, J., Chan, K., & Hamao, Y. (2007). Tick size change and liquidity provision on the Tokyo Stock Exchange. *Journal of Japanese and International Economies, 21*, 173–194.

Ane, T., & Geman, H. (2000). Order flow, transaction clock and normality of asset returns. *The Journal of Finance, 55*, 2259–2284.

Ascioglu, A., Comerton-Forde, C., & McInish, T. H. (2010). An examination of minimum tick sizes on the Tokyo Stock Exchange. *Japan and the World Economy, 22*, 40–48.

Bera, A. K., & Higgins, M. L. (1993). Arch models: properties, estimation and testing. *Journal of Economic Surveys, 7*(4), 305–366.
Berchtold, A. (1999). The double chain Markov model. *Communications in Statistics - Theory and Methods, 28*(11), 2569–2589.
Bessembinder, H. (2000). Tick size, spreads, and liquidity: An analysis of Nasdaq Securities Trading near Ten Dollars. *Journal of Financial Intermediation, 9*, 213–239.
Bessembinder, H. (1997). The degree of price resolution and equity trading costs. *Journal of Financial Economics, 45*, 9–34.
Bessembinder, H. (1999). Trade execution costs on NASDAQ and the NYSE: A post-reform comparison. *The Journal of Financial and Quantitative Analysis, 34*(3), 387–407.
Bessembinder, H. (2003). Trade execution costs and market quality after decimalization. *The Journal of Financial and Quantitative Analysis, 38*(4), 747–777.
Bollen, N. P. B., & Busse, J. A. (2006). Tick size and institutional trading costs: evidence from mutual funds. *The Journal of Financial and Quantitative Analysis, 41*(4), 915–937.
Bouchaud, J. P., & Potters, M. (2009). *Theory of financial risk and derivative pricing: from statistical physics to risk management*, (2nd ed.). Cambridge, UK: Cambridge University Press.
Christie, W. G., & Schultz, P. H. (1994). Why do NASDAQ market makers avoid odd-eighth quotes? *The Journal of Finance, 49*(5), 1813–1840.
Christie, W. G., Harris, J. H., & Schultz, P. H. (1994). Why did NASDAQ market makers stop avoiding odd-eighth quotes? *The Journal of Finance, 49*(5), 1841–1860.
Chung, K. H., Chuwonganant, C., & McCormick, D. T. (2004). Order preferencing and market quality on NASDAQ before and after decimalization. *Journal of Financial Economics, 71*, 581–612.
Clark, P. K. (1973a). A subordinated stochastic process model with finite variance for speculative prices. *Econometrica, 41*(1), 135–155.
Clark, P. K. (1973b). Comments on: a subordinated stochastic process model with finite variance for speculative prices. *Econometrica, 41*(1), 157–159.
Clauset, A., Shalizi, C. R., & Newman, M. E. J.(2009). Power-law distributions in empirical data. *SIAM Review, 51*(4), 661–703.
Cont, R. (2001). Empirical properties of asset returns: stylized facts and statistical issues. *Quantitative Finance, 1*, 223–236.
Curato, G., & Lillo, F. (2013). Modeling the coupled return-spread high frequency dynamics of large tick assets. Retrieved from http://arxiv.org/abs/1310.4539.
Dacorogna, M. M., Gençay, R., Müller, U. A., Olsen, R. B., & Pictet, O. V. (2001). *An introduction to high-frequency finance.* San Diego, California, USA: Academic Press.
Dayri, K., & Rosenbaum, M. (2013). Large tick assets: implicit spread and optimal tick size. Retrieved from http://arxiv.org/abs/1207.6325.
Dayri, K. A., Bacry, E., & Muzy, J. F. (2011). The nature of price returns during periods of high market activity. In F. Abergel, B. K. Chakrabarti, A. Chakraborti & M. Mitra (Eds.), *Econophysics of order-driven markets* (pp. 155–172). Milano, Italy: Springer-Verlag Italia.
Eisler, Z., Bouchaud, J. P., & Kockelkoren, J. (2012). The price impact of order book events: market orders, limit orders and cancellations. *Quantitative Finance, 12*(9), 1395–1419.
Engle, R. F., Focardi, S. M., & Fabozzi, F. J. (2008). ARCH/GARCH models in applied financial econometrics. In F. J. Fabozzi (Ed.), *Handbook of Finance.* New York, NY: John Wiley & Sons.
Gibson, S., Singh, R., & Yerramilli, V. (2003). The effect of decimalization on the components of the bid-ask spread. *Journal of Financial Intermediation, 12*, 121–148.
Gillemot, L., Farmer, J. D., & Lillo, F. (2006). There's more to volatility than volume. *Quantitative Finance, 6*, 371–384.
Goldstein, M. A., & Kavajecz, K. A. (2000). Eights, sixteenths, and market depth: changes in tick size and liquidity provision on the NYSE. *Journal of Financial Economics, 56*, 125–149.
Gopikrishnan, P., Plerou, V., Amaral, L. A. N., Meyer, M., & Stanley, H. E. (1999). Scaling of the distribution of fluctuations of financial market indices. *Physical Review E, 60*(5), 5305–5316.
Hamilton, J. D. (2008). Regime-Switching models. In S. N. Durlauf, & L. E. Blume (Eds.), *The new palgrave dictionary of economics*, 2nd edn. Basingstoke, UK: Palgrave Macmillan.

Harris, L. (1991). Stock price clustering and discreteness. *The Review of Financial Studies, 4*(3), 389–415.
Hautsch, N. (2012). *Econometrics of financial high-frequency data.* Berlin: Springer-Verlag.
He, Y., & Wu, C. (2004). Price rounding and bid-ask spreads before and after decimalization. *International Review of Economics and Finance, 13*, 19–41.
Huang, R. D., & Stoll, H. R. (2001). Tick size, bid-ask spreads, and market structure. *The Journal of Financial and Quantitative Analysis, 36*(4), 503–522.
La Spada, G., Farmer, J. D., & Lillo, F. (2011). Tick size and price diffusion. In F. Abergel, B. K. Chakrabarti, A. Chakraborti & M. Mitra (Eds.), *Econophysics of order-driven markets* (pp. 173–187). Milano, Italy: Springer-Verlag Italia.
Loistl, O., Schossmann, B., & Veverka, A. (2004). Tick size and spreads: The case of Nasdaq's decimalization. *European Journal of Operational Research, 155*, 317–334.
MacKinnon, G., & Nemiroff, H. (2004). Tick size and the returns to providing liquidity. *International Review of Economics and Finance, 13*, 57–73.
Mandelbrot, B., & Taylor, H. M. (1967). On the distributions of stock price differences. *Operations Research, 15*(6), 1057–1062.
Münnix, M. C., Schäfer, R., & Guhr, T. (2010). Impact of the tick-size on financial returns and correlations. *Physica A, 389*, 4828–4843.
Onnela J. P., Töyli, J., Kaski, K. (2009). Tick size and stock returns. *Physica A, 388*, 441–454.
Osborne, M. F. M. (1962). Periodic structure in the Brownian Motion of stock prices. *Operations Research, 10*(3), 345–379.
Plerou, V., Gopikrishnan, P., Amaral, L. A. N., Meyer, M., & Stanley, H. E. (1999). Scaling of the distribution of price fluctuations of individual companies. *Physical Review E, 60*(6), 6519–6529.
Plerou, V., Gopikrishnan, P., & Stanley, H. E. (2005). Quantifying fluctuations in market liquidity: Analysis of the bid-ask spread. *Physical Review E, 71*, 046131-1/046131-8.
Plerou, V., & Stanley, H. E. (2007). Test of scaling and universality of the distributions of trade size and share volume: Evidence from three distinct markets. *Physical Review E, 76*, 046109-1/046109-10.
Ponzi, A., Lillo, F., & Mantegna, R. N. (2009). Market reaction to a bid-ask spread change: A power-law relaxation dynamics. *Physical Review E, 80*, 016112-1/016112-12.
U.S. Securities and Exchange Commission. (2012). *Report to Congress on Decimalization.* Retrieved from http://www.sec.gov/news/studies/2012/decimalization-072012.pdf.
Robert, C. Y., & Rosenbaum, M. (2011). A new approach for the dynamics of ultra-high-frequency data: the model with uncertainty zones. *Journal of Financial Econometrics, 9*(2), 344–366.
Wyart, M., Bouchaud, J. P., Kockelkoren, J., Potters, M., & Vettorazzo, M. (2008). Relation between bid-ask spread, impact and volatility in order driven markets. *Quantitative Finance, 8*(1), 41–57.

Market Shocks: Review of Studies

Mariya Frolova

Abstract This paper gives a brief description of the current state of research on market shocks, presents its main results and denotes problems researchers are faced with. We consider such aspects of shocks analysis as price formation mechanism, origins of market jumps, price–volume relationship, cojumps, empirical description of financial markets around shocks, shocks identification.

Keywords Cojumps • Market shocks • Price formation mechanism • Shocks identification

1 Price Formation Mechanism

The aim to understand the price formation mechanism is not novel. It is well known that price process of any financial instrument follows a stochastic-like path: a price path can include or not a deterministic trend; but in any case the price process is smeared by noise movements. The noise movements are known as market volatility, and they make the price unpredictable. These noise movements can be decomposed into two components: the first component is called regular noise, it represents noise that is frequent but does not bring any abrupt changes, the second component is known as price jumps, it designates rare but very abrupt price movements. The origin of regular noise is in the statistical nature of the markets: any market is a result of the interplay between many different market participants with different incentives, purposes and financial constraints. This interaction of many different agents can be mathematically described as the standard Gaussian distribution (Merton 1976), this assumption allows dealing easy with in mathematical models of the price processes of financial instruments, calculating expectations and establishing various characteristics of financial instruments. The discontinuities in price evolution (price jumps) have been recognized as an essential part of the price time series generated on financial markets. Price jumps can't be fitted by the description of the first noise component and thus have to be modeled

M. Frolova (✉)
Department of Economics, Prognoz Risk Lab, Perm State University, Perm, Russia
e-mail: frolovam@prognoz.ru

A.K. Bera et al. (eds.), *Financial Econometrics and Empirical Market Microstructure*, DOI 10.1007/978-3-319-09946-0_7

on their own, (Merton 1976). But it is worth noting, that the unpredictability of the price movements is not a negative feature, it is rather the nature of financial markets.

Many studies (Andersen et al. 2002; Gatheral 2006) demonstrate that continuous-time models have to incorporate the discontinuous component. Andersen et al. (2002) extend the class of stochastic volatility diffusions by allowing for Poisson jumps of time-varying intensity in returns. However, the problem is the mathematical description of price jumps cannot be easily handled (Pan 2002; Broadie and Jain 2008). The serious problems in the mathematical description of price jumps are very often the reason why price jumps are wrongly neglected. However, the non-Gaussian price movements influence the models employed in finance to estimate the performance of various financial vehicles (Heston 1993; Gatheral 2006). Andersen et al. (2007) conclude that most of the standard approaches in the financial literature on pricing assets assume a continuous price path. Since this assumption is clearly violated in most cases the results tend to be heavily biased.

The literature contains a broad range of ways to classify volatility. Each classification is suitable for an explanation of a different aspect of volatility or an explanation of volatility from a different point of view (see e.g. Harris 2003, where the volatility is discussed from the financial practitioners' points of view). The most important aspect is to separate the Gaussian-like component from price jumps (Merton 1976; Gatheral 2006).

Mathematical finance has developed a class of models that make use of jump processes (Cont and Tankov 2004) and that are used for pricing derivatives and for modeling volatility. Financial econometrics has developed several methods to disentangle the continuous part of the price path from the discontinuous one (Lee and Mykland 2008; Barndorff-Nielsen and Shephard 2006), and the latter is modeled as jumps.

Bormetti et al. (2013) found that, as far as individual stocks are concerned, jumps are clearly not described by a Poisson process, the evidence of time clustering can be accounted for and modelled by means of linear self-exciting Hawkes processes. Clustering of jumps means that the intensity of the point process describing jumps depends on the past history of jumps, and a recent jump increases the probability that another jump occurs. The second deviation from the Poisson model is probably more important in a systemic context. Bormetti et al. find a strong evidence of a high level of synchronization between the jumping times of a portfolio of stocks. They find a large number of instances where several stocks (up to 20) jump at the same time. This evidence is absolutely incompatible with the hypothesis of independence of the jump processes across assets. Authors use Hawkes processes for modeling the dynamics of jumps of individual assets and they show that these models describe well the time clustering of jumps. However they also show that the direct extension of the application of Hawkes processes to describe the dynamics of jumps in a multi-asset framework is highly problematic and inconsistent with data. For this reason, Bormetti et al. introduce Hawkes factor models to describe systemic cojumps. They postulate the presence of an unobservable point process describing a market factor, when this factor jumps, each asset jumps with a given probability, which is different for each stock. In general, an asset can jump also by following an idiosyncratic point process. In order to capture also the time clustering of jumps, they model the point processes as Hawkes processes. Authors show how to estimate this model and discriminate between systemic and idiosyncratic jumps and they claims that the model is able to reproduce both the longitudinal and the cross sectional properties of the multi-asset jump process.

On the opposite, tests applied by Bajgrowicz and Scaillet (2011) do not detect time clustering phenomena of jumps arrivals, and, hence, do not reject the hypothesis that jump arrivals are driven by a simple Poisson process.

The presence of price jumps has serious consequences for financial risk management and pricing. Thus, it is of great interest to describe the noise movements as accurately as possible. Nyberg and Wilhelmsson (2009) discuss the importance of including event risk as recommended by the Basel II accord, which suggests employing a VAR model with a continuous component and price jumps representing event risks.

2 Origins of Market Shocks

It is still not clear what the main source of price jumps is. Price jumps, understood as an abrupt price change over a very short time, are also related to a broad range of market phenomena that cannot be connected to the noisy Gaussian distribution. Researchers agree on the presence of price jumps, but they disagree about the origins. All the explanations are very different in nature. One branch of the literature considers new information as a primary source of price jumps (Lee and Mykland 2008; Lahaye et al. 2009; Cutler et al. 1989). They also show a connection between macroeconomic announcements and price jumps on developed markets. A possible explanation of the source of these jumps says that they originate in the herd behavior of market participants (Cont and Bouchaud 2000; Hirshleifer and Teoh 2003). An illustration of such behavior is a situation when a news announcement is released, and every market participant has to accommodate the impact of that announcement. However, this herding behavior can provide an arbitrage opportunity and can be thus easily questioned. Bajgrowicz and Scaillet (2011) found that majority of news do not cause jumps. One exception is share buybacks announcements, Fed rate news have an important impact but rarely cause jumps. Another finding is that 60 % of jumps occur without any news event. Also authors admit that liquidity pressures are probably another important factor of jumps—for one third of the jumps with no news they found there is unusual behavior in the volume of transactions.

Joulin et al. (2010) and Bouchaud et al. (2004) conclude that price jumps are usually caused by a local lack of liquidity on the market and news announcements have a negligible effect on the origin of price jumps. A hidden liquidity problem is when either the supply or the demand side faces a lack of credit and thus is not able to prevent massive price changes. Madhavan (2000) also claims that the inefficient provision of liquidity caused by an imbalanced market microstructure can cause extreme price movements. Easley et al. (2010) introduced a new metric Volume-

Synchronized Probability of Informed trading (the VPIN) as a real-time indicator of order flow toxicity. Order flow is toxic when it adversely selects market makers, who may be unaware they are providing liquidity at a loss. They find the measure useful in monitoring order flow imbalances and conclude it may help signal impending market turmoil, exemplified by historical high readings of the metric prior to the Flash crash. More generally, they show that VPIN is significantly correlated with future short-term return volatility. In contrast, empirical investigation of VPIN performed by Andersen and Bondarenko (2011) documents that it is a poor predictor of short run volatility, that it did not reach an all-time high prior, but rather after, the Flash crash, and that its predictive content is due primarily to a mechanical relation with the underlying trading intensity.

Filimonov and Sornette (2012) suggests that price dynamics are mostly endogenous and driven by positive feedback mechanisms involving investors' anticipations that lead to self-fulfilling prophecies, as described qualitatively by Soros' concept of "market reflexivity". Filimonov and Sornette introduce a new measure of activity of financial markets that provides a direct access to their level of endogeneity. This measure quantifies how much of price changes are due to endogenous feedback processes, as opposed to exogenous news. They calibrate the self-excited conditional Poisson Hawkes model, which combines exogenous influences with self-excited dynamics, to the E-mini S&P 500 futures contracts traded in the Chicago Mercantile Exchange from 1998 to 2010. They find that the level of endogeneity has increased significantly from 1998 to 2010, with only 70 % in 1998 to less than 30 % since 2007 of the price changes resulting from some revealed exogenous information. Filimonov and Sornette claim that this measure provides a direct quantification of the distance of the financial market to a critical state defined precisely as the limit of diverging trading activity in absence of any external driving. But Hardiman et al. (2013) challenge this study and say that markets are and have always been close to criticality and it is not the result of increased automation of trading. They also note that the scale over which market events are correlated has decreased steadily over time with the emergence of higher frequency trading.

The behavioral finance literature provides other explanations for price jumps. Shiller (2005) claims that price jumps are caused by market participants who themselves create an environment that tends to cause extreme reactions and thus price jumps. Finally, price jumps can be viewed as a manifestation of Black Swans, as discussed by Taleb (2007), where the jumps are rather caused by complex systemic interactions that cannot be easily tracked down. In this view, the best way to understand jumps is to be well aware of them and be ready to react to them properly, instead of trying to forecast them.

Price jumps can also reflect moments when some signal hits the market or a part of the market. Therefore, they can serve as a proxy for these moments and be utilized as tools to study market efficiency (Fama 1970) or phenomena like information-driven trading, see e.g., Cornell and Sirri (1992) or Kennedy et al. (2006). An accurate knowledge of price jumps is necessary for financial regulators to implement the most optimal policies, see Becketti and Roberts (1990) or Tinic (1995).

3 Price–Volume Relationship

The price–volume relationship is one of the most studied in the field of finance when studying price dynamics. One of the oldest models used to study price–volume relationship is the model of Osborne (1959) who models the price as a diffusion process with its variance dependent on the quantity of transaction at that particular moment. Subsequent relevant work can be found in Karpoff (1987), Gallant et al. (1992), Bollerslev and Jubinski (1999), Lo and Wang (2002), and Sun (2003). In general this line of research studies the relationship between volume and some measure of variability of the stock price (e.g., the absolute deviation, the volatility, etc.). Most of these works use models in time, they are tested with low frequency data and the main conclusion is that the price of a specific equity exhibits larger variability in response to increased volume of trades. Engle and Russell (1998) use the Autoregressive Conditional Duration (ACD) model which considers the time between trades as a variable related to both price and volume. Bozdog et al. (2011) study the exception of the conclusion presented in the earlier literature, they do not consider models in time but rather make the change in price dependent on the volume directly. Authors present a methodology of detecting and evaluating unusual price movements defined as large change in price corresponding to small volume of trades. They classify these events as "rare" and show that the behavior of the equity price in the neighborhood of a rare event exhibits an increase in the probability of price recovery. The use of an arbitrary trading rule designed to take advantage of this observation indicates that the returns associated with such movements are significant. Bozdog et al. confirm the old Wall Street adage that "it takes volume to move prices" even in the presence of high frequency trading.

4 Jumps Identification

Before a price jump can be accounted for in an estimation stage, it first has to be identified. Surprisingly, but the literature up to now does not offer a consensus on how to identify price jumps properly. Jumps are identified with various techniques that yield different results.

Generally, a price jump is commonly understood as an abrupt price movement that is much larger when compared to the current market situation. But this definition is too general and hard to define and test. The best way to treat this definition is to define the indicators for price jumps that fit the intuitive definition.

The most frequent approach in the literature is based on the assumption that the price of asset S_t follows stochastic differential equation, where the two components contribute to volatility:

$$dS_t = \mu_t dt + \delta_t dW_t + Y_t dJ_t$$

Table 1 A comparison of two modelling approaches

Jump-diffusion models	Infinite activity models
Must contain Brownian component	Do not necessarily contain Brownian component
Jumps are rare events	The process moves essentially by jumps
Distribution of jump sizes is known	"Distribution of jump sizes is known" do not exist: jumps arrive infinitely often
Perform well for implied volatility smile interpolation	Give a realistic description of the historical price process
Easy to simulate	In some cases can be represented via Brownian subordination, which gives additional tractability

where μ_t is a deterministic trend, δ_t is time-dependent volatility, dW_t is standard Brownian motion and $Y_t dJ_t$ corresponds to the Poisson-like jump process (Merton 1976). The term $\delta_t dW_t$ corresponds to the regular noise component, while $Y_t dJ_t$ corresponds to price jumps, both terms together form the volatility of the market. Based on this assumption for the underlying process, one can construct price jump indicators and theoretically assess their efficiency. Their efficiency, however, deeply depends on the assumption that the underlying model holds. Any deviation of the true underlying model from the assumed model can have serious consequences on the efficiency of the indicators.

Another approach to describe price formation mechanism is to use models with infinite number of jumps in every interval, which are known as infinite activity Levy models. Cont and Tankov (2004) provide detailed description of ways to define a parametric Levy process. Table 1 compares the advantages and drawbacks of these two approaches.

Since the price process is observed on a discrete grid, it is difficult if not impossible to see empirically to which category the price process belongs. The choice is more a question of modelling convenience than an empirical one (Cont and Tankov 2004).

The key role price jumps play in financial engineering triggered interest in the financial econometrics literature, especially how to identify price jumps. Numerous statistical methods to test for the presence of jumps in high-frequency data have been introduced in recent years. Novotný (2010) propose following classification of market shocks filters: the model-independent price jump indicators, which do not require any specific form of underlying price process, and the model-dependent price jump indicators, which assume a specific form of the underlying price process The first group includes such methods as extreme returns, temperature, p-dependent realized volatility, the price jump index, and the wavelet filter, the second—integral and differential indicators based on the difference between the bi-power variance and standard deviation, and the bi-power statistics.

4.1 Model-Independent Price Jump Indicators

1. *Extreme returns indicator*: a price jump occurs at time t if the return at time t is above some threshold. The threshold value can be selected by two ways: it can be selected globally—one threshold value for the entire sample, for example, when the threshold is a given centile of the distribution of returns over the entire data set. Or, it can be selected locally, and consequently, some sub-samples may have different threshold values. A global definition of the threshold allows to compare the behavior of returns over the entire sample, however, the distribution of returns can vary, e.g., the width of the distribution can change due to changes in market conditions, and thus the global definition of the threshold is not suitable to directly compare price jumps over periods with different market conditions. This group is represented by the works of Ait-Sahalia (2004), Ait-Sahalia et al. (2009) and Ait-Sahalia and Jacod (2009a, b). The indicators have well-defined analytic properties; but they do not identify price jumps one by one but rather measure the jumpiness of the given period. These methods are more suitable to assess the jumpiness of ultra-high-frequency data.
2. *Temperature*. Kleinert (2009) shows that high-frequency returns at a 1-min frequency for the S&P 500 and the NASDAQ 100 indices have the property that they have purely an exponential behavior for both the positive as well as negative sides. The distribution can fit the Boltzmann distribution:

$$B(r) = \frac{1}{2T}\exp\left(\frac{-|r|}{T}\right) \tag{1}$$

   where T is the parameter of the distribution conventionally known as the temperature, and r stands for returns. The parameter T governs the width of the distribution; the higher the temperature of the market, the higher the volatility. Kleinert and Chen (2007) and Kleinert (2009) document that this parameter varies slowly, and its variation is connected to the situation on the market.
3. *p-dependent Realized Volatility*. The general definition of the p-dependent realized volatility can be written as:

$$pRV_t^p(r) = \left(\sum_{\tau=t-T+1}^{t} |r_\tau|^p\right)^{\frac{1}{p}} \tag{2}$$

   where the sample over which the volatility is calculated is represented by a moving window of length T (Dacorogna 2001). The interesting property of this definition is that the higher the p is, the more weight the outliers have. Since price jumps are simply extreme price movements, the property of realized volatility can be translated into the following statement: the higher the p is, the more price jumps are stressed. The ratio of two realized volatilities with different p can be thus used as an estimator of price jumps.

4. *Price Jump Index*. The price jump index $j_{T,t}$ at time t (as employed by Joulin et al. 2010) is defined as

$$j_{T,t} = \frac{|r_\tau|}{\frac{1}{T}\sum_{i=0}^{T-1} |r_{t-i}|} \tag{3}$$

where T is the market history employed. Gopikrishnan et al. (1999), Eryigit et al. (2009) and Joulin et al. (2010) take normalized price time series—the normalization differs across these papers—and define the scaling properties of the tails of the distributions. This technique has its roots in Econophysics, it is based on the scaling properties of time series known in physics, see e.g. Stanley and Mantegna (2000).

5. *Wavelet Filter*. The Maximum Overlap Discrete Wavelet Transform (MODWT) filter represents a technique that is used to filter out effects at different scales. In the time series case, the scale is equivalent to the frequency, thus, the MODWT can be used to filter out high frequency components of time series. This can be also described as the decomposition of the entire time series into high- and low-frequency component effects (Gencay et al. 2002). The MODWT technique projects the original time series into a set of other time series, where each of the time series captures effects at a certain frequency scale.

4.2 Model-Dependent Price Jump Indicators

1. *The Difference Between Bi-power Variance and Standard Deviation*
   The method is based on two distinct measures of overall volatility, where the first one takes into account the entire price time movement while the second one ignores the contribution of the model-dependent price jump component. Barndorff-Nielsen and Shephard (2004a) discuss the role of the standard variance in the models where the underlying process follows Eq. (1): the standard variance captures the contribution from both the noise and the price jump process unlike the realized variance, which does not take into account the term with price jumps. It is called the realized bi-power variance. The difference between the standard and the bi-power variance can be used to define indicators that assess the jumpiness of the market. Generally, there are two ways to employ bi-power variance: the differential approach and the integral approach.

   1.1 *The Differential Approach*. The standard variance is defined as

$$\widehat{\sigma}_t^2 = \frac{1}{T-1}\sum_{\tau=t-T}^{t-1}\left(r_\tau - \frac{1}{T}\sum_{i=0}^{T-1} r_{t-i}\right)^2 \tag{4}$$

The bi-power variance is defined according to Barndorff-Nielsen and Shephard (2004b) as

$$\widehat{\widehat{\sigma}}_t^2 = \frac{1}{T-2} \sum_{\tau=t-T+2}^{t-1} |r_\tau| \, |r_{\tau-1}| \tag{5}$$

The higher the ratio $\widehat{\sigma}_t^2 / \widehat{\widehat{\sigma}}_t^2$, the more jumps are contained in the past T time steps back.

1.2 *The Integral Approach.* The integral approach is motivated by the work of Pirino (2009). The integral approach employs the two cumulative estimators for the total volatility over a given period. The first one is the cumulative realized volatility estimator defined as

$$RV_{day} = \sum_{day} (r_\tau)^2 \tag{6}$$

The second estimator is the bi-power cumulative volatility estimator defined in an analogous way:

$$BPV_{day} = \frac{\pi}{2} \sum_{day} |r_\tau| \, |r_{\tau-1}| \tag{7}$$

Analogously the ratio of the two cumulative estimators RV_{day}/BPV_{day} can be considered as a measure of the relative contribution of price jumps to the overall volatility over the particular period.

2. *Bi-power Test Statistics.* The bi-power variance can be used to define the proper statistics for the identification of price jumps one by one. This means testing every time step for the presence of a price jump as defined in Eq. (1). These statistics were developed by Andersen et al. (2007) and Lee and Mykland (2008) and are defined as

$$L_t = \frac{r_\tau}{\widehat{\widehat{\sigma}}_t^2} \tag{8}$$

Following Lee and Mykland, the variable ξ is defined as

$$\frac{\max_{\tau \subset A_n} |L_\tau| - C_n}{S_n} \to \xi \tag{9}$$

Where A_n is the tested region with n observations and the parameters are defined as

$$C_n = \frac{(2 \ln n)^{\frac{1}{2}}}{c} - \frac{\ln \pi + \ln(\ln n)}{2c(2 \ln n)^{\frac{1}{2}}} \tag{10}$$

$$S_n = \frac{1}{c(2\ln n)^{\frac{1}{2}}} \tag{11}$$

$$c = \frac{\sqrt{2}}{\sqrt{\pi}} \tag{12}$$

The variable ξ has in the presence of no price jumps the cumulative distribution function $P(\xi \leq x) = \exp(e^{-x})$. The knowledge of the underlying distribution can be used to determine the critical value ξ_{CV} at a given significance level. Whenever ξ is higher than the critical value ξ_{CV}, the hypothesis of no price jump is rejected, and such a price movement is identified as a price jump. In contrast, when ξ is below the critical value, we cannot reject the null hypothesis of no price jump. Such a price movement is then treated as a noisy price movement. These statistics can be used to construct a counting operator for the number of price jumps in a given sample. However, the main disadvantage of bi-power variation-based methods lie in the sensitivity of the intraday volatility patterns, which leads to a high rate of jump misidentification.

Jiang and Oomen (2008) modified this approach. They suggest to calculate Swap Variance as:

$$SwV = 2\sum_{i=2}^{n} R_i - r_i$$

where

$$R_i = \frac{P_i - P_{i-1}}{P_i}$$

$$P_i = \exp(p_i)$$

$$r_i = p_i - p_{i-1}$$

Jiang and Oomen claim that employing swap variance further amplifies the contribution coming from price jumps and thus makes the estimator less sensitive to intraday variation. The Jiang–Oomen statistics is defined as

$$JO = \frac{nBV}{\sqrt{\Omega SwV}}\left(1 - \frac{RV}{SwV}\right)$$

JO is asymptotically equal to $z \sim N(0,1)$ and tests the null hypothesis that a given window does not contain any price jump. The indicator for a price jump is defined as those price movements for which $JOt - 1 \leq \Phi - 1(\alpha)$ and $JOt > \Phi - 1(\alpha)$. The authors claim that their test is better than the one based on bi-power variation since it amplifies the discontinuities to a larger extent, as they show with a comparative analysis using Monte Carlo simulation. The amplification of discontinuities tends to suppress the effects of intraday volatility patterns.

To define extreme events on tick scale Nanex methodology can be applied. This methodology defines down (up) shock if stock had to tick down (up) at least 10 times before ticking up (down)—all within 2 s and the price change had to exceed 0.8 %. Tick means a price change caused by trade(s).

Hanousek (2011) performed an extensive simulation study to compare the relative performance of many price-jump indicators with respect to false positive and false negative probabilities. The results suggest large differences in terms of performance among the indicators: in the case of false positive probability, the best-performing price-jump indicator is based on thresholding with respect to centiles, in the case of false negative probability, the best indicator is based on bipower variation. The differences in indicators is very often significant at the highest significance level, which further supports the initial suspicion that the results obtained using different price-jump indicators are not comparable.

Another problem specific for any statistical filter is spurious detection. The problem is that performing the tests for many days simultaneously results in conducting multiple testing, which by nature leads to making a proportion of spurious detections equal to the significance level of the individual tests (Bajgrowicz and Scaillet 2011). Bajgrowicz and Scaillet (2010) treat the problem of the spurious identification of price jumps by adaptive thresholds in the testing statistics. The problem with most of the price-jump indicators lies in what model they are built upon and there is the need for robustness of each filter when dealing with price jumps. Bajgrowicz and Scaillet (2011) developed a method to eliminate spurious detections that can be applied very easily on top of most existing jump detection tests, a Monte Carlo study shows that this technique behaves well infinite sample. Applying this method on high-frequency data for the 30 Dow Jones stocks over the 3-year period between 2006 and 2008, authors found that up to 50 % of days selected initially as containing a jump were spurious detections. Abramovich et al. (2006) introduce the data adaptive thresholding scheme based on the control of the false discovery rate (FDR). FDR control is a recent innovation in simultaneous testing, which ensures that at most a certain expected fraction of the rejected null hypothesis correspond to spurious detections. Bajgrowicz and Scaillet (2010) use the FDR to account for data snooping while selecting technical trading rules. The choice of which threshold to use: universal or FDR, depends on the application. If the main purpose of research is the probability of a jump conditional on a news release, the FDR threshold is more appropriate as it reduces the likelihood of missing true jumps. If the goal is to study what kind of news cause jumps, it is better to apply the universal threshold in order to avoid looking vainly for a news when in fact the detection is spurious.

5 Cojumps

Documenting the presence of cojumps and understanding their economic determinants and dynamics are crucial for a risk measurement and management perspective.

Bajgrowicz and Scaillet (2011) did not detect cojumps affecting all stocks simultaneously for the sample including high-frequency data for the 30 Dow Jones stocks over the 3-year period between 2006 and 2008, which supports the assumption in Merton (1976) that jump risk is diversifiable and thus does not require a risk premium.

Other empirical studies of cojumps include Bollerslev et al. (2008) who examine the relationship between jumps in a sample of 40 large-cap U.S. stocks and the corresponding aggregate market index. To more effectively detect cojumps authors developed a new cross product statistic, termed the cp-statistic, that directly uses the cross-covariation structure of the high-frequency returns and examines cross co-movements among the individual stocks. Employing this statistic allows to detect many modest-sized cojumps. Cross product statistic defined by the normalized sum of the individual high-frequency returns for each within-day period:

$$cp_{t,j} = \frac{1}{2n\,(n-1)} \sum_{i=1}^{n-1} \sum_{l=i+1}^{n} r_{i,t,j} r_{l,t,j}$$

where j = 1, ..., M, M—total number of observations in a day, n—number of stocks. The cp-statistic provides a direct measure of how closely the stocks move together.

Lahaye et al. (2009) investigated cojumps between stock index futures, bond futures, and exchange rates in the relation with news announcements, they found that exchange rates experience frequent but relatively small jumps because they are subject to news from two countries and because they probably experience more idiosyncratic liquidity shocks during slow trading in the 24-h markets. Forex jumps tend to be smaller than bond or equity jumps because national macro shocks produce much smaller changes in expected relative fundamentals between currencies. Equity and bond market cojumps are much more strongly associated with news releases than foreign exchange cojumps. But also authors admit that most of news does not cause jumps. A generic announcement only produces an exchange rate jump about 1–2 % of the time and a bond or equity jump only about 3–4 % of the time.

By investigating a set of 20 high cap stocks traded at the Italian Stock Exchange, Bormetti et al. (2013) found that there is a large number of multiple cojumps, i.e. minutes in which a sizable number of stocks displays a discontinuity of the price process. As mentioned above, they show that the dynamics of these jumps is not described neither by a multivariate Poisson nor by a multivariate Hawkes model, which are unable to capture simultaneously the time clustering of jumps and the high synchronization of jumps across assets. Authors introduce a one factor model approach where both the factor and the idiosyncratic jump components are described by a Hawkes process. They develop a robust calibration scheme which is able to distinguish systemic and idiosyncratic jumps and show that the model reproduces very well the empirical behaviour of the jumps of the Italian stocks.

6 Empirical Description of Markets Around Shocks

A broad range of research works tries to give an empirical description for price jumps and analyze their statistical properties and the behaviour of market quantities around such events.

The attempt to compare shocks on different time scales is relatively little explored. Fan and Wang (2007), for example, used wavelets to identify jumps on multiple time scales, but the method is not used to compare shocks on different scales, but to detect shocks using a multiscale tool. On the other hand the attempt to investigate shocks and pre- and aftershock market behaviour is not novel. Lillo and Mantegna studied the relaxation dynamics of the occurrence of large volatility after volatility shocks (Lillo and Mantegna 2004), Zawadowski et al (2004) examined the evolution of price, volatility and the bid-ask spread after extreme 15 min intraday price changes on the New York Stock Exchange and the NASDAQ, Ponzi et al. (2009) studied possible market strategies around large events and they found that the bid-ask spread and the mid-price decay very slowly to the normal values when conditioned to a sudden variation of the spread. Sornette found that the implied variance of the Standard and Poor's 500 Index after the Black Monday decays as a power law with log-periodic oscillations (Sornette et al. 1996).

Mu et al. (2010) study the dynamics of order flows around large intraday price changes using ultra-high-frequency data from the Shenzhen Stock Exchange. They find a significant reversal of price for both intraday price decreases and increases with a permanent price impact. The volatility, the volume of different types of orders, the bid-ask spread, and the volume imbalance increase before the extreme events and decay slowly as a power law, which forms a well established peak. They also study the relative rates of different types of orders and find differences in the dynamics of relative rates between buy orders and sell orders and between individual investors and institutional investors. There is evidence showing that institutions behave very differently from individuals and that they have more aggressive strategies. Combing these findings, they conclude that institutional investors are more informed and play a more influential role in driving large price fluctuations.

Novotný (2010) tries to determine if there is any increase in market volatility and any change in the behaviour of price jumps during the recent financial crisis. He employs data on 16 highly traded stocks and one Exchange Traded Fund (ETF) from the North American exchanges found in the TAQ database from January 2008 to July 2009. It was found that the overall volatility significantly increased in September 2008 when Lehman Brothers filed for bankruptcy protection, the periods immediately after this announcement reveal significantly higher levels of volatility. However, the ratio between the regular noise and price jump components of volatility does not change significantly during the crisis. The results suggest individual cases where the ratio increases as well as decreases.

Conclusion

The literature review suggests that the mathematical description of price jumps cannot be easily handled, there is no general approach to model price jumps, but it would be wrong to neglect their presence since market shocks are the essential part of price time series and have serious consequences for pricing models and financial risk management. Thus market shocks have to be the subject for further research and analysis.

References

Abramovich, F., Benjamini, Y., Donoho, D. L., & Johnstone, I. M. (2006). Adapting to unknown sparsity by controlling the false discovery rate. *The Annals of Statistics, 34*(2), 584–653.

Ait-Sahalia, Y. (2004). Disentangling diffusion from jumps. *Journal of Financial Economics, 74*, 487–528.

Ait-Sahalia, Y., Cacho-Diaz, J., & Hurd, T. (2009). Portfolio choice with jumps: A closed form solution. *Annals of Applied Probability, 19*, 556–584.

Ait-Sahalia, Y., & Jacod, J. (2009a). Testing for jumps in a discretely observed process. *Annals of Statistics, 37*, 184–222.

Ait-Sahalia, Y., & Jacod, J. (2009b). Estimating the degree of activity of jumps in high frequency data. *Annals of Statistics, 37*, 2202–2244.

Andersen, T. G., Bollerslev, T., & Diebold, F. X. (2002). Parametric and nonparametric volatility measurement. In Y. Ait-Sahalia & L. P. Hansen (Eds.), *Handbook of financial econometrics*. Amsterdam: North-Holland. Available at http://www.ssc.upenn.edu/~fdiebold/papers/paper50/abd071102.pdf.

Andersen, T., Bollerslev, T., & Diebold, F. (2007). Roughing it up: Including jump components in the measurement, modeling, and forecasting of return volatility. *Review of Economics and Statistics, 89*(4), 701–720.

Andersen, T., & Bondarenko, O. (2011). *VPIN and the flash crash*. Available at http://ssrn.com/abstract=1881731.

Bajgrowicz, P., & Scaillet, O. (2010). *Detecting spurious jumps in high-frequency data*. Available at SSRN http://ssrn.com/abstract=1343900.

Bajgrowicz, P., & Scaillet, O. (2011). *Jumps in high-frequency data: Spurious detections, dynamics, and news*. Swiss Finance Institute, Occasional Paper Series N 11–36.

Barndorff-Nielsen, O. E., & Shephard, N. (2004a). *Measuring the impact of jumps on multivariate price processes using bipower variation*. Discussion paper, Nu¢ eld College, Oxford University.

Barndorff-Nielsen, O. E., & Shephard, N. (2004). Power and bipower variation with stochastic volatility and jumps. *Journal of Financial Econometrics, 2*, 1–37.

Barndorff-Nielsen, O., & Shephard, N. (2006). Econometrics of testing for jumps in financial economics using bipower variation. *Journal of Financial Econometrics., 4*, 1–30.

Becketti, S., & Roberts, D. J. (1990). Will increased regulation of stock index futures reduce stock market volatility? *Economic Review, Federal Reserve Bank of Kansas City* (November Issue), 33–46.

Bollerslev, T., & Jubinski, D. (1999). Equity trading volume and volatility: Latent information arrivals and common long-run dependencies. *Journal of Business & Economic Statistics, 17*, 9–21.

Bollerslev, T., Law, T. H., & Tauchen, G. (2008). Risk, jumps, and diversification. *Journal of Econometrics, 144*(1), 234–256.

Bormetti, G., Calcagnile, L. M., Treccani, M., Corsi, F., Marmi, S., & Lillo, F. (2013, 25 January). *Modelling systemic cojumps with Hawkes factor models*. arXiv:1301.6141v1[q-fin.ST]

Bouchaud, J. -P., Kockelkoren, J., & Potters, M. (2004). *Random walks, liquidity molasses and critical response in financial markets*. Science & Finance (CFM) working paper archive 500063, Science & Finance, Capital Fund Management.

Bozdog, D., Florescu, I., Khashanah, K., & Wang, J. (2011). Available at http://papers.ssrn.com/sol3/papers.cfm?abstract_id=2013355.

Broadie, M., & Jain, A. (2008). The effect of jumps and discrete sampling on volatility and variance swaps. *International Journal of Theoretical and Applied Finance, 11*(8), 761–797.

Cont, R., & Bouchaud, J.-P. (2000). Herd behavior and aggregate fluctuations in financial markets. *Macroeconomic Dynamics, 4*, 170–196.

Cont, R., & Tankov, P. (2004). Non-parametric calibration of jump–diffusion option pricing models. *Journal of Computational Finance, 7*(3), 1–49.

Cornell, B., & Sirri, E. R. (1992). The reaction of investors and stock prices to insider trading. *Journal of Finance, 47*(3), 1031–1060.

Cutler, D. M., Poterba, J. M., & Summers, L. H. (1989). What moves stock prices? *Journal of Portfolio Management, 15*, 4–12.

Dacorogna, M. M. (2001). *An introduction to high-frequency finance*. San Diego: Academic.

Easley, D., Lopez de Prado, M., & O'Hara, M. (2010). *The microstructure of 'Flash Crash'*. Available at http://ssrn.com/abstract=1695041.

Engle, R. F., & Russell, J. R. (1998). Autoregressive conditional duration: A new model for irregularly spaced transaction data. *Econometrica, 66*, 1127–1162.

Eryigit, M., Cukur, S., & Eryigit, R. (2009). Tail distribution of index fluctuations in world market. *Physica A, 388*, 1879–1886.

Fama, E. (1970). Efficient capital markets: A review of theory and empirical work. *Journal of Finance, 25*, 383–417.

Fan, J., & Wang, Y. (2007). Multi-scale jump and volatility analysis for high-frequency financial data. *Journal of the American Statistical Association, 102*, 1349–1362.

Filimonov, V., & Sornette, D. (2012). *Quantifying reflexivity in financial markets: Towards a prediction of flash crashes*. Available at arXiv:1201.3572v2[q-fin.ST]. 17 April 2012.

Gallant, A. R., Rossi, P. E., & Tauchen, G. E. (1992). Stock prices and volume. *The Review of Financial Studies, 5*, 199–242.

Gatheral, J. (2006). *Volatility surface: A practitioner's guide*. New Jersey: Wiley.

Gencay, R., Selcuk, F., & Whitcher, B. (2002). *An introduction to wavelets and other filtering methods in finance and economics*. San Diego: Elsevier.

Gopikrishnan, P., Plerou, V., Nunes Amaral, L. A., Meyer, M., & Stanley, H. E. (1999). Scaling of the distribution of fluctuations of financial market indexes. *Physical Review E, 60*(5), 5305–5316.

Hanousek, J. (2011). *The identification of price jumps*. Working Paper Series 434 (ISSN 1211–3298).

Hardiman, S., Bercot, N., & Bouchaud, J. -P. (2013). *Critical reflexivity in financial markets: A Hawkes process analysis*. Available at arXiv:1302.1405v1[q-fin.ST]. 6 February 2013.

Harris, L. (2003). *Trading and exchanges: Market microstructure for practitioners*. New York: Oxford University Press.

Heston, S. L. (1993). A closed-form solution for options with stochastic volatility with applications to bond and currency options. *The Review of Financial Studies, 6*(2), 327–343.

Hirshleifer, D., & Teoh, S. H. (2003). Herd behaviour and cascading in capital markets: A review and synthesis. *European Financial Management, 9*(1), 25–66.

Jiang, G., & Oomen, R. (2008). Testing for jumps when asset prices are observed with noise: A swap variance approach. *Journal of Econometrics, 144*(2), 352–370.

Joulin, A., Lefevre, A., Grunberg, D., & Bouchaud, J. -P. (2010). *Stock price jumps: News and volume play a minor role*. Resource document. http://arxiv.org/pdf/0903.0010.pdf. Accessed 3 Oct 2010.

Karpoff, J. (1987, March). The relation between price change and trading volume: A survey. *Journal of Financial and Quantitative Analysis*, 109–126.

Kennedy, D. B., Sivakumar, R., & Vetzal, K. R. (2006). The implications of IPO underpricing for the firm and insiders: Tests of asymmetric information theories. *Journal of Empirical Finance, 13*(1), 49–78.

Kleinert, H. (2009). *Path integrals in quantum mechanics, statistics, polymer physics, and financial markets* (5th ed.). Berlin: World Scientific.

Kleinert, H., & Chen, X. J. (2007). Boltzmann distribution and market temperature. *Physica A, 383*(2), 513–518.

Lahaye, J., Laurent, S., & Neely, C. J. (2009). *Jumps, cojumps and macro announcements*. Working Paper of Federal Reserve Bank of St. Louis, No 2007-032, Revised Version.

Lee, S. S., & Mykland, P. A. (2008). Jumps in financial markets: A new nonparametric test and jump dynamics. *Review of Financial Studies, 21*(6), 2535–2563.

Lillo, F., & Mantegna, R. (2004). Dynamics of a financial market index after a crash. *Physica A, 338*, 125–134.

Lo, A. W., & Wang, J. (2002). *Trading volume: Implications of an intertemporal capital asset price model* (pp. 1–23). Advances in Economic Theory: Eighth World Congress.

Madhavan, A. (2000). Market microstructure: A survey. *Journal of Financial Markets, 3*(3), 205–258.

Merton, R. C. (1976). Option pricing when underlying stock returns are discontinuous. *Journal of Financial Economics, 3*(1–2), 125–144.

Mu, G. -H., Zhou, W. -X., Chen, W., & Kertesz, J. (2010). *Order flow dynamics around extreme price changes on an emerging stock market*. Resource document. http://arxiv.org/pdf/1003.0168.pdf. Accessed 28 Feb 2010.

Novotný, J. (2010). *Were stocks during the financial crisis more jumpy: A comparative study*. Electronic copy available at http://ssrn.com/abstract=1692331.

Nyberg, P., & Wilhelmsson, A. (2009). Measuring event risk. *Journal of Financial Econometrics, 7*(3), 265–287.

Osborne, M. F. M. (1959). Brownian motion in the stock market. *Operations Research, 7*(2), 145–173.

Pan, J. (2002). The jump-risk premia implicit in options: Evidence from an integrated time-series study. *Journal of Financial Economics, 63*, 3–50.

Pirino, D. (2009). Jump detection and long range dependence. *Physica A, 388*, 1150–1156.

Ponzi, A., Lillo, F., & Mantegna, R. (2009). Market reaction to a bid-ask spread change: A power-law relaxation dynamics. *Physical Review E, 80*, 016112.

Shiller, R. J. (2005). *Irrational exuberance*. Princeton: Princeton University Press.

Sornette, D., Johansen, A., & Bouchaud, J. P. (1996). Stock market crashes, precursors and replicas. *Journal of Physics I France, 6*, 167–175.

Stanley, H. E., & Mantegna, R. N. (2000). *An introduction to econophysics*. Cambridge: Cambridge University Press.

Sun, W. (2003). *Relationship between trading volume and security prices and returns* (MIT LIDS Technical Report 2638). February 2003 Area Exam.

Taleb, N. (2007). *The black swan: The impact of the highly improbable*. New York: Random House.

Tinic, S. M. (1995). Derivatives and stock market volatility: Is additional government regulation necessary? *Journal of Financial Services Research, 9*(3–4), 351–362.

Zawadowski, A. G., Kertesz, J., & Andor, G. (2004). Large price changes on small scales. *Physica A, 344*, 221–226.

The Synergy of Rating Agencies' Efforts: Russian Experience

Alexander Karminsky

Abstract We examine the synergy of the credit rating agencies' efforts. This question is important not only for regulators, but also for commercial banks if the implementation of the internal ratings and the advanced Basel Approach are discussed. We consider Russian commercial banks as a good example where proposal methods might be used. Firstly, a literature overview was supplemented with an analysis of the activities of rating agencies in Russia. Secondly, we discussed the methods and algorithms of the comparison of rating scales. The optimization task was formulated and the system of rating maps onto the basic scale was obtained. As a result we obtained the possibility of a comparison of different agencies' ratings. We discussed not only the distance method, but also an econometric approach. The scheme of correspondence for Russian banks is presented and discussed. The third part of the paper presents the results of econometric modeling of the international agencies' ratings, as well as the probability of default models for Russian banks. The models were obtained from previous papers by the author, but complex discussion and synergy of their systematic exploration were this paper's achievement. We consider these problems using the example of financial institutions. We discuss the system of models and their implementation for practical applications towards risk management tasks, including those which are based on public information and a remote estimation of ratings. We expect the use of such a systemic approach to risk management in commercial banks as well as in regulatory borders.

Keywords Econometric model • Mapping • Rating • Rating scale • Risk management

JEL Classification G21, G24, G32

The work is partially supported by the International Laboratory of Quantitative Finance, NRU HSE, RF government grant, ag. 14.A12.31.0007.

A. Karminsky (✉)
Department of Finance, National Research University Higher School of Economics, IIEPD MGIMO-U, Moscow, Russia
e-mail: karminsky@mail.ru

A.K. Bera et al. (eds.), *Financial Econometrics and Empirical Market Microstructure*, DOI 10.1007/978-3-319-09946-0_8

1 Introduction

Ratings have been an essential tool for risk evaluation for more than a century and their range of use is still growing. Ratings transform a great volume of information into the rating agencies' opinion on the current financial stability and risk of an entity. They represent the result of a complex assessment of separate companies or single financial instruments (further named as entities). An increasing number of banks, especially those from emerging markets, have become a part of the rating systems in recent years, and the expectation that banks and other entities are going to be rated has become conventional. Rating costs are relatively low for both the issuers and the investors, but the percentage of all banks and companies with ratings is still not large. Moreover, there are no widely accepted instruments to compare rating estimations by different agencies.

Previous research has shown that ratings are *important for many reasons*, including: regulatory rules, as well as the Basel Accords, asset management and investors for portfolio allocations, government and market regulation covenants for investments and participation at financial tenders and auctions, information for fixed income and equity markets, and so on.

We should also mention that interest in resolving these issues is still increasing. The development of approaches based on internal ratings systems under the Basel II Accord (Basel 2004) has a practical interest for internal ratings and their models that would help to predict the credit ratings of banks using only freely accessible public information, especially for developing markets. The topic has received increased attention in connection with the global crisis that began in 2007 and the implementation of Basel III (Basel 2010). The regulation of rating agencies' activities was one of the main topics of the G20 meeting in Moscow in February 2013 (G20 2013).

The key goals of this research are to develop methods of comparison and to compare the bank ratings of the main rating agencies from different points of view. We focus on the synergy of the common use of the ratings of an entity estimated by different agencies, as well as cooperated internal ratings in this integration process. We also consider previous ratings and the probability of default models of different entities to extend the sphere of influence of rating methods for risk management.

For this purpose we executed an analysis of the connected literature, as well as the dynamics of the process of setting ratings to Russian banks (Sect. 2), considering different methods and algorithms for the comparison of ratings (see Sect. 3). Particular attention is devoted to the rating business in Russia and the comparative analysis of ratings of Russian banks that has been rapidly developing and redeveloping in recent years and has involved substantial efforts by the rating agencies.

Later on in Sects. 4 and 5 we discuss the rating model system, which has been obtained in previous papers from the synergy position. We briefly discuss the structure and parameters of the databases, the type of econometric models (order and binary choice), the financial and macroeconomic indicators for the models, and the comparison of the main international ratings connected with Russian financial institutions. Conclusions are provided in last section.

2 Comparison of the Ratings: Literature and Practice Overview

The process of rating assignment is similar for different international rating agencies. Frequently, agencies publish their methodologies. However, they do not include detailed information, but rather general directions for rating assessment.

The basic problem for using credit ratings by regulatory bodies and commercial banks is the comparability of the ratings from different agencies. From a practical point of view it is important to compare ratings. So the question is how a relationship between the rating scales can be found when different levels of defaults and expected losses are established.

2.1 Rating Comparisons in the Literature

Among the first papers aimed to compare the ratings of different agencies was the one by Beattie and Searle (1992). Long-term credit ratings were gathered from 12 international credit rating agencies (CRA) that used similar scales. The sample of differences between the pairs of ratings for the same issuer was found. Around 20 % of the pairs in that sample involved differences in excess of two gradations. That may be explained by differing opinions about the financial stability of the issuers, as well as by different methodologies used by the rating agencies. But the average difference between ratings of the main international agencies S&P and Moody's was insignificant.

Cantor and Packer (1994) compared Moody's ratings of the international banks with the ratings of nine other rating agencies. It was found that the differences were greater on average than those discussed earlier. The average rating difference among the biggest international and three Japanese rating agencies was nearly three gradations.

The CRAs sometimes explain this effect in terms of a conservative approach when dealing with an unrequested rating because they do not have as much information about a company with which they have a rating contract, as they would with a company that has entered into a rating agreement. Poon (2003) empirically concluded that unrequested ratings were lower on average than the requested ratings, and found that the effect could be explained as self-selection.

The questions connected with the desire of issuers to use rating shopping to obtain the best ratings were developed to overcome the difficulties to apply ratings for regulatory aims (Cantor and Packer 1994; Karminsky and Peresetsky 2009).

A lot of studies have analyzed the reasons for differences in ratings from different agencies rather than constructing a mapping between the different scales. Liss and Fons (2006) compared the national rating scales supported by Moody's with its global rating scale.

Ratings have also been compared in Russia by some authors (Hainsworth et al. 2012), according to Russian bank ratings connected both national and international agencies. Matovnikov (2008) looked at the relationship between the gradations of rating scales and the total assets and capital of banks. Hainsworth used an iterative application of linear regressions to find mappings between the rating scales of all the credit rating agencies.

A wide array of literature on rating modeling uses econometric models; for example, for bank ratings (Caporale et al. 2010; Iannotta 2006; Peresetsky and Karminsky 2011). Typical explanatory variables from publicly available sources have been defined for models of ordered choice. Examining changes in rating gradation over time for a limited sample of international CRAs was fulfilled.

The selection of the explanatory variables is an important step for the elaboration of such models. Firstly, quantitative indicators that are employed by the rating agencies may be examined (see, for example, Moody's 2007), as well as non-confidential indicators that have previously been employed by other researchers. Typical informative indicators are connected with the CAMELS classification and include the size of the company, its profitability, stability, liquidity, and structure of the business, as expressed through companies' balance-sheet figures. In recent years, the use of such factors as state support for banks or companies, and support from the parent company or group of companies has also become more frequent.

Secondly, the use of macroeconomic indicators has become popular recently (Carling et al. 2007; Peresetsky and Karminsky 2011). Among the most common indicators there are inflation index, real GDP growth, industrial production growth and oil prices, and changes in the foreign exchange cross-rates of currencies for export-oriented countries. Because of the correlation between the majority of macroeconomic indicators they may be used mostly separately. Thirdly, the potential efficiency of market indicator exploration (Curry et al. 2008) for public companies should be mentioned. It should also be noted that alternate indicators may be informative for developing and developed markets.

At the Higher School of Economics and the New Economic School in Moscow there has been research on modeling the ratings of international credit rating agencies in Russia (Peresetsky et al. 2004; Karminsky et al. 2005; Peresetsky and Karminsky 2011). These studies have focused on finding economic and financial explanatory factors, that affect ratings, and on comparing the ratings of international agencies.

2.2 *Dynamic of the Rating Agencies Activities in Russia*

The growth of the number of Russian agencies ratings has been significant in recent years. Four Russian rating agencies achieved registration in the Russian Ministry of Finance as well as three international ones. Due to this fact, the question of the integration of these agencies' efforts and comparison of their rating scales is important. As for now we have nearly 700 ratings for banks only. We observed a threefold growth in 5 years (2006–2011). We also see that the number of ratings given by Russian agencies is roughly similar to the international agencies' ratings (Karminsky et al. 2011b).

Despite the comparative growth in the number of ratings, the rating methods are largely unclear, and expertise plays a significant role. This hinders the usage of ratings for risk evaluation and decision-making even at the state level. It is the reason for interest in the creation of internal ratings and model ratings.

Our long-term goal is to research the possibility of forecasting company ratings based solely on publicly available information, including indicators from international financial reports and market conditions on stock exchanges.

3 Comparison of Ratings: Methods and Algorithms

The rating process has some problems, such as

- A relatively small number of updated communicative ratings.
- Difficulties of comparison of estimation between different rating agencies.
- Absence of any integrative effect from available competitive estimations of independent agencies.
- A demand for extended usage on independent rating estimations primarily owing to modeling techniques.

We aim to achieve a comparison capability of independent estimations of different ratings. In this way the elaboration and development of the approaches and methods are especially urgent because of synergy opportunities connected with the limitations mentioned above. For these aims the Joint Rating Environment (JRE) was introduced, and included a selection of basic rating scales, the building of a mapping system of external and internal ratings to a base scale, and the common usage of all rating estimations for every class of issuer or financial instrument.

We used statistical approaches to calculate the distance between different ratings for the same entities. Also we selected a basic scale, in which we proposed to measure the difference between ratings, and proposed to use mapping between rating scales, while our aim is to find functional approximations of such maps.

Econometric approaches were studied in the paper (Ayvazyan et al. 2011). In this method, firstly, the econometric order choice models for every CRA were determined. Then the correspondence between latent variables for the model for the basic CRA and every other CRA model in polynomial form was estimated. These gave an opportunity to determine the mapping of every CRA scale to the basic scale at last.

The main points of distance algorithm for the rating scales' comparison include not only the methodology of agency-scales mapping, principles and criteria for

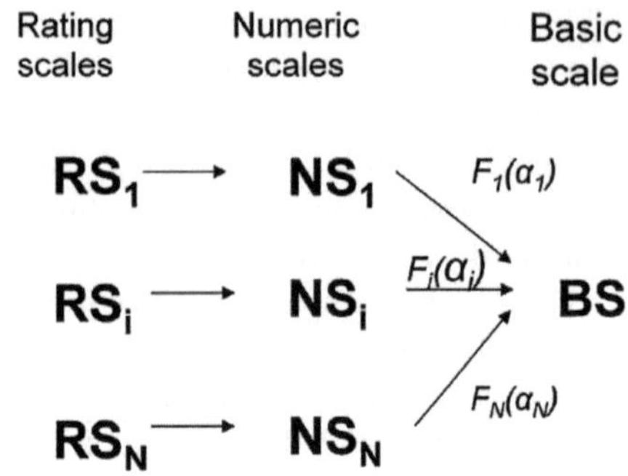

Fig. 1 The system of scale mappings

comparison of rating scales, but also the choice of an optimization algorithm, the construction of a comparison scheme and a table, the principles of result auditing during that time and so on (Hainsworth et al. 2012).

In this paper Moody's rating scale is used as a basic scale, but the results must be practically invariant to the choice. The system of mapping, which was presented in Fig. 1, was established. In this figure the first group of mapping deals with the correspondence between the rating and numerical scales, which is reasonable because of the rating's orderliness. The mappings to the basic scale

$$F_i\ (\alpha_i) :\ NS_i\ \rightarrow BS$$

for every rating scale R_i were parameterized, and our aim is to find the vectors α_i for each scale $i = 1, \ldots, N$, where N—the number of the scales.

We have considered some parameterization of mappings $F_i(\alpha_i) = a_{i1} * f_i(R_i) + a_{i2}$, using functions $f_i(R_i)$ from some classes and a vector of parameters of the map $\alpha_i = (a_{i1},\ a_{i2})$. At this step we have formulated the task of the parametric optimization problem. We used a square measurement between rating images in this research:

$$\min_{\{\alpha_i, i=1,\ldots N\}} \sum_{Q} \left(F_{i1}\left(R_{i1jt}, \alpha_{i1}\right) - F_{i2}\left(R_{i2jt}, \alpha_{i2}\right)\right)$$

Above we mean that

Q—the set of combinations of points over time
$q = \{$quarter t, bank j, the rating of the basic agency R_{i1jt}, the rating of the other agency $R_{i2jt}\}$;
F_{i1} and F_{i2}—the maps for *i1* and *i2* scales as defined above.

During the research we compare linear, power and logarithmic function classes f_i, which were used for the evaluation of map dependences.

An additional analysis of the default statistics for Moody's and S&P gives us an opportunity to use a priority logarithmic approximation, which we use in this paper for empirical analysis. It must also be mentioned that for the previous problem we

Table 1 Table of parameters for bank scale mappings in a logarithmic model specification

Rating scale	a_{i1}	a_{i2}
Moody's (Russian scale)	0.254	2.202
Standard and poor's	0.916	0.146
Standard and poor's (Russian scale)	0.265	2.113
Fitch ratings	0.749	0.594
Fitch ratings (Russian scale)	0.213	2.162
AK&M	0.269	2.491
Expert RA	0.373	2.329
RusRating	0.674	1.016
National rating agency	0.163	2.474
Number of estimations	*3,432*	
Pseudo-R^2	*0.902*	

Italic texts were connected with statistical summaries of the tables.

could have used econometric program packages such as eViews or STATA because of the use of the quadratic criteria (the experiments with other criteria showed the robustness of the comparison results).

We provided this analysis for both Russian and international data. For the Russian data we had a sample for a time span of 20 quarters (from 1Q 2006 till 4Q 2010), as well as the data for periods until 2012 in other examples. We have collected data from three international agencies (Moody's, S&P and Fitch) on both international and national scales, as well as from four Russian agencies (AK&M, NRA, RusRating and Expert RA). This sample has included 7,000+ pairs of ratings for 370 Russian banks with any rating during this time span.

The result of the optimization task decision is presented in Table 1.

The results derived from this can be presented both in scheme (Fig. 2) and table interpretations. At this point we have constructed a scale correspondence, which may be used in practice for regulatory and risk management purposes.

It should be mentioned that the correspondence between international agencies on traditional scales are not identical, and we can compare the difference between these agencies with the Russian banks.

It also should be noted that the results included in the scheme are stable. We have compared the results not only with a different base scale, but also with two different methods such as distance and econometric methods. The results obtained give us the opportunity to acquire comparable estimations of entities for both regulation and risk management aims.

For the international banks' models an accurate forecast was generated in nearly 40 % of cases. The forecasting power may be estimated by mistakes on the part of the models, which in the case of no more than two grades gave a probability of 1–2 %. These results were comparable with previous models, but extended to three international rating agencies simultaneously.

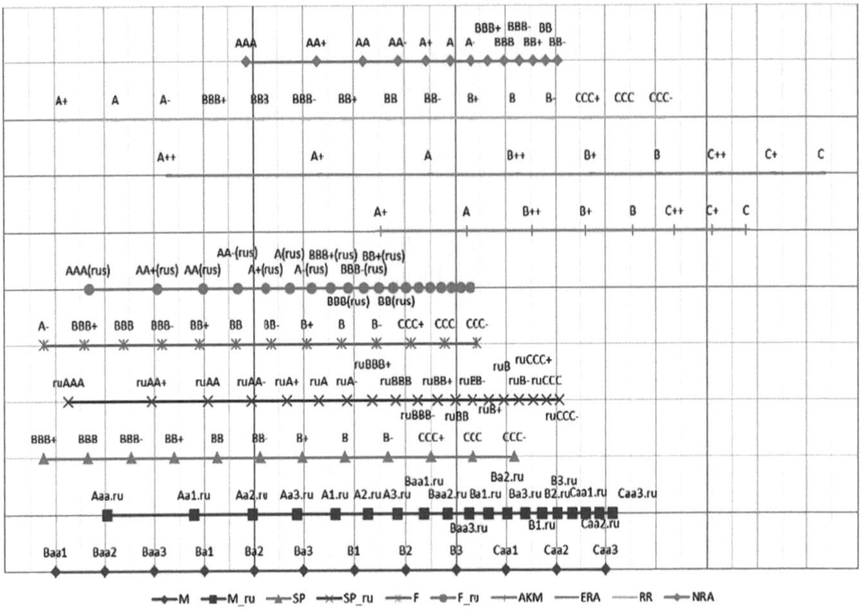

Fig. 2 Scheme of correspondence for CRA scales working in Russia

The signs for all the models were almost equal, and could be easily explained from a financial point of view. Coefficient sign analysis allowed us to make the following conclusions:

- The size of the bank is positive for a rating level increase, also as capital ratio and asset profitability as the retained earnings to total assets ratio.
- Such ratios as debt to asset and loan loss provision to total assets have a negative influence on the rating grade.
- Macro variables are also important for understanding the behavior of bank ratings, and are presented with a negative sign for the corruption index and inflation.

We also constructed models for Russian bank ratings using a Russian data base and have concluded that the influence of financial indicators is mainly the same (Vasilyuk et al. 2011).

4 Modeling of Ratings and the Probability of Default Forecast Models

A lot of research is devoted to the difference in the ratings of the main international CRAs. They provide adjustments of explanatory financial and macroeconomic variables on the new horizon analysis dependence of ratings on their affiliation to specific groups of countries, their degradation over time, lags between dependence and independence variables, etc.

Firstly, econometric rating modeling needs comprehensive and well-organized data. Secondly, the class of econometric models and principles of their verification should be selected. A modern risk management system based on best practice is the next important component. Finally, such a system needs domestic experience data that would take into account the specifics of a country.

In this section we systemize the practice of research of such models for banks, corporations and countries in Russian bank applications. Additionally we will discuss the opportunities of the probability of default models in the case of Russia. We use the existing experience of such research, which was obtained and published in previous works. In this paper we try to understand how this knowledge may be accumulated in the JRE system.

4.1 Models and Data for Bank Ratings

Here, and further in this section, ordered probit/logit econometric models were used to forecast rating grades (for example, see Peresetsky and Karminsky (2011)). Numeric scales for ratings were also used as a result of the mappings mentioned in Fig. 1. For the main international CRAs, nearly 18 corporate rating grades were used.

The original databases for different classes of entity were used. There were two different databases used separately for banks for both international and Russian ones. The first database was obtained from Bloomberg data during the period 1995–2009. The database includes 5,600+ estimations for 551 banks from 86 countries. The data contains the banks from different countries including more than 50 % from developed and 30 % from developing countries. Russian banks are also included in the sample and form nearly 4 %.

The second database was constructed from the data for Russian banks according to Russian financial reporting. It contains 2,600+ quarterly estimations from 2006 until 2010 for 370 Russian banks.

We carried out model choices from different points of view for three agencies simultaneously. We determined which financial explanatory variables were the most informative ones. Then we considered quadratic models, using macro, market and institutional variables, as well as dummies. We used a rating grade as a dependent variable where the lower numbers were associated with a better rating. So a positive sign in the coefficient related to a negative influence on the ratings, and vice versa.

You can see the chosen models for international banks in Table 2 (Karminsky and Sosyurko 2010).

For the international bank models, an accurate forecast was generated in nearly 40 % of the cases. The forecasting power may be estimated by the mistakes of the models, which in the case of no more than two grades gave the probability of

Table 2 Bank rating models: international banks

Variable	Influence	S&P—issuer credit	Fitch—issuer default	Moody's—bank deposits	Moody's—BFSR
Ln (assets)	+	−0.523***	−0.561***	−0.545***	−0.383***
Equity capital/total assets	+	−3.012***	−1.945***	−2.758***	−1.607***
Equity capital/risk weighted assets	+	0.045***	0.014*	0.028***	
Loan loss provision/average assets	−	42.763***	37.284***	19.188***	12.245***
Long-term debt/total assets	−	0.008*	0.017**	0.023***	0.020***
Interest expenses/interest income	−	0.353***	0.277***	0.294***	0.171***
Retained earnings/total assets	+	−9.841***	−5.063***	−1.404*	−2.345***
Cash and near cash items/total liabilities	−	2.303***	1.814***	1.985***	1.917***
Corruption index	−	−0.408***	−0.356***	−0.383***	−0.316***
Annual rate of inflation	−	0.038***	0.020**	0.028***	−0.009*
Exports/imports	+	−0.584***	−0.400***	−0.559***	−0.017
GDP	+	−4.40***	−4.40***	−12.20***	−15.80***
Pseudo R^2		*0.293*	*0.266*	*0.295*	*0.192*
Number of estimations		*1,804*	*1,985*	*1,787*	*1,897*

Notes: *, **, *** represent 10%, 5%, 1% levels of significant, respectively.
Italic texts were connected with statistical summaries of the tables.

1–2 %. These results were comparable with the previous models, but extended to three international rating agencies simultaneously.

The signs for all the models were almost equal and could be easily explained from a financial point of view. Coefficient sign analysis allowed us to make the following conclusions:

- The size of the bank is positive for a rating level increase, as are capital ratio and asset profitability as the retained earnings to total assets ratio.
- Such ratios as debt to asset and loan loss provision to total assets have a negative influence on the rating grade.
- Macro variables are also important for understanding the behavior of bank ratings, and are presented with a negative sign for the corruption index and inflation.

We also constructed the models for Russian banks ratings using a Russian database, and have concluded that the influence of financial indicators is mainly the same (Vasilyuk et al. 2011).

4.2 Models of Corporations and Sovereigns

The sample of corporations included information from different industries (oil and gas, utilities, retail, telecom, etc.) and countries. We considered the rated companies from these industries which also had financial and market indicators. Financial explanatory variables included such group indicators as size of company, capitalization, assets, management, efficiency, and liquidity. Among the macro indicators it stands out on the corruption perception index by Transparency International. While among market indicators the volatility of the market prices stands out. We also added industry classification dummies, as well as such factors as groups of countries and a company's affiliation.

We used both the agencies' and Bloomberg data for this sample. Financial indicators were selected for 30+ countries during 2000–2009 for 211 corporations. Our database included nearly 1,800 estimations (non-balance panel) for three international rating agencies; S&P, Fitch and Moody's ratings.

Order probit model parameters are presented in Table 3. We do not have the opportunity to use all the explanatory variables. You can see the best models, which differed in profitability indicators (Karminsky 2010).

The signs for all three models are equal, and have a good explanation from a financial point of view. As for its interpretation, a positive sign of coefficient relates

Table 3 Comparison of corporate rating models for international CRA

Variable	S&P	Fitch	Moody's
LN (market capital)	−0.692***	−0.806***	−0.691***
Sales/Cash	0.00004***	−0.00051	−0.00049
EBIT/interest expenses	−0.0017***	0.0006	−0.0054***
LT debt/capital	0.006***	0.011***	0.019***
Retained earnings/capital	−1.107***	−0.581**	−1.230***
Volatility (360d)	0.012***	0.013***	0.016***
Corruption perception index	−0.217***	−0.088***	−0.088
Chemicals	−0.235***	0.381***	−0.182
Metal and mining	0.322***	1.317***	0.947***
Pseudo-R^2	*0.215*	*0.220*	*0.276*
Number of observations	*1,362*	*423*	*339*
$\|\Delta\| = 0$	*40.6 %*	*34.3 %*	*42.5 %*
$\|\Delta\| \leq 1$	*87.7 %*	*87.7 %*	*87.0 %*

Notes: *, **, *** represent 10%, 5%, 1% levels of significant, respectively.
Italic texts were connected with statistical summaries of the tables.

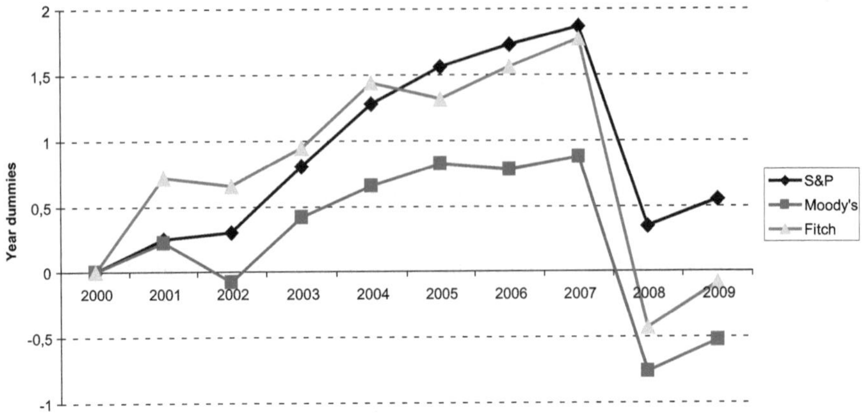

Fig. 3 Procyclicity of corporate ratings: year dummies dynamics

to a negative influence on rating, and vice versa, because of the fact that the scale mapping choice should be taken into account. From this model we can make the following conclusions:

- The size of the company, asset profitability and the EBITDA to interest expenses ratio have a positive influence on the rating level. A ratio such as LT Debt to Capital has a negative influence on the rating grade.
- Industry dummies are significant. We can see that companies from the utility and oil and gas industries have higher ratings.
- Market variables are also important for understanding the behavior of companies, for example, the corruption index has a negative influence.

Time has an important influence as well. We used a system of dummies during the years 2000–2009 to understand the impact of methodology and crisis. Most of the dummies are significant. We can see in Fig. 3 that all the agencies have the same procyclicity connected with the crisis of 1998 and 2008.

The main explanatory variables for sovereign rating models may be classified into 6 groups of quantitative variables such as: bank characteristics, economic growth, international finance, monetary policy, and public finance and stock market characteristics. In our research 30+ parameters from all groups were analyzed.

We also used dummies for regions, financial crisis type and indicators of corruption (CPI index). Our sample included nearly 1,500 estimations for 100+ countries during the 1991–2010 periods. We dealt with Moody's bank ceiling ratings as a sovereign rating proxy. The models are presented in Karminsky et al. (2011a).

We derived a strong association of sovereign ratings with economic growth, the public sector, monetary policy, the banking sector, the foreign sector, stock market variables and geographical regions. The forecast accuracy of the models is higher for investment-level grades than for speculative-level grades.

The majority of working explanatory variables for higher-investment ratings consists of the financial sector variables and GDP per capita. The majority of working explanatory variables for speculative-grade ratings includes budget deficit, inflation growth rate, export-to-import ratio and GDP per capita.

4.3 Probability of Default Models

Here, and later in the paper, the default is understood as one of the following signals for its registration:

- A bank's capital sufficiency level falls below 2 %.
- The value of a bank's internal resources drops lower than the minimum established at the date of registration.
- A bank fails to reconcile the size of the charter capital and the amount of internal resources.
- A bank is unable to satisfy the creditors' claims or make compulsory payments.
- A bank is subject to sanitation by the Deposit Insurance Agency or another bank.

We propose a forecast probability of default (PD) model, which is based on the relationship between banks' default rates and public information. We have constructed a quarterly bank-specific financial database on the basis of Mobile's information from 1998 to 2011: data in accordance with Russian Financial Reporting Standards, taken from bank Balance sheets and Profit and Loss statements.

During a 14-year period there were 467 defaults in compliance with our definition, as well as 37 bank sanitations. The quarterly database created has a good coverage of default events and the banking sector. We have applied a binary choice logistic model to forecast default probability. The maximum likelihood approach is used to estimate the model. The sample was split into two parts: "1998–2009"—to estimate models, and "2010–2011"—to test the predictive power of the models.

Financial ratios used as explanatory variables were determined from the literature review and common sense. They were tested on their separating power between bankrupt and healthy banks, as well as being divided into blocks according to the CAMELS methodology. We have also employed non-linearities in our model and found the optimal lag on financial ratios.

- Macroeconomic variables are highly correlated, and there were only two variables used in order to account for the effect of the macroeconomic environment on bank performance: quarterly GDP growth rates and the Consumer Price Index. We also controlled for the impact of the following on a bank's default probabilities:
- Monopoly power of a bank on the market (with the Lerner index).
- Its participation in a Deposit insurance system (with a dummy variable).
- The territorial location of the bank's operational activity (Moscow or regional)

Our key findings (Karminsky et al. 2012) were that:

- Banks with extremely high and low profitability have higher default rates due to their impact on the default probability of the profit-to-assets ratio (poor and risky banks).
- Banks with a higher proportion of corporate securities in assets carry a higher risk of a price crash on the market.
- Lower turnover on correspondent accounts in comparison with total assets increases the probability of default (a bank's potential inability to make payments).
- Banks with a considerable number of bad debts are less stable.

Additionally, a growing consumer price index increases a bank's default probability:

- Inflation reduces the real return on loans.
- Depositors are able to withdraw money and deposit it into the bank again at a higher interest rate or spend it.

We have also found that banks with a higher monopoly power are financially stable. Moscow-based banks have higher PDs on average.

We have found no evidence that a bank's participation in the Deposit insurance system influences its PD. The explanation is that the set of System participants is too diversified. The out-of-sample prediction performance of the model (for 2010–2011) is prominent: over 60 % of bank failures were correctly classified with a moderately sized risk group.

At the same time, the developed model underestimated the default probabilities for 2009. This result reveals some unrecorded channels that significantly increased the risks during the period of the recent financial crisis.

5 System of Models and Synergy of Rating Estimations

Previously we considered the capabilities which were given to us by rating mappings and models. Later we will discuss the synergy of these approaches as instruments of the Joint Rating Environment system (JRE-system). Such a system may be used for risk management in commercial banks; its main components for financial institutions are presented in Fig. 4.

The main part of such a system is the correspondence between rating scales, including the connection with internal ratings. They provide the opportunity to compare different ratings, as well as to use a comparable estimation of ratings received by several models. The synergy of such estimations gives a basic scale by independent risk weightings.

The system of models brings to the IRB Approach some possibilities, among which there may indications such as

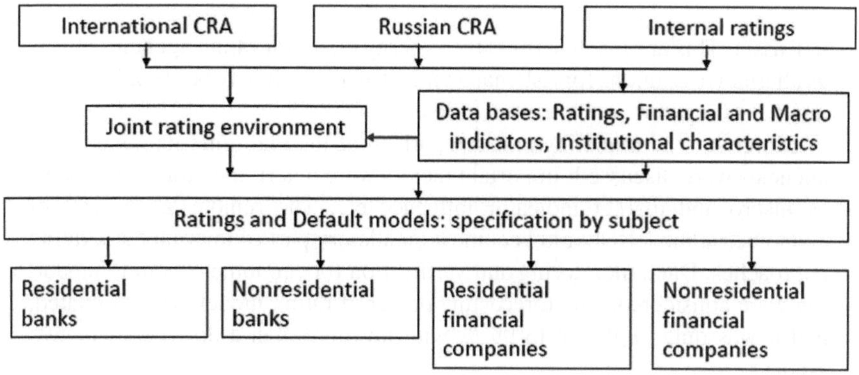

Fig. 4 Rating model system for financial institutions

- A basic scale established for the development and practical usage of econometric rating models within the IRB-approach for Russian and international rating agencies.
- Rating scale comparison methods defined for different agencies including external and internal rating reconciliation.
- Rating estimation forecasting approach and banking risk measurement dependent on internal and external factors.
- Rating forecasting for financial and non-financial companies which have no rating.
- Implementation of an econometric modeling system which requires:
  - Structured databases (data warehouse).
  - Support for all life cycle stages of models.
  - Monitoring, data gathering and the integration problem solution.

Of course such systems may be constructed for all types of entity, which were indicated at the specified risk management system according Basel II (Basel 2004, 2010). The details should be discussed for every bank or regulator separately. The discussion of these details is beyond the scope of this paper and may be done later.

Conclusion

We considered some methods of rating system construction, including a comparison of different rating estimates and modeling ratings for unrated entities.

The mapping of rating scales was introduced as the foundation for the comparison of rating scales using a distance method. We proposed this method for all the international and national agencies, which were recognized

(continued)

in Russia. This approach permits the synergy effect for rating agencies efforts as alternative opinions for risk management analysis. It may be combined with internal ratings for an increase in efficiency.

Moreover, the modeling and comparison of the main international rating agencies were discussed. Important factors were determined for such models as macro and market indicator influence etc. The remote assessment of econometric models should become a mandatory part of internal bank rating approaches. Data, monitoring and verification for econometric rating modeling were considered. The forecasting power of rating models was estimated, and it was quite high (up to 99 % with no more than a divergence of two grades).

Besides the bank rating models, the system should include corporate, sovereign and bond rating models. Some of them were presented in the paper, also as principles of their creations and main findings.

Bank and government financial regulators may be perspective users of the proposed methods. They can use such methods for the synergy of rating estimations.

References

Ayvazyan, S., Golovan, S., Karminsky, A., & Peresetsky, A. (2011). The approaches to the comparison of rating scales. *Applied Econometrics, 3*, 13–40 (in Russian).

Basel. (2004). *International convergence of capital measurement and capital standards. A revised framework*. Basel: Bank for International Settlements, Basel Committee on Banking Supervision.

Basel. (2010). *Basel III: A global regulatory framework for more resilient banks and banking systems*. Basel: Bank for International Settlements, Basel Committee on Banking Supervision.

Beattie, V., & Searle, S. (1992). Bond ratings and inter-rater agreement. *Journal of International Securities Markets, 6*, 167–172.

Cantor, R., & Packer, F. (1994). The credit rating industry. *FRBNY Economic Policy Review*, 1–26.

Caporale, G. M., Matouse, R., & Stewart, C. (2010). EU banks rating assignments: Is there heterogeneity between new and old member countries? *Review of International Economics, 19*(1), 189–206.

Carling, K., Jacobson, T., Linde, J., & Roszbach, K. (2007). Corporate credit risk modeling and the macroeconomy. *Journal of Banking and Finance, 31*, 845–868.

Curry, T., Fissel, G., & Hanweck, G. (2008). Is there cyclical bias in bank holding company risk ratings? *Journal of Banking & Finance, 32*, 1297–1309.

G20. (2013). New rules on credit rating agencies (CRAs), European Commission, MEMO, Brussels, 16 January 2013. http://europa.eu/rapid/press-release_MEMO-13-13_en.htm.

Hainsworth, R., Karminsky, A., & Solodkov, V. (2012). *Arm's length method for comparing rating scales*. Working Paper WP BRP 01/FE/2012. Higher School of Economics.

Iannotta, J. (2006). Testing for Opaqueness in the European banking industry: Evidence from bond credit ratings. *Journal of Financial Service Researches, 30*, 287–309.

Karminsky, A. (2010). Rating model opportunities for emerging markets. In *Proceedings of the International Scientific Conference "Challenges for Analysis of the Economy, the Businesses, and Social Progress"*. Szeged: University Press.

Karminsky, A., & Peresetsky, A. (2009). Ratings as measure of financial risk: Evolution, function and usage. *Journal of the New Economic Association, 1–2*, 86–104.

Karminsky, A., & Sosyurko, V. (2010). Comparative analyses of rating models generation. *Financial analyses: problems and solutions, 14*(38), 2–9 (in Russian).

Karminsky, A., Peresetsky, A., & Petrov, A. (2005). Ratings in the economics: Methodology and practice. In A. M. Karminsky (Ed.), *Finance and statistics*. Moscow (in Russian).

Karminsky A., Kiselev, V., & Kolesnichenko, A. (2011a). Modeling sovereign ratings: Estimators, models, forecasting. In *EBES 2011 Conference – Zagreb*. EBES Abstract Book, Istanbul.

Karminsky, A., Polozov, A., & Ermakov, S. (2011b). *Encyclopedia of ratings: Economics, society, sport. Public house "Economic and Life"* (in Russian).

Karminsky, A., Kostrov, A., & Murzenkov, T. (2012). *Comparison of default probability models: Russian experience*. Working Paper WP BRP 06/FE/2012 FE. Higher School of Economics.

Liss, H., & Fons, J. (2006, December). *Mapping Moody's national scale ratings to global scale ratings, Moody's rating methodology*.

Matovnikov, M. (2008). How to authorize the credit rating agencies to assess the creditworthiness of banks. *Money and Credit, 12* (in Russian).

Moody's. (2007). *Bank financial strength ratings: Moody's investors service, global methodology*.

Peresetsky, A., & Karminsky, A. (2011). Models for Moody's bank ratings. *Frontiers in Finance and Economics, 1*, 88–110.

Peresetsky, A., Karminsky, A., & van Soest, A. (2004). Modeling the ratings of Russian banks. *Economics and Mathematical Methods, 40*(4), 10–25 (in Russian).

Poon, W. P. H. (2003). Are unsolicited credit ratings biased downward? *Journal of Banking and Finance, 27*, 593–614.

Vasilyuk, A., Karminsky, A., & Sosyurko, V. (2011). *A system of bank rating models for IRB approach: Comparison and dynamics*. Working Paper WP7/2011/07, National Research University Higher School of Economics (in Russian).

Spread Modelling Under Asymmetric Information

Sergey Kazachenko

Abstract Bid–ask spread is a key measure of pricing efficiency in a microstructure framework. Today there is no universal model of spread formation that includes all three factors of transaction costs, inventory risk (losses in case of a changing value of a stored asset) and information asymmetry that influence the behaviour of traders and market-makers. Empirical evaluations of these three components of spread are very contradictory (Campbell et al., The econometrics of financial markets. University Press, Princeton, 1997; Easley and O'Hara, Microstructure and asset pricing. In: George MC, Milton H, Rene HS (eds) Handbook of the economics of finance. Elsevier, Amsterdam, pp 1022–1047, 2003). In our work, after the introduction of the additional uncertainty about the real asset value, we propose an algorithm of bid–ask spread formation for the market-maker, based on classical model of Glosten and Milgrom (J Financ Econ 14:71–100, 1985). Our modification allows us to the reproduce intertemporal spread dynamics under asymmetric information and limited inventory risk of a market-maker.

Keywords Asymmetric information • Bid–ask spreads • Glosten–Milgrom • Inventory risk • Market microstructure • Price formation

JEL Classification G14, D47, D82

1 Introduction

Today there is some controversy about the asset pricing process between the microstructure approach and macro models of finance theory. In history, we can find a similar period of misunderstanding between the micro and macro economies (Ball and Romer 1990). The experience of restoring the integrity of economic theory draws attention to the efficient market hypothesis (Fama 1970). We can

S. Kazachenko (✉)
Faculty of Economics, National Research University Higher School of Economics, Perm, Russia
e-mail: kazachenko.serg@gmail.com

A.K. Bera et al. (eds.), *Financial Econometrics and Empirical Market Microstructure*, DOI 10.1007/978-3-319-09946-0_9

assume that the key "macro request" to microstructure analysis is a quantitative measure of information efficiency (measure of market price deviation from fair price) or mechanism designed to regulate the asset pricing process, which provides given characteristics that suits assumptions of macro models. For instance, Agarwal and Wang (2007) stated that there is an explanation of the high descriptive power paradox of the empirical three-factor Fama–French model (Fama and French 1993), which arises because transaction costs were not taken into account. In this case, the spread acts as an indirect measure of information efficiency of the asset pricing (Roll 1984).

Bid–ask spread depends on three key factors: transaction costs (Roll 1984), inventory risk (Stoll 1978) and information asymmetry (Glosten and Milgrom 1985). Today there is no universal model of spread formation that includes all three factors. Generally, when authors have modelled spread formation, they took into account only one factor, as discussed above. The impact of other factors is limited. Empirical estimations of these three factors are controversial (Campbell et al. 1997; Easley and O'Hara 2003).

The mechanism of information allocation and incorporation into the market price plays a key role in bid–ask spread formation. The American scientists L. Glosten and P. Milgrom provided in 1985 the basic research in this area. The authors, hereafter referred to as GM, constructed their model to show the influence of information asymmetry on bid–ask spread. That is why they introduced strict assumptions:

- Uniqueness of informational event and common knowledge of the moment when informed traders receive information about real asset value
- Knowledge of possible real asset value ($\underline{V}$ and $\overline{V}$, lower higher price)
- Fixed volume for one transaction
- Traders cannot refuse to perform a transaction
- Informed traders have no power over price manipulation
- The market-maker has no need to account for inventory risk
- The market-maker has zero profit and losses
- Authors exclude competition between market-makers

The key point in obtaining a complex model of bid–ask spread formation, based on the GM model, is the accounting of inventory risk. Straight incorporation of inventory risk in a bid–ask spread yields explosive growth of spread and price. In our work, we attempt to find such a relaxation of assumptions of the GM model that allows the market-maker to implement simultaneous control of inventory costs and costs from adverse selection and, at the same time, keep the key features of the GM model (i.e. the martingale property of prices and intertemporal dynamics of spread). In our study, we do not include transaction costs.

In our work, we have made following changes in assumptions of the GM model:

- We introduced uncertainty of market-maker's expectations about real asset value (the market-maker has no knowledge about expecting higher $\overline{V}$ and lower $\underline{V}$ prices. Instead, he/she knows only the range $\left[\underline{V}_{MM}; \overline{V}_{MM}\right]$, where real asset value is located).

- Informed traders make errors, but they still know about the exact expected value of real asset value ($\overline{V}$ or $\underline{V}$).
- Informed traders can refuse to perform non-profitable transactions.
- We added some statistical functions to analyze inventory risk of the market-maker and its financial result.

Other basic assumptions of the GM model remain unchanged. The proposed modification of the original assumption was influenced by studies (Das 2005; Zachariadis 2012; Gerig and Michayluk 2010). The logic of proposed changes to the GM model is as follows:

- A necessary condition of write-off of the market-maker's inventory costs is dilution of an informed trader's monopoly by introducing informational uncertainty for informed traders. After that, informational uncertainty for the market-maker must also be introduced.
- The correct solution for the informed trader's informational uncertainty problem assumes introduction of the learning mechanism. However, in our study, we restricted ourselves to the introduction of a simplified version of informed traders' information uncertainty, which involves the consideration of a certain percentage of mistakes made by informed traders.
- We accepted that the informed trader could refuse a non-profit transaction when profit from expected operation (purchase or sale) generates a loss.
- We introduced an algorithm that allows the market-maker to correct bid–ask spread by taking into account the refusals of informed traders.

Comparative analysis of the numerical example of the proposed modification shows that the speed of incorporation of information decreases, which creates opportunities for the market-maker to control some inventory risk and adverse selection risk during bid–ask spread formation.

The rest of the paper is organized as follows. In the first chapter, we provide a review of studies concerning GM model modifications and distinguishing features of our extension. In the second chapter, we describe and analyze two stages of GM model modifications: market-maker's uncertainty about real asset value and errors made by informed traders. In the third chapter, we conduct a comparative analysis of results of modelling the basic GM model and the modified GM model. The findings are attached.

Our violation of modification logic is connected with conservation of research chronology, when we first searched for a solution for the market-maker strategy, as the most complicated stage of GM model modification.

2 Distinguishing Features of Glosten and Milgrom Model Modification

There are numerous studies devoted to analysis of changing or relaxing assumptions of the GM model. Back and Baruch (2004) investigate relations between two major models of market mircostructure: the GM model and the Kyle model (Kyle 1985). The authors show that, under certain conditions, the equilibrium of the GM model converges into one of the equilibrium states of the Kyle model. Thus, an opportunity arises to introduce the concepts of a strategic informed trader, i.e. volume and classical characteristics of market microstructure (tightness, depth and resiliency) in bid–ask spread models. However, the authors emphasize that they managed to construct equilibrium in the GM model only for a special case and in numerical form. It should be noted that most GM and Kyle models studies use the same limitations. Takayama (2013) confirms this tendency in his 2013 paper: research of microstructure price dynamics and information disclosure is not complete, because there is no closed-form solution for equilibrium in a GM model; moreover, it is not yet known if the equilibrium is unique in the Kyle model analytical solution.

Questions about a market-maker's existence and regulation, and necessity and conditions of competition, are closely related with the issue of High Frequency Trading (HFT). HFT replaces classic market intermediaries by providing liquidity to markets. Today, when there is no valid constraint on HFT activity, they have a number of advantages over classic market-makers: information processing and decision-making speed, instant arbitrage on many financial markets and the exclusive right to stop trading at any moment, because they do not have any commitment to maintain liquidity or pricing stability. Gerig and Michayluk (2010) tried to update the GM model by adding multiple assets and introducing HFT into the list of market participants. The authors conclude that the bid–ask spread on a market with a large share of uninformed traders is lower than in the classic model. The opposite situation is observed in a market with a high share of informed traders: bid–ask spread is higher than in the classic model. Thus, HFT activity increases informed traders' transaction costs. After adding elasticity of liquidity traders' demand, Gerig and Michayluk concluded that HFT helps to increase trading volume and generally decreases the transaction costs of other market participants.

Zachariadis (2012) studied the issues of information allocation in time and between market participants. In the GM model, information is distributed evenly and simultaneously between informed traders. Zachariadis (2012) modified the GM model by reducing the difference between informed and uninformed traders. This renders the GM model more realistic, because in reality every market participant has information about the real value of an asset, which changes over time. Thus, information efficiency of asset pricing is not constant over time and does not depend on the ratio of noise and informed traders. The author suggested giving every market participant the ability to learn new information about the real value of an asset from price, spread and volume dynamics, and showed that in spite of eliminating pure noise trading, the main conclusions of the GM model are still correct.

Das (2005) studied the GM model modification when the market-maker has no information about possible real asset value ($\underline{V}$ or $\overline{V}$). At the same time, the market-maker knows the exact time when information that could change the asset price comes to market. Informed traders are still the same, as in the standard GM model. The market-maker is forced to learn real asset value from the actions of market participants (buying, selling). To do that, the author suggests a numerical algorithm for explicitly computing approximate solutions to the expected-value equations for setting prices in an extension of the GM model. Moreover, Das (2005) trained the market-maker in inventory control. Empirical analysis of the artificial time series, obtained during the author's modification of the GM model and real market data, sufficiently differ from one another.

The main and distinguishing features of the GM model modifications discussed above and our work are described in Table 1.

When GM (1985) discussed assumptions about the market-maker's zero profit, they accepted that a specialist accumulates inventory risk.[1] The long-term asset market is rising; thus, according to GM (1985), the market-maker will accumulate shorts on the rising market and will not be able to sell them without losses. The authors argue that the market-maker may remain at break even in the case of competition between market-makers. In GM's opinion, the addition of competition between market-makers must yield the Nash equilibrium (Nash 1951) and so market-makers will remain at break-even. However, competition between market-makers is not always possible. For instance, NYSE assigns only one market-maker for each asset and GM's assumption about market-maker's break-even cannot be satisfied.

Gerig and Michayluk (2010) created grounds for the possible relaxation of GM's assumption about market-makers' competition by taking into account HFT influence. Competition between market-makers, needed to create the Nash equilibrium, to some extent can be replaced with competition between the market-maker and HFT, or only between HFT. This hypothesis needs additional verification and goes beyond the scope of our work.

In contrast to straight incorporation of inventory costs into spread, as Das (2005) did, in our work the introduction of information uncertainty for market participants about real asset value allows us to change dynamic characteristics of the market-maker's inventory costs.

[1]For the GM model, we simulated straight incorporation of inventory costs into the bid-ask spread and this leads to explosive growth of spread and price. Program code and results are available upon the readers' request.

Table 1 Distinguishing features of our GM model modification

Paper	Features and assumptions
Kyle (1985)	In our GM model modification (unique informational event with known for arrival time of information participants and market-makers), we do not use volume and market microstructure characteristics (tightness, depth, resiliency). There is no chance for informed traders to build strategies either
Gerig and Michayluk (2010)	Only one market-maker is used and HFT introduction is not analysed. Moreover, no account is made of the consequences of competition between market-makers or the impact of HFT activity
Zachariadis (2012)	Uneven informational allocation is simulated when the market-maker does not know the possible real asset value ($\underline{V}$ or $\overline{V}$) and informed traders receive information about real asset value with error. The share of informed traders is constant over time. There is no trader who changes from informed trader to noise trader (or vice versa) according to received information during trade
Das (2005)	We use an intuitively simple algorithm to help the market-maker find real asset value, while Das's algorithm is very hard for the market-maker to use, because it takes a very long time to compute the next step We introduce no assumption about the normal distribution of real asset value Noise traders cannot refuse to engage in buy or sell operations when they meet with the market-maker. Informed traders can skip operations, but only if they are not profitable. Informed traders are prohibited from making timing-refusals with a view to profit We do not introduce a market-maker's inventory control function

3 Unknown Real Asset Value and Informed Traders' Errors During the Trading Process

Our GM model extension is divided into two stages. During the first stage, the market-maker loses its knowledge about possible real asset value, while conditions for informed and uninformed traders remain the same. During the second stage, in addition to uncertainty for the market-maker about possible real asset value, we introduce errors of informed participants.

The disruption of the modification sequence is motivated by conservation of research chronology, when in the beginning our task was to search for a market-maker's strategy, since it was the hardest stage of GM model modification.

The first stage of modification is a relaxation of the assumption about the market-maker's knowledge of possible real asset value ($\underline{V}$ and $\overline{V}$). This is the same as in the work of Das (2005). In our study, informed traders can refuse unprofitable transactions. Thus, the market-maker takes into account these actions and, with some additional algorithms, reduces the range in which the real asset value is located.

Indeed, original assumptions of GM about the market-maker's knowledge of possible real asset value moves the GM model away from reality. In the real asset market, especially short-term and medium-term markets, market-makers do not have time to calculate possible real asset value in the event of the arrival of new information. Therefore, the assumption that the market-maker knows only the range in which the real asset value is located seems more realistic.

In contrast to GM's medium-term market, our modification characterizes bid–ask spread formation on the short-term market, where information arrives frequently and at different times, (Gerig and Michayluk 2010). Market is still pure dealership, so all orders are market orders. There are informed traders, uninformed traders and only one market-maker. The trading process is separated into T periods. In every period there is only one deal.

At the beginning of trade, informed traders receive real asset value V, which will be publically known at the moment of time T. The market-maker knows that real asset value is located in the range from $\underline{V}_{MM}$ until $\overline{V}_{MM}$, i.e. $V \in [\underline{V}_{MM}; \overline{V}_{MM}]$. Limits of this range are used by the market-maker to calculate bid and ask price instead of values $\underline{V}$ and $\overline{V}$ of the standard GM model. The market-maker sets the bid and asks price, using knowledge of $\underline{V}_{MM}$, $\overline{V}_{MM}$, estimated share of informed traders μ and direction of price movement δ. In the standard GM model, the δ parameter determines the probability of the real asset value equalling the higher or lower price. In our modification the δ parameter determines the probability that real asset value will be above or below mid-range $[\underline{V}_{MM}; \overline{V}_{MM}]$.

Subsequently, based on the actions of market participants (buy, sell, refusal of a transaction) the market-maker corrects the higher or lower limit of the range $[\underline{V}_{MM}; \overline{V}_{MM}]$ so that one of these limits becomes equal to the real asset value.

After the market-maker has established bid and ask, a random trader observes quotes and makes a decision: to buy or sell the asset or refuse the transaction. We assume that only informed traders can refuse a transaction and they can make a refusal only from non-profitable transactions. There is no timing refusal in our model. Uninformed traders must make a deal at any price.

As in the GM model, we assume that in every moment of time, there is only one transaction and trading volume is limited to one block of assets, e.g. one share.

Informed traders are prohibited from performing manipulative strategies, due to random selection of traders. They cannot evaluate how many times they will participate in trading. Thus, we exclude volume and price manipulation from our modification.

Informed participants buy if real asset value is higher than the ask price of the market-maker. If real asset value is lower than bid price, informed traders sell the asset. Finally, if real asset value is located between bid and ask, informed traders refuse the transaction. In official terms, informed traders' actions can be described by the formula (Eq. (1)).

$$Informed \rightarrow \begin{cases} Sell,\ if\ V < Bid \\ Refuse,\ if\ Bid < V < Ask \\ Buy,\ if\ V > Ask \end{cases}, \tag{1}$$

where *V*—real asset value, *Bid*, *Ask*—market-maker's quotes to sell and buy, *Sell*, *Buy*, *Refuse*—informed traders' actions.

After each transaction or refusal of a transaction, the market-maker reviews the bid and ask prices. A specialist knows that a refusal can only be made by an informed trader when real asset value is between bid and ask quotes. Therefore, the market-maker takes a refusal as a signal to correct the bid and ask quotes in a special way.

After initial designations, we have constructed an event tree for the first stage modification. For the full event tree, please see the appendix. The part of the event tree that corresponds with the interaction of the market-maker and informed trader, when real asset value is located below mid-range $\left[\underline{V}_{MM}; \overline{V}_{MM}\right]$, is shown on Fig. 1.

$Ask < \underline{V}_{MM}$ is the event when the ask is lower than the low limit of the range, where, according to the market-maker's suggestions, the real asset value is located. *AL* is the possibility of event $Ask < \underline{V}_{MM}$. $Bid < \underline{V}_{MM}$ is the event when the bid is lower than the low limit of the range, where, according to the market-maker's suggestions, real asset value is located. *BL*—is the possibility of event $Bid < \underline{V}_{MM}$. Events $Bid > \overline{V}_{MM}$, $Ask > \overline{V}_{MM}$ and their possibilities can be described in the same manner.

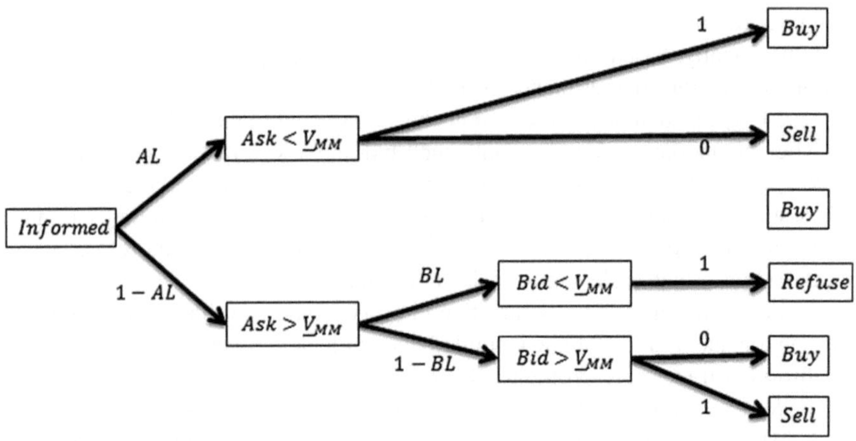

Fig. 1 Part of the event tree of the first stage modification

The difference between event trees of the standard GM model and our modification of the first stage consists of the informed trader's behaviour. The informed trader, as before, uses knowledge of real asset value to maximize his/her profit from every transaction. Consequently, he/she will sell if real asset value is lower than the bid. This happens when $Bid > \overline{V}_{MM}$, or $Bid > \underline{V}_{MM}$ and $Ask > \underline{V}_{MM}$ take place at one time. Moreover, the informed trader will buy if the ask is lower than real asset value. This happens when $Ask < \underline{V}_{MM}$ or $Bid < \overline{V}_{MM}$ and $Ask < \overline{V}_{MM}$ take place at one time. In the classic GM model, $V = \overline{V}$ real asset value is above or equal to ask during the entire trading process. In our modification, bid–ask spread can be above or below real asset value. Thus, informed traders, by making buys and sells, give signals to the market-maker about spread location relative to real asset value.

Moreover, in our first stage modification, it is possible that real asset value is located within bid and ask. This happens when two conditions take place at one time: $Bid < \overline{V}_{MM}$ and $Ask > \overline{V}_{MM}$ or $Ask > \underline{V}_{MM}$ and $Bid < \underline{V}_{MM}$. In this case, the informed trader has no incentives to trade, because he/she will incur losses, which he/she cannot accept due to his/her utility function. Therefore, the informed trader refuses the transaction. Consequently, refusal from trade is a signal to the market-maker that real asset value lies between bid and ask.

We calculated the probability of traders' buys and sells depending upon possible events and placed them in Table 2.

As in the GM model, in our first stage modification, to calculate ask and bid, we use formulas (Eqs. (2) and (3)).

$$Ask = E\left[V \middle| Buy\right] = \underline{V}_{MM} P\left[\underline{V}_{MM} \middle| Buy\right] + \overline{V}_{MM} P\left[\overline{V}_{MM} \middle| Buy\right] \tag{2}$$

$$Bid = E\left[V \middle| Sell\right] = \underline{V}_{MM} P\left[\underline{V}_{MM} \middle| Sell\right] + \overline{V}_{MM} P\left[\overline{V}_{MM} \middle| Sell\right] \tag{3}$$

Table 2 Probability of traders' buys and sells

Target price	Type of trader	Location relative to *Ask*	Location relative to *Bid*	Type of deal	Possibility
$\underline{V}_{MM}$	*Inf*	$Ask < \underline{V}_{MM}$	–	*Buy*	$AL\mu\delta$
$\underline{V}_{MM}$	*Inf*	$Ask < \underline{V}_{MM}$	–	*Sell*	0
$\underline{V}_{MM}$	*Inf*	$Ask > \underline{V}_{MM}$	$Bid > \underline{V}_{MM}$	*Buy*	0
$\underline{V}_{MM}$	*Inf*	$Ask > \underline{V}_{MM}$	$Bid > \underline{V}_{MM}$	*Sell*	$(1-BL)(1-AL)\mu\delta$
$\underline{V}_{MM}$	*Uninf*	–	–	*Buy*	$0.5(1-\mu)\delta$
$\underline{V}_{MM}$	*Uninf*	–	–	*Sell*	$0.5(1-\mu)\delta$
$\overline{V}_{MM}$	*Inf*	–	$Bid > \overline{V}_{MM}$	*Buy*	0
$\overline{V}_{MM}$	*Inf*	–	$Bid > \overline{V}_{MM}$	*Sell*	$BH\mu(1-\delta)$
$\overline{V}_{MM}$	*Inf*	$Ask > \overline{V}_{MM}$	$Bid < \overline{V}_{MM}$	*Buy*	$(1-AH)(1-BH)$ ** $\mu(1-\delta)$
$\overline{V}_{MM}$	*Inf*	$Ask > \overline{V}_{MM}$	$Bid < \overline{V}_{MM}$	*Sell*	0
$\overline{V}_{MM}$	*Uninf*	–	–	*Buy*	$0.5(1-\mu)(1-\delta)$
$\overline{V}_{MM}$	*Uninf*	–	–	*Sell*	$0.5(1-\mu)(1-\delta)$

Using Bayes' rule and probability from Table 2, we derive final formulas for bid and ask calculation. For final formulas of probability from Eqs. (2) and (3), please see the appendix.

After each transaction, the market-maker should review his/her suggestions about δ and the probability of real asset value falling above or below mid-range $\left[\underline{V}_{MM};\overline{V}_{MM}\right]$. Through δ_t we denote the probability of real asset value being located relative to mid-range $\left[\underline{V}_{MM};\overline{V}_{MM}\right]$ after transaction (buy or sell) in step t. For example, if during t period the trader bought an asset from the market-maker, then δ_t will be calculated through the following formula (Eq. (4)).

$$\delta_t\left(Buy_t\right) = \frac{\delta_{t-1}\left(AL\mu + 0.5\left(1-\mu\right)\right)}{\delta_{t-1}\left(AL\mu + 0.5\left(1-\mu\right)\right) + \left(1-\delta_{t-1}\right)\left(\left(1-AH\right)\left(1-BH\right)\mu + 0.5\left(1-\mu\right)\right)} \tag{4}$$

A similar expression can be written for $\delta_t(Sell_t)$.

We did not attempt to construct an analytical form of equilibrium, because during our review of literature we generally found that for GM model modifications, the solution can be written only in numerical form. To simulate reduction of bid–ask spread after the trader's refusal of a transaction, we have chosen a simple and intuitive bisection method. After refusal, the market-maker determines the direction of the last transaction and corrects the corresponding limit of the range $\left[\underline{V}_{MM};\overline{V}_{MM}\right]$ with half the size of the spread. For instance, if the last transaction was closer to the lower limit of range $\left[\underline{V}_{MM};\overline{V}_{MM}\right]$, then correction can be calculated by the formula (Eq. (5)).

$$\underline{V}_{MM}(t) = \underline{V}_{MM}\left(t-1\right) + \left(Ask_t - Bid_t\right)/2, \tag{5}$$

where $\underline{V}_{MM}(t)$ is the new value of the lower limit of the range, $\underline{V}_{MM}\left(t-1\right)$ is previous value of lower limit of range.

For a higher limit of range correction will be (Eq. (6)).

$$\overline{V}_{MM}(t) = \overline{V}_{MM}\left(t-1\right) - \left(Ask_t - Bid_t\right)/2 \tag{6}$$

In addition to the formulas of bid–ask spread, we have calculated the inventory accumulation of the market-maker and his/her financial result to study inventory risk. Inventory accumulation of the market-maker is calculated through formula (Eq. (7)).

$$rsNT_t = \sum_{k=1}^{t} trade_k,\ t \in \left[1, T\right], \tag{7}$$

where $rsNT_t$ is total specialist inventory after t transactions, T is the whole number of operations that the market-maker made before public disclosure of real asset value; $trade_k$ is the quantitative result of the transaction that reflects changes in the specialist's cash. The quantitative result of the transaction is calculated through formula (Eq. (8)).

$$trade_k = \begin{cases} 1, if\ Sell \\ -1, if\ Buy \end{cases}, \quad (8)$$

where *Sell*, *Buy* are the market-maker's sells or buys of an asset.

The financial result of the market-maker is the difference between committed transactions and inventory liquidation at current price (Eq. (9)).

$$rsPT_t = \sum_{k=1}^{t} Price_k * trade_k - rsNT_t * Price_t, t \in [1, T], \quad (9)$$

where $rsPT_t$ is the financial result of the specialist after committing t transactions; $Price_t$ is price of t transaction.

Thus, our first stage of modification can be presented as follows:

- Unit of the market-maker's valuations and decisions (we use Eqs. (2)–(6)).
- Unit of traders' valuations and decisions (we use probabilities at the end of nodes of the event tree).
- Unit of statistical computations (we use Eqs. (7)–(9)).

During the second stage of modification, to simulate information uncertainty of informed traders, we introduce a determined share of informed traders' errors. Moreover, we assume that additional information uncertainty of informed traders will help the market-maker to reduce inventory during the trading process.

Before the beginning of trade, the informed trader assumes that real asset value will be equal to $\underline{V}$ or $\overline{V}$. At the beginning of trading, the informed trader receives a signal that future real asset value will be $\overline{V}$ with the probability π_I. Consequently, the probability that real asset value will be $\underline{V}$ is equal to $(1 - \pi_I)$. The informed trader takes into account this signal during transactions and trades in the direction of $\overline{V}$ with the probability π_I and in the direction of $\underline{V}$ with the probability of $(1 - \pi_I)$.

Part of the tree that corresponds to the event when real asset value $V = \overline{V}$ is located above mid-range $\left[\underline{V}_{MM}; \overline{V}_{MM}\right]$ is shown on Fig. 2. For the full event tree, see the appendix.

Designations on Fig. 2 are equal to Fig. 1.

Probability of buys and sells of the second-stage modification can be calculated similarly to the probabilities in Table 2.

Formulas of bid and ask calculation can be derived using the same algorithm as in the first-stage modification, taking into account the probability of informed traders' errors.

The market-maker corrects limits of range $\left[\underline{V}_{MM}; \overline{V}_{MM}\right]$ using the same algorithm as in the first-stage modification (Eqs. (5) and (6)). After correction of the limit, the market-maker reviews bid and ask.

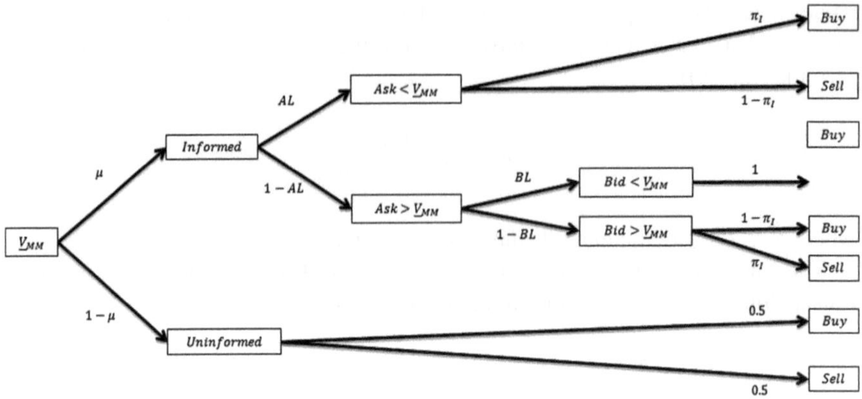

Fig. 2 Part of the event tree of the second-stage modification

4 Simulation Results

Our numerical simulation includes the GM model, first-stage modification (the market-maker's uncertainty about real asset value) and second-stage modification (simultaneous uncertainty of the market-maker and the informed trader about real asset value).

We made only one trial simulation of model modification to study changes in inventory risk, price, spread and the market-maker's financial result.[2] This simulation is only the first attempt and we recognize the need for further simulations to test our results.

We performed the simulation according to the following conditions: $\underline{V} = 10\$$ (low value), $\overline{V} = 20\$$ (high value), $\delta = 0.5$ (starting possibility of $V = \underline{V}$), $\mu = 0.2$ (share of informed traders), $T = 1{,}000$ (time periods). At the beginning of the trading period, informed traders receive a signal that real asset value will be $V = \overline{V} = 20\$$.

For the first-stage modification, we added conditions that $\underline{V}_{MM} = 13\$$, $\overline{V}_{MM} = 22\$$ (the range where the market-maker assumes the location of real asset value). Under these conditions at the beginning of trading, the informed traders receive a signal that real asset value will be $V = \overline{V} = 20\$$ and the market-maker learns the real asset value by taking into account the traders' actions. For the second-stage modification, we introduce an additional parameter: the probability of informed traders' error, $\pi_I = 0.85$.

---

[2]We made our simulation using a computer program, which is written in R. We can provide the program code upon the reader's request. You can send your request to the author via e-mail. We have a code to simulate the GM model for straight incorporation of inventory risk into bid-ask spread, and the first and the second stages of modification.

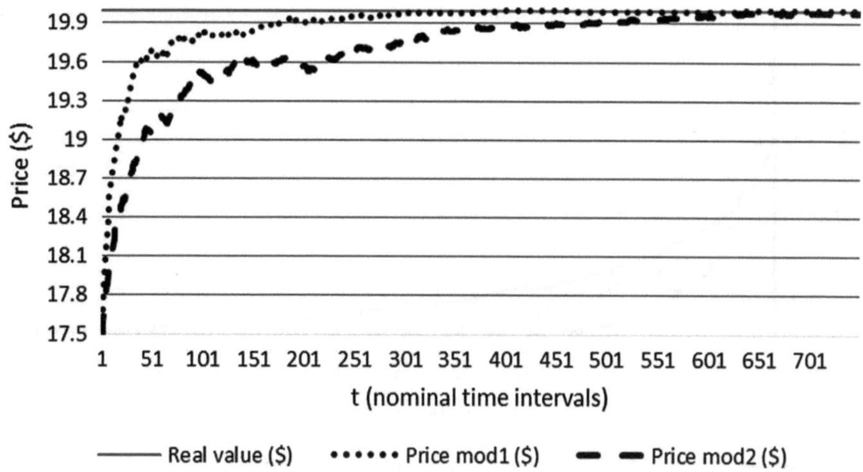

Fig. 3 Price change for the first stage (*Price mod1*) and the second stage of (*Price mod2*) modification

We repeated our tests 500 times. After that, we took the averaged value for each observed variable.

Price change graphs for the first- and second-stage modifications are shown in Fig. 3.

Simulation results prove that the market-maker manages to determine real asset value in spite of the absence of knowledge about higher and lower prices ($\overline{V}$ or $\underline{V}$). Thus, we manage to maintain the key properties of the GM model (incorporation of informed trader information in market price) even when the market-maker has no information about possible real asset value.

One can note that the quantity of trades needed for market price to become equal to real asset price significantly increases from the first stage to the second stage of modification. This result is quite predictable, because informed traders start to make errors and trade in the opposite direction from real asset value. Thus, the share of buys increases more slowly and the market-maker gradually changes bid and ask quotes.

Graphs of bid–ask spread change for the GM model for the first- and second-stage modification are shown in Fig. 4.

With an increase in number of transactions and refusals, the spread in our modifications tends toward the lower. Consequently, knowledge of possible real asset value ($\underline{V}$ or $\overline{V}$) is not a necessary assumption for the market-maker to find real asset value during the trading process. The spread decreases more slowly in the first-stage modification than in the GM model. This can be explained by increased uncertainty for the market-maker about real asset value. The spread in the second-stage modification decreases even more slowly, because actions of informed traders bring less information through errors.

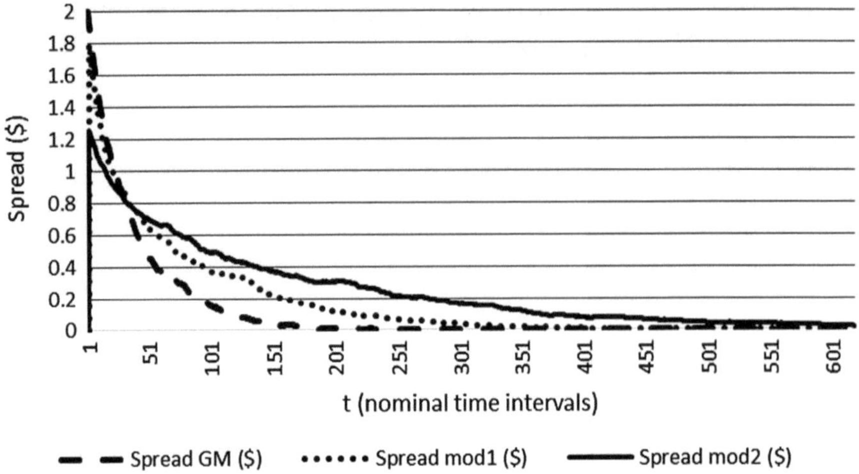

Fig. 4 Bid–ask spread change for GM model (*Spread GM*) the first (*Spread mod1*) and the second (*Spread mod2*) stage modification

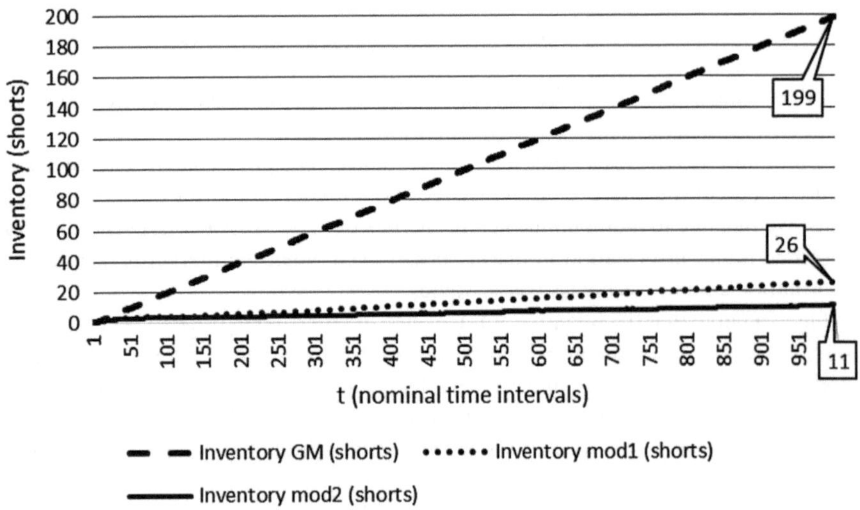

Fig. 5 Market-maker's inventory change for GM model (*Inventory GM*) with the first-stage (*Inventory mod1*) and the second-stage (*Inventory mod2*) modification

Graphs of the market-maker's inventory change for the GM model, for the first- and the second-stage modifications, are shown in Fig. 5:

Due to the decrease of price learning speed, inventory accumulation becomes smoother and the total volume significantly reduces in both the first and the second stages of modification in contrast to the GM model. It should be noted that we have no inventory control function, as did Das (2005). However, we reserve the right to introduce an additional inventory control function and to include the remaining inventory costs into the spread.

Graphs of a market-maker's financial result for the GM model, and for the first and the second stages of modification, are shown in Fig. 6:

Market-maker's financial result for the first stage significantly increases in comparison with the GM model. The rate of profit growth is proportional to spread decrease. We can conclude that our algorithm of bid–ask spread correction is very attractive to the market-maker, because he/she earns profits. On the other hand, this result shows that our algorithm is non-optimal, because the market-maker must earn zero profit and has no losses. Due to the increased period of spread decrease in the second-stage modification, the market-maker earns even bigger profits than in the first-stage modification.

It should be noted that a positive financial result is not the purpose of our work. Without additional study, we cannot affirm that this financial result is the outcome of relaxation assumptions of the GM model. In any event, GM's proposition about the introduction of competition between market-makers should deprive them from profits. Even in the absence of competition between market-makers, conclusions Gerig and Michayluk (2010) show that some aspects of the competition effect are created by HFT, which gradually drive out market-makers. Thus, we returned to our starting point. However, we made a spiral motion instead of a circular motion, because we achieved significant approximation of reality of the GM model.

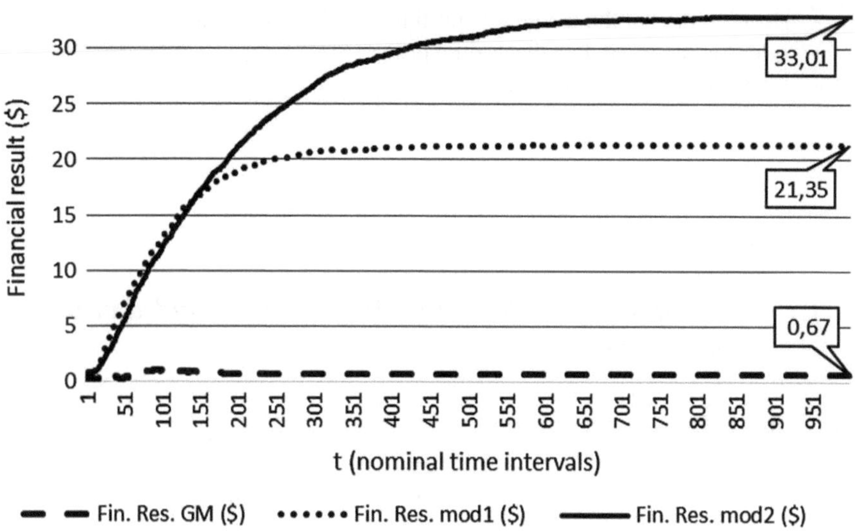

Fig. 6 Financial result for GM model (*Fin. Res. GM*), with the first (*Fin. Res. mod1*) and the second (*Fin Res. mod2*) stages of modification

Conclusion
In our work, after the introduction of the additional uncertainty about the real asset value, we propose an algorithm of bid–ask spread formation for the market-maker, based on classical model of Glosten and Milgrom (1985). Our modification allows us to reproduce the intertemporal spread dynamics under the asymmetric information and limited inventory risk of a market-maker.

Our modification of the GM model is divided in two steps. During the first step of modification we introduced uncertainty of market-maker's expectations about real asset value. The market-maker knows only the range, where real asset value is located and learns it only by actions of traders. The second step introduces information uncertainty of informed traders. We modelled this process by introduction of a certain percentage of errors of informed traders during transactions.

Results, obtained after the modification of the GM model, allow us to make a step closer in building complex model of bid–ask spread formation. This model will include all three factors of spread formation: transaction costs, inventory risk and information asymmetry.

Our study can be extended in the following areas:

- Development of a model with the introduction of uncertainty about frequency and arrival time of new information for traders and the market-maker
- Development of the learning algorithm of informed traders
- Introduction of HFT in the trading process
- Incorporation of transaction costs

A.1 Appendix

A.1.1 Calculation of Probabilities, Which Are Included in Bid and Ask Quotes

According to Bayes' rule:

$$P\left[\underline{V}_{MM}\middle|Buy\right] = \frac{P\left[Buy\middle|\underline{V}_{MM}\right]P\left[\underline{V}_{MM}\right]}{P\left[Buy\right]}$$

$$P\left[\overline{V}_{MM}\middle|Buy\right] = \frac{P\left[Buy\middle|\overline{V}_{MM}\right]P\left[\overline{V}_{MM}\right]}{P\left[Buy\right]}$$

Full probabilities of making buy and sell can be described by formulas:

$$P\,[Buy] = AL\mu\delta + 0.5\,(1-\mu)\,\delta + (1-AH)\,(1-BH)\,\mu\,(1-\delta) + 0.5\,(1-\mu)\,(1-\delta)$$

$$P\,[Sell] = (1-BL)\,(1-AL)\,\mu\delta + 0.5\,(1-\mu)\,\delta + BH\mu\,(1-\delta) + 0.5\,(1-\mu)\,(1-\delta)$$

Integrating full probabilities into formulas, obtained according to Bayes' rule, yields final formulas for calculating probabilities, which are included in bid and ask quotes. For example, the probability that real asset value asset is equal to lower limit of range $\left[\underline{V}_{MM}; \overline{V}_{MM}\right]$ can be calculated through this formula:

$$P\left[\underline{V}_{MM}\middle|Buy\right] = \frac{\delta\,(AL\mu + 0.5\,(1-\mu))}{\delta\,(AL\mu + 0.5\,(1-\mu)) + (1-\delta)\,((1-AH)\,(1-BH)\,\mu + 0.5\,(1-\mu))}$$

Conditional probabilities $P\left[\overline{V}_{MM}\middle|Buy\right]$, $P\left[\underline{V}_{MM}\middle|Sell\right]$ and $P\left[\overline{V}_{MM}\middle|Sell\right]$ can be calculated in a similar way.

A.1.2 Event Tree for the First-Stage GM Model Modification

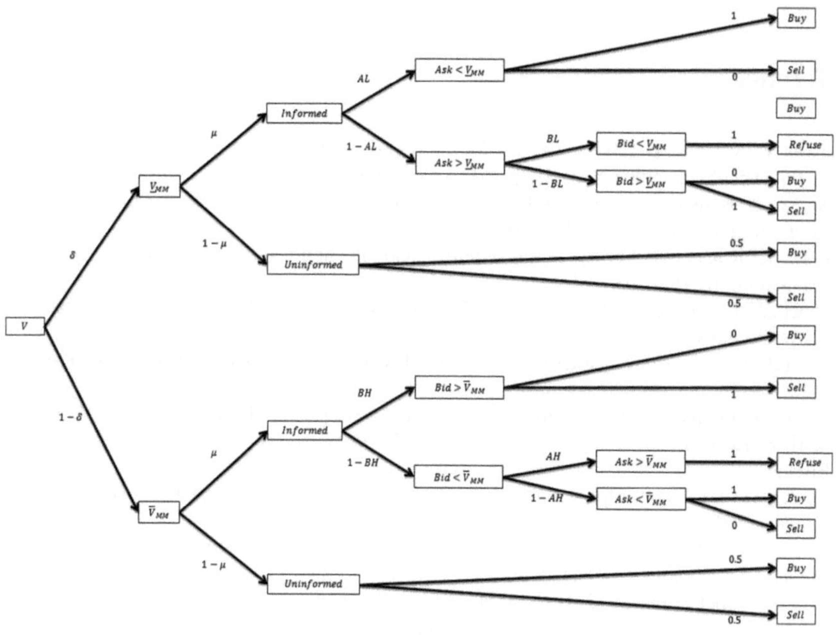

A.1.3 Event Tree for the Second-Stage GM Model Modification

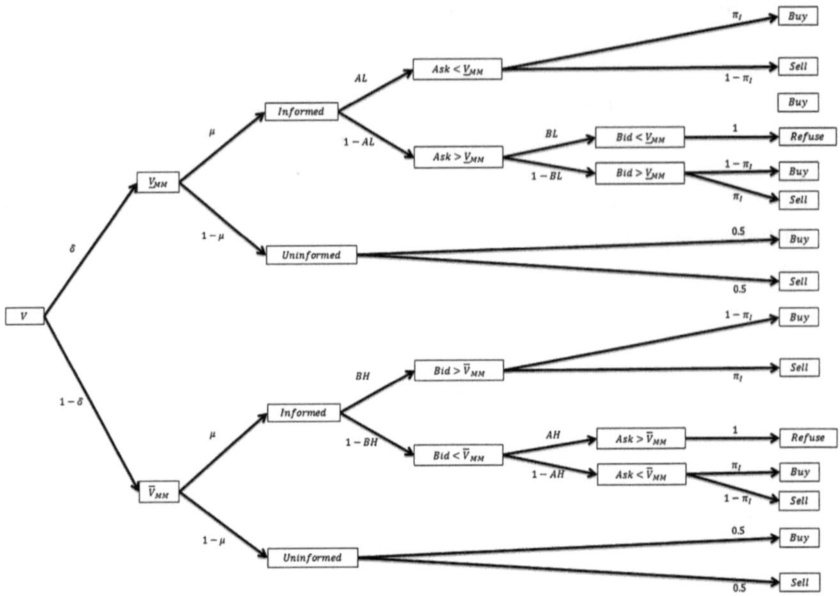

Acknowledgments The author would like to thank Igor Zakharov for his helpful comments, suggestions and support throughout the work on this paper.

References

Agarwal, V., & Wang, L. (2007). *Transaction costs and value premium*. CFR Working Paper, No. 07-06. http://hdl.handle.net/10419/57763. Accessed 9 Aug 2013.

Back, K., & Baruch, S. (2004). Information in securities markets: Kyle meets Glosten and Milgrom. *Econometrica, 72*, 433–465.

Ball, L., & Romer, D. (1990). Real rigidities and the non-neutrality of money. *Review of Economics, 57*, 183–203.

Campbell, J., Lo, A., & MacKinlay, A. (1997). *The econometrics of financial markets*. Princeton: University Press.

Das, S. (2005). A learning market-maker in the Glosten–Milgrom model. *Quantitative Finance, 5*(2), 169–180.

Easley, D., & O'Hara, M. (2003). Microstructure and asset pricing. In M. C. George, H. Milton, & H. S. Rene (Eds.), *Handbook of the economics of finance* (pp. 1022–1047). Amsterdam: Elsevier.

Fama, E. (1970). Efficient capital markets: A review of theory and empirical work. *The Journal of Finance, 25*(2), 383–417.

Fama, E., & French, K. (1993). Common risk factors in the returns on stocks and bonds. *Journal of Financial Economics, 33*(1), 3–56.

Gerig, A., & Michayluk, D. (2010). *Automated liquidity provision and the demise of traditional market making*. Retrieved 9 Aug 2013 from http://arxiv.org/pdf/1007.2352v1.pdf.

Glosten, L., & Milgrom, P. (1985). Bid, ask and transaction prices in a specialist market with heterogeneously informed traders. *Journal of Financial Economics, 14*, 71–100.
Kyle, A. (1985). Continuous auctions and insider trading. *Econometrica, 53*, 1315–1336.
Nash, J. (1951). Non-cooperative games. *The Annals of Mathematics, 54*(2), 286–295.
Roll, R. (1984). A simple measure of the effective bid/ask spread in an efficient market. *Journal of Finance, 39*, 1127–1139.
Stoll, H. R. (1978). The supply of dealer services in securities markets. *Journal of Finance, 33*(4), 1133–1151.
Takayama, S. (2013). *Price manipulation, dynamic informed trading and uniqueness of equilibrium in a sequential trade model*. Retrieved 9 Aug 2013 from http://www.kier.kyoto-u.ac.jp/~game/2010/100520Takayama.pdf.
Zachariadis, K. (2012). *A baseline model of price formation in a sequential market*. London School of Economics. Retrieved 9 August 2013 from http://papers.ssrn.com/sol3/papers.cfm?abstract_id=1661825.

On the Modeling of Financial Time Series

Aleksey Kutergin and Vladimir Filimonov

Abstract This paper discusses issues related to modeling of financial time series. We discuss so-called empirical "stylized facts" of real price time-series and the evolution of financial models from trivial random walk introduced by Louis Bachelier in 1900 to modern multifractal models, that nowadays are the most parsimonious and flexible models of stochastic volatility. We focus on a particular model of Multifractal Random Walk (MRW), which is the only continuous stochastic stationary causal process with exact multifractal properties and Gaussian infinitesimal increments. The paper presents a method of numerical simulation of realizations of MRW using the Circulant Embedding Method and discuss methods of its calibration.

Keywords Circulant Embedding Method • Estimation of parameters • Financial time series • Multifractal Random Walk • Numerical simulations • Stylized facts

1 Introduction

Financial modeling is being one of the most actively evolved topics of quantitative finance for many decades. Having a numerous practical applications especially in the fields of derivative pricing or risk management, it is of an extreme interest of academic research as well. Financial markets are a global social system in which many agents make decisions, every minute being exposed to risk and uncertainty. Participants interact with each other trying to make profit, taking into account not only recent news and internal market events but also action of other participants as well. However the result of such complex behavior is reduced to a small set of entities, by far the most important of which is the price of some given asset.

A. Kutergin (✉)
Prognoz Risk Lab, Perm, Russian Federation
e-mail: kutergin@prognoz.ru

V. Filimonov
Department of Management, Technology and Economics ETH Zürich, Zürich, Switzerland
e-mail: vfilimonov@ethz.ch

A.K. Bera et al. (eds.), *Financial Econometrics and Empirical Market Microstructure*, DOI 10.1007/978-3-319-09946-0_10

The first attempt to describe observable assets price dynamics was made by Bachelier in 1900 with his seminal work "Théorie de la spéculation" (Bachelier 1900). Bachelier suggested that asset prices follows random walk. In other words, increments (returns) of the price are independent identically distributed (iid) random variables which he suggested to have Gaussian probability distribution. Having many merits, such simple model is not able to fully account for complex interaction of many random factors underling the price formation process. However, analytical tractability due to simple construction and underlying Gaussian distribution, allowed to construct on top of the random walk process many financial theories, such as Black and Scholes option pricing theory or Markowitz portfolio theory.

Rebounding from the naive random walk, the evolution of the financial modeling has brought a variety of models that are aimed to describe the complex statistical properties of real price time series, summarized in the so-called "stylized facts". One of the most widely used is the "conditional volatility" models such as ARCH/-GARCH family, that model volatility as an autoregressive process on the past values of volatility and returns as well. Another class of models represent volatility as some stochastic process. The most interesting subclass of these "stochastic volatility" models are so-called multifractal models. Despite having simple construction, multifractal models are able to represent most of the empirically observed "stylized facts". In this paper we focus on particularly one multifractal model, namely on the Multifractal Random Walk (MRW) that was proposed by Bacry, Delour and Muzy in 2000 (Bacry et al. 2001), which is the only continuous stochastic stationary causal process with exact multifractal properties and Gaussian infinitesimal increments. We describe the procedure of numerical simulation of the realization of MRW process and discuss issues related to estimation of parameters.

2 Stylized Facts

Though the empirical analysis of asset price time series has been performed for more than half a century, only the development of computerized trading in 1980s allowed to record enough data for robust statistical analysis. Recent two decades of evolution of IT infrastructure opened new horizons for empirical finance by bringing huge amount of high-frequency data, numerical tools and computational power for analysis. Nowadays every large exchange record terabytes of high-frequency data at every trading session.

The study of these new financial datasets results in a number of empirical observations quantified with robust statistical methods. Interestingly, some of these statistical laws were found to be common across many types of sufficiently liquid instruments on many markets. These common statistical laws, discovered independently by many researchers, were called "stylized facts" of financial time series (Cont 2001; Bouchaud and Potters 2000; Lux 2009). Here we present a non-exhaustive list of most important of "stylized facts":

1. **Absence of linear autocorrelation** in assets returns except short intra-day time scales, where effects of market microstructure plays substantial role. Absence of linear autocorrelation is perfectly described by the naive model of Bachelier (1900), though this is almost the only "stylized fact" that this model could reproduce. For this reason "stylized facts" could be viewed as a collection of facts that differ real price dynamics from the random walk. Despite its simplicity, absence of linear autocorrelation (which could be stated in other words as absence of linear predictability) of financial returns plays extremely important role in financial modeling. This observation was embedded in the so-called *no arbitrage hypothesis* and *Efficient Market Hypothesis (EMH)* (Fama 1970, 1991), that in a very broad sense claim impossibility of obtaining excess returns (more than a risk-free rate) without being exposed to risk.
2. **Long memory in volatility.** Despite the absence of linear autocorrelation in signed price returns, autocorrelation function of absolute (or squared) returns decays very slowly and are statistically significant even on scales of hundreds days for intra-day returns.
3. **Heavy tails in probability distribution of asset returns.** The Bachelier's random walk model assume probability density function (pdf) of iid increments (returns) to be Normal. However empirical analysis of real financial time series at many time scales show that pdf of asset returns is very skewed, having narrow peak and tails decaying as a power law with exponent γ in the range $2 \lesssim \gamma \lesssim 4$ for intraday and daily time scales. Such "heavy tails" of asset returns distribution accounts for the presence of extreme events (large positive or negative returns) in real time series in contrast to the idealized Gaussian random walk model.
4. **Aggregational Gaussianity.** The distribution function of the returns is not the same at different time scales and exponent γ of the tail of pdf depends in fact on the scale over which returns are calculated and increase with increase of this time scale. With moving from intraday returns towards weekly or monthly returns the distribution slowly converges towards Gaussian distribution and for quarterly or annual returns one typically can not reject the null hypothesis of normal distribution. In a way this "stylized fact" is a direct consequence of the Central Limit Theorem and the fact that exponent γ was never found smaller than 2, which ensures finite variance of returns.
5. **Volatility clustering.** Presence of long memory in volatility and heavy tails of returns merged in an interesting observation that of sufficient irregularity of returns time series. Its typical pattern has periods of high volatility, which are followed by periods of low volatility, and vice versa. In other words, volatility bursts tend to group into clusters.
6. **Multifractal properties.** Above properties (heavy tails, absence of autocorrelation, long-range memory in volatility) are observed at various time scales, implying scale invariance of financial time series. More specifically, financial time series are found to exhibit so-called multifractal properties (we discuss it in details in Sect. 4) (Muzy et al. 2000; Arneodo et al. 1998b; Calvet and Fisher 2002; Liu et al. 2008).

7. **Leverage effect.** One important observation of the financial time series is absence of time-reversal symmetry. In other words, statistical properties of time series in direct time and reversed time are different. Leverage effect describes particular aspect of time-reversal asymmetry in terms of correlation function between returns and volatility

$$L(\tau) = \frac{\mathrm{E}\left[r_{t+\tau}^2, r_t\right]}{\mathrm{E}\left[r_t^2\right]^{3/2}}. \quad (1)$$

which is negative for $\tau >= 0$ and decay to zero with $\tau \to \infty$ and almost zero for $\tau < 0$ (Bouchaud et al. 2001). In other words, past returns are negatively correlated with future volatility, but past values of volatility do not correlate with future signed returns, satisfying the absence of arbitrage hypothesis.
8. **Gain-loss asymmetry.** Leverage effect is tightly linked with another breaking of symmetry in behavior of asset prices. In stocks, indices and their derivatives one could observe large drawdowns in prices but not equally large drawups (this is typically not true for exchange rates that are highly symmetrical in price movements). Moreover, it typically takes longer time to reach a gain of a certain value than a loss of a same value (Siven and Lins 2009).
9. **Volume-volatility correlation.** Most of the statistical properties of volatility could be observed in series of trading volume as well. Moreover, trading volume is correlated with all measures of volatility.
10. **Extreme events.** Finally, more than 400 years of history of financial markets have shown that bubbles and crashes are not exceptions and observed extremely often in different markets. Statistical properties of such extreme events are non-trivial and differ from statistical properties of the financial time series in normal regimes (Sornette 2003).

3 Brief Review of Financial Time-Series Models

As discussed above, the naive random walk model is too simple to describe the complexity of price dynamics. The fractional Brownian motion (Mandelbrot and Van Ness 1968), which is the natural extension of the random walk that accounts for long memory, can not be directly applied for modeling due to the presence of memory both in volatility and signed returns (violation of the "no arbitrage hypothesis"). The most obvious way to account for the absence of linear autocorrelation, but preserve structure in volatility is to separate noise term from volatility term in the equation for returns in the following multiplicative manner:

$$r_t = \xi_t \sigma_t, \quad (2)$$

where ξ_t structure-less represents iid innovations (often considered to be Gaussian) and σ_t represents volatility of the process. For $\sigma_t \equiv \sigma_0 =$ const one recovers the simple random walk model. Depending on the structure of σ_t models are typically classified into two groups: *stochastic volatility* models,[1] where σ_t is modeled as an independent from r_t and ξ_t stochastic process, and *conditional volatility* models where σ_t is defined as a functional form of the past values of r_τ and ξ_τ (for $\tau < t$).

One of the most well-known models of conditional volatility family are Autoregressive Conditional Heteroscedasticity model (ARCH) (Engle 1982) and Generalized Autoregressive Conditional Heteroscedasticity (GARCH) (Bollerslev 1986) models. They are successfully used for reproducing volatility clustering and non-trivial (but though sufficiently short) memory in volatility. The ARCH/GARCH models gave birth to the whole family that accounts for more than 50 different models (see review in Bollerslev (2010)), the most popular of which are: t-GARCH with innovations having t-Student distribution that reproduces heavy tails of returns distribution; EGARCH (Exponential GARCH) and T-GARCH (Threshold GARCH) that model leverage effect; FIGARCH (Fractional Integrated GARCH), MS-GARCH (Markov Switching GARCH) and LM-GARCH (Long Memory GARCH) that account for long memory in volatility and some others. Without spending time on discussion of all of them we suggest a number of reviews and handbooks with details, such as (Engle 2001; Aït-Sahalia and Hansen 2009; Zakoian and Francq 2010). Being very flexible with respect to modifications, ARCH/GARCH family is constrained with its autoregressive form and does not allow to easily and parsimoniously combine different stylized facts within one model. However due to its simplicity and sufficient robustness the whole family is being very widely used nowadays.

In the present paper we focus on another class of models—stochastic volatility models, and in particular at its subclass of so-called multifractal models. The theory of multifractal random processes started with rethinking and generalization of cascade models that was introduced by Richardson (1961) and Kolmogorov (1941, 1962). Being proposed to model velocity in turbulence, they reflect the fact, that in turbulent gas or fluid flow energy is transferred from large-scale vortices to small-scale vortices by cascades where structures at different scales are similar to each other (resulting in *self-similarity* of the whole system). Similar cascade structures for returns at different time scales were observed at financial markets as well (Ghashghaie et al. 1996), and the idea of self-similar cascades were used in several models, most successful of which are Multiplicative Cascades Model (MCM) (Breymann et al. 2000) and Markov Switching Multifractal (MSM) model (Calvet and Fisher 2008, 2004). When MCM model successfully reproduced heavy tails of returns distribution, long memory in absolute returns and volatility clustering, practical application of this model is limited due to nontransparent

[1] We must notice that typically stochastic volatility models are defined not within the framework of Eq. (2), but as an extension of stochastic differential equation of the geometric Brownian motion. Strictly speaking, these equations do not always have solution in form of (2).

parametrization and absence of robust method of parameters estimation. In contrast, when calibration of the MSM model is relatively simple and sufficiently robust, the model describes heavy tails and long memory only in the limit of infinite number of components. However despite these drawbacks and unclear economic underpinning and a rather artificial discrete hierarchical structure, MSM model was shown to be much better in terms of volatility forecasting than GARCH and some of its siblings (Calvet and Fisher 2004).

The multifractal random walk (MRW) (Bacry et al. 2000, 2001) is the only continuous stochastic stationary causal process with exact multifractal properties and Gaussian infinitesimal increments. Being first introduced within stochastic volatility framework (2), later it was shown to have also exact cascade representation (Bacry et al. 2008; Bacry and Muzy 2003). The exact multifractality comes with a cost of a delicate tuning to a critical point associated with logarithmic decay of the correlation function of the log-increment up to an integral scale. As a consequence, the moments of the increments of the MRW process become infinite above some finite order, which depend on the intermittency parameter of the model. The extension of MRW—Quasi-Multifractal (QMF) model (Saichev and Filimonov 2008, 2007; Saichev and Sornette 2006)—was free of these drawbacks. Rather than insisting on the exact multifractal properties, QMF model described process that was approximately multifractal within the finite range of scales. This approximation makes the model more flexible and removes above contradictions. Being very successful in reproducing many stylized facts, most of multifractal models failed to describe leverage and gain-loss asymmetry effects. To account for asymmetry effects, the Skewed MRW (Pochart and Bouchaud 2002) explicitly introduce the negative correlations between returns and volatility. More parsimonious way was implemented in the so-called Self-Excited Multifractal (SEMF) model (Filimonov and Sornette 2011), which describes the self-reinforcing feedback mechanism (explicit dependence of the future returns on the dynamics of past returns) in a manner similar to autoregressive models.

4 Multifractal Formalism for Stochastic Processes

Original definition of fractal was proposed by Mandelbrot with respect to sets. He defined fractal as a mathematical set with fractal dimension is strictly larger than its topological dimension (Mandelbrot 1975, 1982). Later he extends this definition, calling a fractal any kind of self-similar structure (Mandelbrot 1985).

For the stochastic processes the notion of fractality is based on the definition of self-affine processes—processes that keep statistical properties under any affine transformations. Being more strict, a stochastic process $X = \{X(t);\ t \geqslant 0; X(0) = 0\}$ is called *self-affine*, if for $\forall\, c > 0$ and time moments $t_1, \ldots, t_k \geq 0$, the following expression holds:

$$\{X(c\,t_1), \ldots, X(c\,t_k)\} \stackrel{d}{=} \{c^H X(t_1), \ldots, c^H X(t_k)\}, \tag{3}$$

where H is a constant named *self-affine index* and symbol "$\stackrel{d}{=}$" stands for equality in distribution. For stochastic processes with stationary increments $\delta_l X(t) = X(t+l) - X(t)$ the self-affinity is usually defined not using the multivariate probability distribution functions as in (3) but via moments of increments:

$$M_q(l) = \mathrm{E}[|\delta_l X(t)|^q] = \mathrm{E}[|X(t+l) - X(t)|^q], \tag{4}$$

where $\mathrm{E}[\ldots]$ stands for averaging over ensemble of realizations. Functional form, which describes the dependency of moments as a function of q, plays the key role in the determination of scale invariance properties. If all moments of increments $M_q(l)$ can be represented in a power law form:

$$M_q(l) = K_q l^{\zeta_q}, \tag{5}$$

where K_q and ζ_q depends only on q, then the process $X(t)$ is said to have *multifractal properties*. The functional dependency of scale indices ζ_q on order q ($\zeta_q = f(q)$) is called a *multifractal spectrum*, and its form defines the self-similarity properties of the process: stochastic process $X(t)$ is said to have *monofractal properties* if ζ_q is a linear function of q ($\zeta_q = qH$). If ζ_q is a nonlinear function of q then the process is said to have *multifractal properties*. Typical examples of monofractal processes are random walks, their genralisation of fractional Brownian motion and Lévi flights. The examples of multifractal processes were discussed in previous section.

We need to mention, that multifractal properties cannot be maintained for arbitrary small or arbitrary large scales l. For strictly convex or strictly concave function ζ_q the interval of scales is bounded either from below or above correspondingly (Mandelbrot et al. 1997). Alternatively, function ζ_q may have both convex, concave and linear part (like in the QMF model (Filimonov 2010)) and there exists strictly bounded interval of scales

$$\tau \leqslant l \leqslant L, \tag{6}$$

where scale index ζ_q in (5) has nonlinear dependency with respect to q. Analogically to turbulence theory, interval (6) is called inertial. Similarly to the theory of turbulence, scale τ is called *scale of viscosity* and scale L is an *integral scale*. Finally, it can be shown analytically that multifractal spectrum of the strictly nondecreasing stochastic process has to satisfy following condition (Filimonov 2010):

$$\zeta_q \geqslant 1, \qquad \text{when} \qquad q > 1. \tag{7}$$

5 Multifractal Random Walk Model

5.1 Model Description

As discussed above, continuous Multifractal Random Walk (MRW) (Bacry et al. 2000, 2001) process is the only continuous stochastic stationary causal process with exact multifractal properties and Gaussian infinitesimal increments. It is defined as a continuous limit

$$X(t) = \lim_{\Delta t \to 0} X_{\Delta t}[t] \tag{8}$$

of the discrete random process of following type

$$X_{\Delta t}[t] = \sum_{k=1}^{t/\Delta t} \delta_{\Delta t} X_{\Delta t}[k\Delta t] = \sum_{k=1}^{t/\Delta t} \xi_{\Delta t}[k]\, e^{\omega_{\Delta t}[k]}, \tag{9}$$

where $X_{\Delta t}[0] = 0$; $\xi_{\Delta t}[k]$ is iid Gaussian noise with zero mean and variance $\sigma^2 \Delta t$ and $\omega_{\Delta t}[k]$ is another Gaussian process uncorrelated with the first one ($\mathrm{Cor}[\xi_{\Delta t}[i], \omega_{\Delta t}[j]] = 0,\ \forall i, j$). In financial applications process (8) can be interpreted as the process for logarithm of price and process $\delta_{\Delta t} X_{\Delta t}[k\Delta t]$ can be viewed as the process for log-returns. Process $\omega_{\Delta t}[k]$ can also be considered as a log-volatility.

According to the definition, $\omega_{\Delta t}[k]$ has zero mean and logarithmically decaying covariance function

$$\mathrm{Cov}[\omega_{\Delta t}[k_1], \omega_{\Delta t}[k_2]] = \lambda^2 \ln \rho_{\Delta t}[|k_1 - k_2|], \tag{10}$$

where

$$\rho_{\Delta t}[k] = \begin{cases} \frac{L}{(|k|+1)\Delta t}, & |k| \leqslant \frac{L}{\Delta t} - 1; \\ 1, & |k| > \frac{L}{\Delta t} - 1. \end{cases}$$

One can see that here the support of the autocorrelation function of process $\omega_{\Delta t}[k]$ is bounded from above by the value of integral scale L.

Process (8) for scales $l \leqslant L$ has strict multifractal properties and parabolic multifractal spectrum

$$\zeta_q = \left(\frac{1}{2} + \lambda^2\right) q - \frac{\lambda^2}{2} q^2. \tag{11}$$

For scales $l \geqslant L$ process (8) has monofractal properties and spectrum $\zeta_q = q/2$, which is identical to the spectrum of regular Brownian motion with Hurst exponent $1/2$.

As discussed, exact multifractal properties comes along with two significant shortcomings. First, the variance of the process $\omega_{\Delta t}[k]$

$$\mathrm{E}\left[\omega_{\Delta t}[k]^2\right] = \lambda^2 \ln \frac{L}{\Delta t} \tag{12}$$

is infinite in the limit

$$\lim_{\Delta t \to 0} \mathrm{E}\left[\omega_{\Delta t}[k]^2\right] = \lim_{\Delta t \to 0} \lambda^2 \ln \frac{L}{\Delta t} = \infty. \tag{13}$$

In order to obtain convergence of variance, the mean value of $\omega_{\Delta t}[k]$ was allowed to decrease logarithmically

$$\mathrm{E}\left[\omega_{\Delta t}[k]\right] = -\mathrm{Var}\left[\omega_{\Delta t}[k]\right] = -\lambda^2 \ln \frac{L}{\Delta t}. \tag{14}$$

Such modification provided the "physical meaning" for the variance, but made it difficult to interpret the meaning of producing noise $\omega_{\Delta t}[k]$ in applications. The second issue is that multifractal spectrum does not exist for high orders (namely, for orders $q > 1/\lambda^2$). Nevertheless, MRW model is flexible enough and has relatively straightforward way of calibration that is discussed below.

5.2 *Numerical Simulation of the MRW Process*

The bottleneck of numerical simulation of the MRW process (8) is simulation of logarithmically correlated noise $\omega_{\Delta t}[k]$. Simulation of the discrete Gaussian noise process with given autocorrelation function (covariance matrix) is a well-known problem and is subjected to a trade-off: exact simulation processes usually requires a lot of computation resources, and fast algorithms typically provide only approximated solution. The most known exact simulation method is based on the Cholesky or LU-decomposition of the covariance matrix into lower- and upper-triangle matrices (Davis 1987). Though this method is very efficient for short time-series, having computational complexity of $\mathcal{O}(N^2)$ it is not suitable for simulation of long (e.g. $N > 10^3$) realizations. Much more efficient decomposition algorithm is the so-called *Circulant Embedding Method (CEM)* (Dietrich and Newsam 1997) that is based on the Fast Fourier Transform (FFT) and thus has complexity of $\mathcal{O}(N \log N)$.

Consider $N \times N$ covariance matrix $\mathbf{R}$ with elements $R_{p,q} = r[|p-q|] = r[k]$ for $k = 1, \ldots, N-1$, where $r[k]$ is the required covariance matrix and in our case is given by (10). CEM consist in embedding of matrix $\mathbf{R}$ into a larger $2M \times 2M$ matrix $\mathbf{S}$ (where $M \geqslant N-1$). The optimal case of $M = N-1$ is called the minimal embedding. The first row of $\mathbf{S}$, denoted by s, consists of two parts of length $N-1$ each: the original first row of $\mathbf{R}$ following with the first row of $\mathbf{R}$ in reverse order:

$$\begin{aligned} s\,[k] &= r\,[k]\,, & k &= 0\,,\ldots,N-1\,, \\ s\,[2M-k] &= r\,[k]\,, & k &= 1\,,\ldots,N-2. \end{aligned} \tag{15}$$

Since matrix $\mathbf{S}$ is circulant, any matrix extracted along its diagonal is a copy of $\mathbf{R}$. One of the properties of circulant matrices tells that matrix $\mathbf{S}$ can be decomposed into a product $\mathbf{S} = \mathbf{FDF}^H$, where $\mathbf{F}$ is a matrix of discrete Fourier transform coefficients, and $\mathbf{F}^H$ is the conjugate transpose of $\mathbf{F}$. The matrix $\mathbf{D}$ is diagonal, and elements along the diagonal can be obtained by the discrete Fourier transform of first row or column of $\mathbf{S}$: $\tilde{s} = \mathbf{F}s$.

Let us construct a vector $\mathbf{y} = \mathbf{FD}^{1/2}\mathbf{x}$, where $\mathbf{x}$ is a complex vector, having iid $N(0, \mathbf{I})$ real and imaginary parts. It was shown (Dietrich and Newsam 1997) that any vector of size N, extracted from either real or imaginary part of $\mathbf{y}$ has covariance matrix exactly equal to $\mathbf{R}$. It should be noted that necessary and sufficient condition of existence of auxiliary vector $\mathbf{y}$ is the matrix $\mathbf{S}$ to be nonnegative definite (nonnegative eigenvalues s_m of the matrix $\mathbf{S}$). Though for strictly positive definite Toeplix matrices $\mathbf{R}$ the existence of nonnegative circulant embedding was proven (Dembo et al. 1989), there is no general recipe and the necessary and sufficient condition should be tested for any specific covariance matrix $\mathbf{R}$.

Without providing a general proof that circulant matrix $\mathbf{S}$ for covariance matrix given by (10) is nonnegative definite, while performing simulations we have tested numerically eigenvalues $\mathbf{S}$ for all used parameter sets. According to the property of circulant matrices, eigenvalues s_m of matrix $\mathbf{S}$ are equal to the discrete Fourier transform of the first row of $\mathbf{S}$:

$$s_m = \sum_{k=0}^{N-1} r_k \exp\left(2\pi i \frac{mk}{2(N-1)}\right) + \sum_{k=N}^{2(N-1)-1} r_{2(N-1)-k} \exp\left(2\pi i \frac{mk}{2(N-1)}\right),$$

where $i^2 = -1$ and $r_k = \mathrm{Cov}\left[\omega_{\Delta t}[i], \omega_{\Delta t}[i+k]\right]$ are given by (10). After the rearrangement and substitution of (10) we obtain:

$$s_m = \lambda^2 \left[\log \frac{L}{\Delta t} + (-1)^m \log \frac{L}{\Delta t(N-1)} + 2\sum_{k=1}^{N-2} \log \frac{L}{\Delta t(k+1)} \cos\left(\pi \frac{mk}{N-1}\right)\right]. \tag{16}$$

As one can see, λ^2 is a multiplicative coefficient in (16) and the sign of s_m is fully determined by the relation between $L/\Delta t$ and N. For any used combination of $L/\Delta t$ and N we have tested that values (16) are nonnegative: $s_m \geq 0$.

The CEM algorithm for minimal embedding case could be summarized in the following steps (Dietrich and Newsam 1997):

1. Evaluate the first row of R for lags from 0 to N − 1;
2. Form the first row s of the circulant matrix S;
3. Compute FFT from s;
4. Compute square root from results of 3;

5. Generate complex iid N(0,1) vector x of length 2(N − 1);
6. Multiply result from step 4 by x;
7. Evaluate y via inverse FFT of the result of previous step;
8. Extract vector of length N from real part of y;
9. Another vector could be extracted from real part of y;
10. To generate additional realizations, proceed to step 5.

The computational complexity of this algorithm is $\mathcal{O}(N \log N)$ compared to $\mathcal{O}(N^2)$ for the Cholesky-based method. Additionally, the memory requirements is $\mathcal{O}(N)$ for CEM compared to $\mathcal{O}(N^2)$ for Cholesky decomposition. This allows to generate extremely long ($N = 10^{20}$) realizations of MRW random process using standard Core i5 4Gb RAM computer.

5.3 *Statistical Properties of MRW Process*

In order to demonstrate distinctive feature of MRW process, one can compare its realization with realization of geometric Brownian motion (original random walk model of Bachelier), which sample increments and path are shown in Figs. 1 and 2. Sample realizations of increments and path of MRW are shown in Figs. 3 and 4.

Comparing Fig. 3 with Fig. 1, one can notice significant differences in the way, which each process goes. When dynamics of increments of random walk (Fig. 1) are very regular and one can not observe large deviations from the mean value, the dynamics of MRW (Fig. 3) is much more intermittent, one can easily spot volatility clustering and large excursions (extreme events).

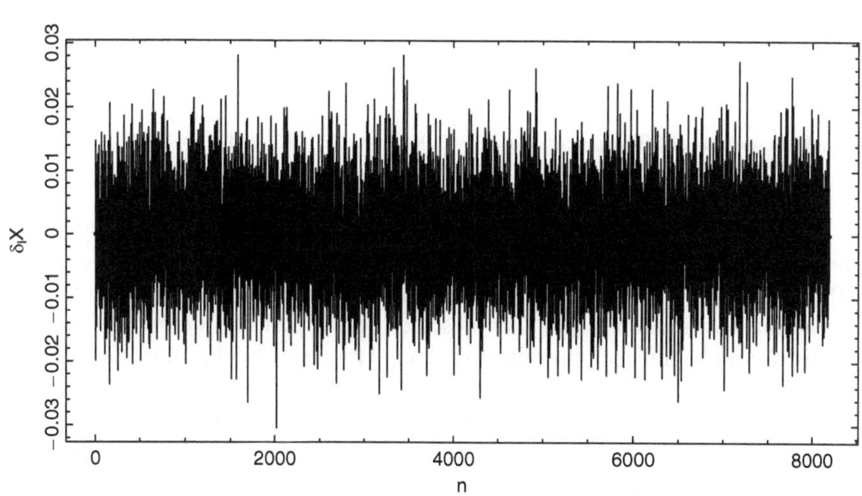

Fig. 1 Increments of geometrical Brownian motion for $\sigma = 0.0078$

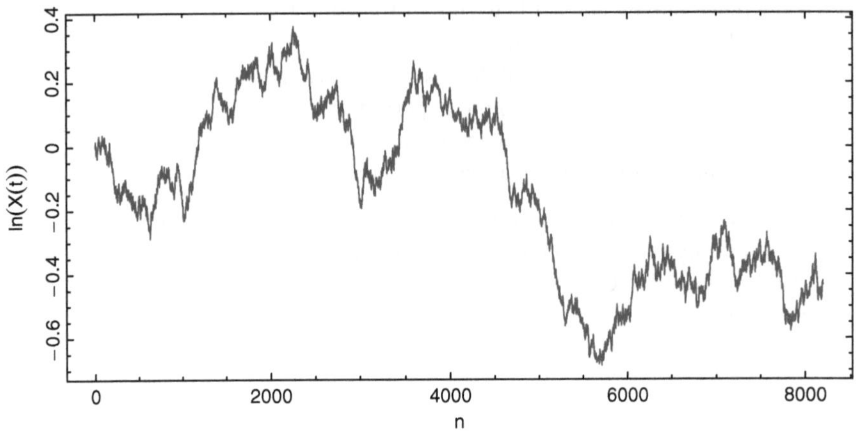

Fig. 2 Path of geometrical Brownian motion for $\sigma = 0.0078$

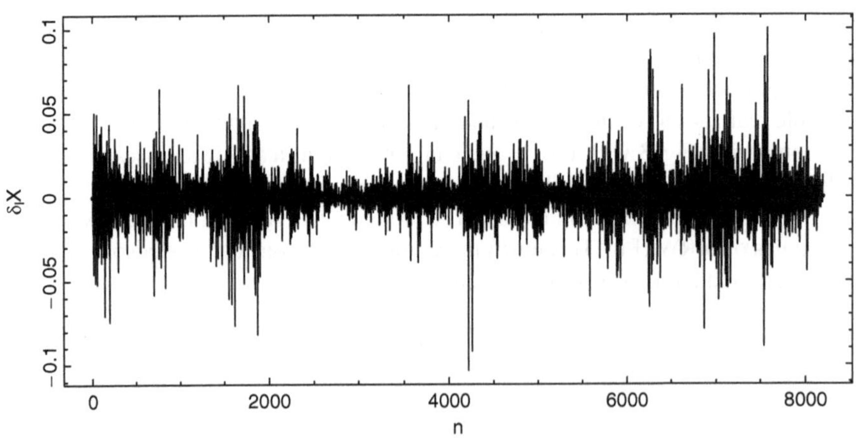

Fig. 3 Increments of MRW process for $\lambda^2 = 0.06$, $\sigma = 7.5 \cdot 10^{-5}$ and $L = 1024$

Presence of the heavy tails of pdf for MRW can be shown more clearly with the ranking plot (see Fig. 5) for various aggregation level. The interval of scales $10^{-3} \leqslant \delta_{\Delta t} X_{\Delta t}\,[t] \leqslant 1$ illustrates the tails of pdf which decay much slower than for the normal distribution that is also presented on the plot for comparison. In other words, the probability of observing extremely large increment (return) for MRW is much larger than for the normal distribution where the probability of observing a value larger than three-four standard deviations is essentially zero.

Figure 5 also illustrates another stylized fact, namely—aggregational gaussianity. One can see from the Fig. 5, that slope of the tail line tends to the slope of the tail line for normally distributed data, when aggregation level (which is defined as a number of consecutive increments of initial MRW process that are summed to obtain single

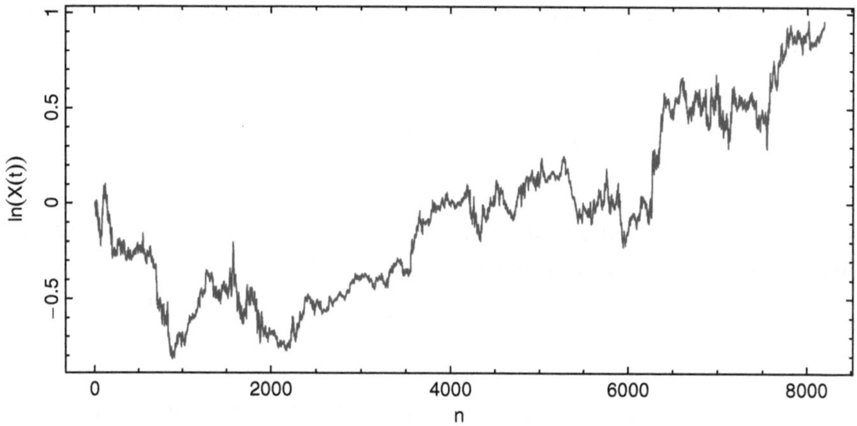

Fig. 4 Path of MRW process for $\lambda^2 = 0.06$, $\sigma = 7.5 \cdot 10^{-5}$ and $L = 1024$

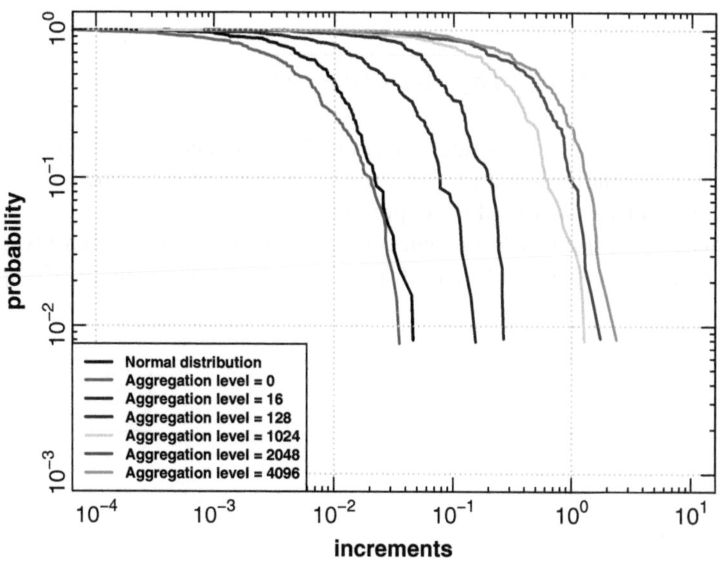

Fig. 5 Ranking plot for the increments of the MRW process for $\lambda^2 = 0.06$, $\sigma = 7.5 \cdot 10^{-5}$, $L = 1024$ and length of realization $N = 2^{20}$. Normally distributed (iid) sample has equal length and simulated for mean value μ and standard deviation σ that are equal to the sample mean and standard deviation of the realization of MRW process

increment of aggregated process) rises. For instance, for aggregation level equals 4096 tail of the distribution converges to Gaussian distribution.

Volatility clustering, that one can observe in Fig. 3 is a result of the presence of long memory in volatility. In order to quantify it we have considered four different measures of the volatility. The first one is the simplest squared values of returns (increments). Second is the definition of volatility as a standard deviation of returns in a rolling window of size n_t:

$$\sigma_t = \sqrt{\frac{1}{n_t - 1} \sum_{i=1}^{n_t} (r_i - \bar{r})^2}. \tag{17}$$

Third is the widely-used volatility estimator as a Exponentially-Weighted Moving Average (EWMA), which can be defined as

$$\sigma_t = \sqrt{\lambda\, \sigma_{t-1}^2 + (1 - \lambda)\, r_{t-1}^2}, \tag{18}$$

where $\lambda \in [0, 1]$ is the rate of decay of the exponential weight within time window. Finally, we have also considered Müller estimator of the volatility (Müller 2000) which is similar to the EWMA, but involves recursion both of lagged and current squared returns:

$$\sigma_{t_n}[\tau] = \sqrt{\mu\, \sigma_{t_{n-1}}^2 [\tau] + (\upsilon - \mu)\, r_{t_{n-1}}^2 + (1 - \upsilon)\, r_{t_n}^2}, \tag{19}$$

where $\alpha = (t_n - t_{n-1})/\tau$ is the rate of decay of the exponential weight; $\mu = e^{-\alpha}$ is an exponential weight itself and $\upsilon = (1 - \mu)/\alpha$. Autocorrelation functions computed for above estimators of volatility are presented in Fig. 6.

As one can see from the Fig. 6, autocorrelation of all proxies of volatility is significantly non zero in a very wide range (of the lags up to 1000 and more). Compared to

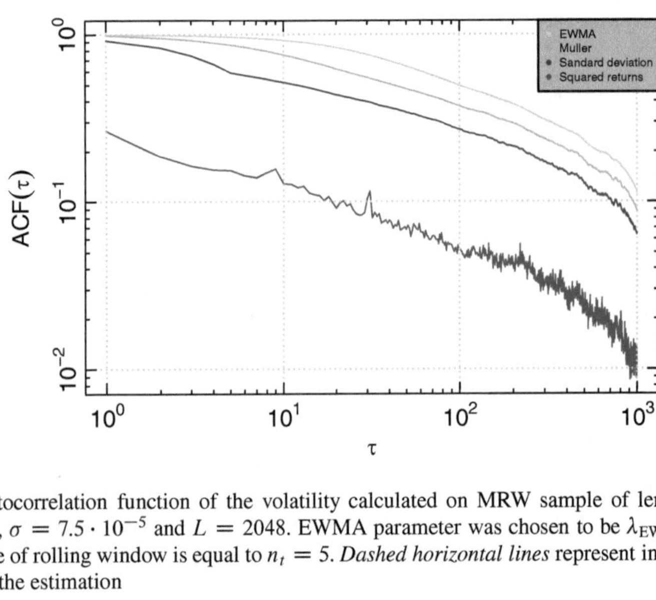

Fig. 6 Autocorrelation function of the volatility calculated on MRW sample of length 2^{17} for $\lambda^2 = 0.06$, $\sigma = 7.5 \cdot 10^{-5}$ and $L = 2048$. EWMA parameter was chosen to be $\lambda_{\mathrm{EWMA}} = 0.94$ and the size of rolling window is equal to $n_t = 5$. *Dashed horizontal lines* represent insignificance interval of the estimation

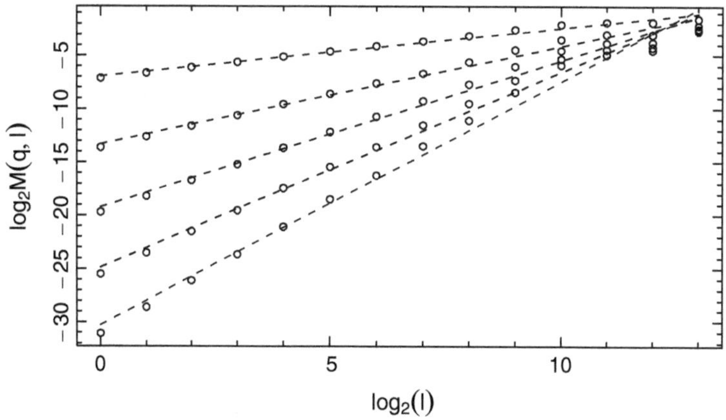

Fig. 7 Log-log plot of moments of increments (4) calculated using the MRW sample of length 2^{17} for $\lambda^2 = 0.06$, $\sigma = 7.5 \cdot 10^{-5}$, $L = 2048$ and $q = 1, 2, 3, 4, 5$. *Dashed lines* correspond to linear fit of dependency (4)

autocorrelation for squared returns, the rate of decay of autocorrelations for standard deviation, EWMA and Müller estimator decay much slower due to the fact, that above the three estimators perform recursive procedures for volatility computation. These recursion-based estimators capture the features of volatility behavior better than squared returns.

In order to illustrate the scale invariance in simulated MRW sample, one have to consider moments of increments of the realization (4). As described above, the presence of scale invariance is qualified with the power law behavior of the the moments (4). As one can see from the Fig. 7 this holds for the analyzed MRW process, as the absolute moments $M_q(l)$ for all q has linear or close to linear (for $q = 5$) form in log-log scale, which tells about the presence of power law dependency in the ordinary scale.

5.4 *Calibration of the Model*

One of the most important issues for the practical applications is the estimation of the three unknown parameters of MRW model (σ, λ, L) with the real data.

The parameter σ can be estimated using the scaling relation for the variance of the increments of the MRW process:

$$\mathrm{Var}\left[\delta_{\Delta t} X_{\Delta t}\left[k\right]\right] = \sigma^2 \, \Delta t, \tag{20}$$

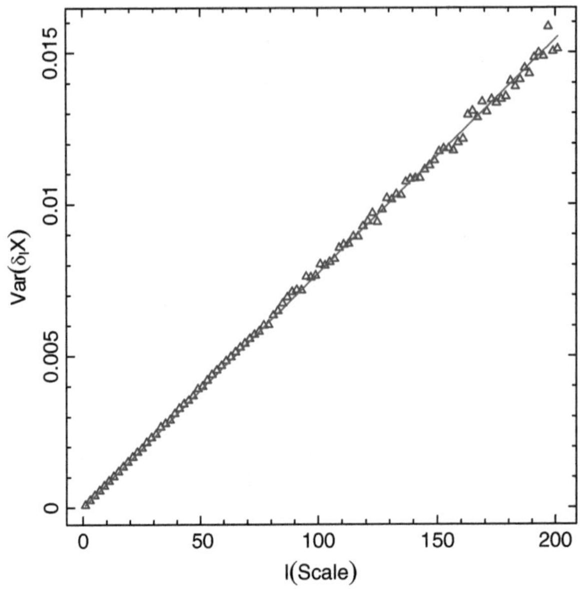

Fig. 8 Calibration of σ^2 based on MRW sample of length 2^{19} for $\lambda^2 = 0.06$, $\sigma = 7.5 \cdot 10^{-5}$ and $L = 2048$. *Triangles* represent empirical observations and *red line* corresponds to the linear regression (20). The estimated $\hat{\sigma}^2$ equals to $7.7237 \cdot 10^{-5}$

where Δt is the scale of log-returns (e.g, 1-, 5-, 10-, 20-min etc.). The parameter σ^2 can be then estimated with the linear regression of the $\mathrm{Var}\,[\delta_{\Delta t} X_{\Delta t}\,[k]]$ on Δt as it is shown in Fig. 8.

Estimation of intermittency coefficient λ and integral scale L is much more complicated, because they define the unobserved log-volatility process $\omega_{\Delta t}\,[k]$. In Bacry et al. (2001) it is shown that the magnitude correlation function

$$C_p\,(\tau, l) = \mathrm{E}\,[|\delta_\tau X\,[k + l]\,|^p, |\delta_\tau X\,[k]\,|^p]\,, \tag{21}$$

where l is lag and $\tau \ll L$ is a scale of the log-returns, in the limit of small scales τ has the following asymptotic behavior

$$C_p\,(\tau, l) \sim K_p^2 \left(\frac{\tau}{L}\right)^{2\zeta_p} \left(\frac{l}{L}\right)^{-\lambda^2 p^2}, \tag{22}$$

where the pre-factor K_p is defined as

$$K_{2p} = L^p \sigma^{2p}\,(2p-1)!! \int_0^1 du_1 \ldots \int_0^1 du_p \prod_{i<j} |u_i - u_j|^{-4\lambda^2}.$$

Analyzing the limit of (22) when $p \to 0$ one can find the following approximate relation (Bacry et al. 2001):

$$C(\tau, l) \sim -\lambda^2 \ln\left(\frac{l}{L}\right). \tag{23}$$

In other words, magnitude correlation function, for small enough τ, has similar behavior to the correlation function of underlying log-volatility process $\omega_{\Delta t}[k]$. Thus, regressing $C(\tau, l)$ on $\log l$ one can estimate the parameter λ^2. Finally, the integral scale L can be obtained as the scale l after which autocorrelation function (23) is indistingushable from noise.

Figure 9 illustrates fits of the λ^2 and L using relation (23). Measures of the slope and intercept of $C_\tau(l) \sim \ln(l)$ provide good estimate of respectively λ^2 and L, though the estimation of the integral scale L is typically worse in comparison with estimation of λ^2. The algorithm of determining L could be summarized as follows:

1. Set the size of small rolling window;
2. Scan values of magnitude function within rolling window;
3. Stop scanning if all elements within rolling window belong to the interval of insignificance;
4. Set L for the index of the middle point of rolling window at its last position.

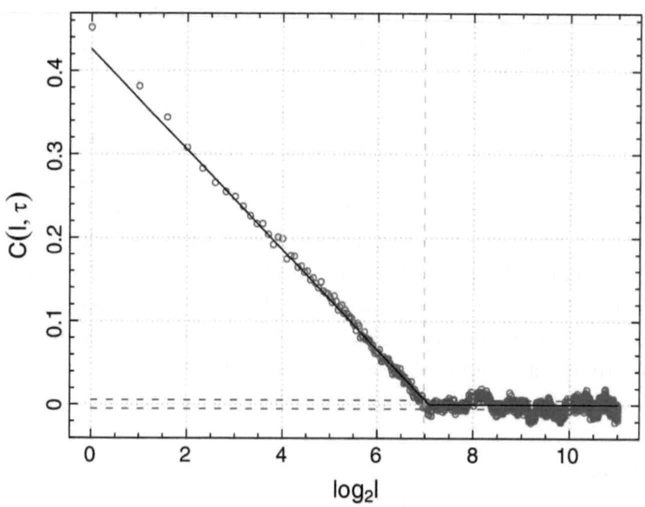

Fig. 9 Magnitude correlation function $C(\tau, l)$ increments of MRW process sample of length 2^{21} for $\lambda^2 = 0.06$, $\sigma = 7.5 \cdot 10^{-5}$, $L = 2048$ and $\tau = 15$. *Black solid line* represents linear regression (23). *Horizontal dashed lines* represent insignificance interval and *vertical dashed line* denotes estimated value of L. The estimated $\hat{\lambda}^2$ equals to 0.0623 and estimated $\hat{L}$ is 1905

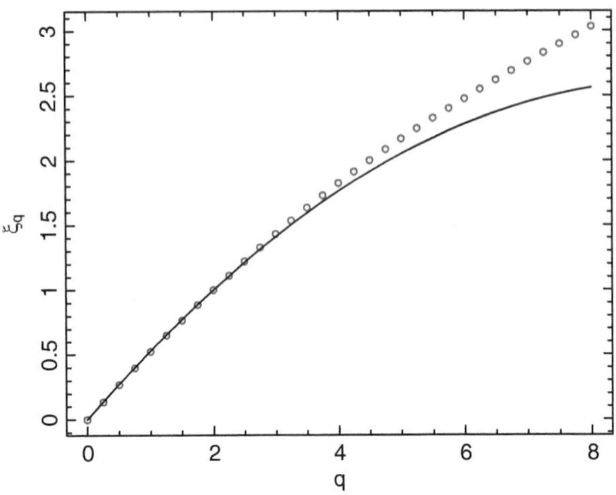

Fig. 10 Estimated multifractal spectrum ζ_q for the realization of MRW process of length 2^{21} for $\lambda^2 = 0.06$, $\sigma = 7.5 \cdot 10^{-5}$ and $L = 2048$. *Solid black line* corresponds to theoretical spectrum (11)

However, this algorithm strongly depends on the choice of the rolling window size τ and requires additional validation of the results.

Alternative way of estimation of intermittency coefficient λ^2 involves estimation of the multifractal spectrum $\zeta_q = f(q)$ of the process. Given the analytical expression (11) one can then estimate λ^2 with the least squares estimator. Straightforward estimation of ζ_q requires calculation of moments of increments $M_q(l)$ as a function of scale l using the definition (4) and then regressing $\log M_q(l)$ on $\log l$ for different values of q, implying relation (5). Results of estimation of the multifractal spectrum for MRW process are presented in Fig. 10. One can see good agreement of the empirical spectrum with theoretical prediction up to orders of $q = 6$. The divergence of analytical and theoretical spectrum for higher values of q results from the insufficient sample size. Alternative methods of estimation of multifractal spectrum are based on the wavelet transform—so-called Wavelet Transform Modulus Maxima (WTMM) (Arneodo et al. 1998a) and detrended fluctuation analysis: Multifractal Detrended Fluctuation Analysis (MF-DFA) (Kantelhardt et al. 2002) and Multifractal Detrended Moving Average (MF-DMA) (Gu and Zhou 2010). However it should be noted that all these methods are subjected to the bias for large values of q and in real cases due to short observed realizations are not efficient with respect to estimation of λ^2.

Conclusions

Concluding topics discussed above, MRW model allows to capture six main stylized facts (absence of linear autocorrelation, volatility clustering, long memory in volatility, heavy tails in probability distribution, aggregational Gaussianity and multifractal scaling). Moreover some modifications of the

(continued)

MRW such as Skewed MRW (Pochart and Bouchaud 2002) accounts also for the leverage effect and most likely for related gain-loss asymmetry. MRW model has an effective procedure for numerical simulation and relatively robust method of calibration, and thus is a prominent candidate for the option pricing applications, using the straightforward method of Monte-Carlo simulations.

References

Aït-Sahalia, Y. & Hansen, L., (Eds.). (2009). Handbook of financial econometrics. Amsterdam: North Holland.

Arneodo, A., Bacry, E., & Muzy, J.-F. (1998a). Random cascades on wavelet dyadic trees. *Journal of Mathematical Physics, 39*(8), 4142–4164.

Arneodo, A., Muzy, J.-F., & Sornette, D. (1998b). "Direct" causal cascade in the stock market. *The European Physical Journal B - Condensed Matter and Complex Systems, 2*(2), 277–282.

Bachelier, L. (1900). Théorie de la spéculation. *Annales scientifiques de l'École normale supérieure, 3*(17), 21–86.

Bacry, E., Delour, J., & Muzy, J.-F. (2000). A multivariate multifractal model for return fluctuations. arXiv:cond-mat/0009260.

Bacry, E., Delour, J., & Muzy, J.-F. (2001). Multifractal random walk. *Physical Review E, 64*(2), 456, 026103C.

Bacry, E., Kozhemyak, A., & Muzy, J.-F. (2008). Log-Normal continuous cascades: aggregation properties and estimation. Application to financial time-series. arXiv:q-fin/0804.0185.

Bacry, E. & Muzy, J.-F. (2003). Log-infinitely divisible multifractal processes. *Communications in Mathematical Physics, 236*(3), 449–475.

Bollerslev, T. (1986). Generalized autoregressive conditional heteroskedasticity. *Journal of Econometrics, 31*(3), 307–327.

Bollerslev, T. (2010). Glossary to ARCH (GARCH). *Volatility and Time Series Econometrics, 28*, 137–164.

Bouchaud, J.-P., Matacz, A., & Potters, M. (2001). Leverage effect in financial markets: the retarded volatility model. *Physical Review Letters*, 87(22), 228701+.

Bouchaud, J.-P. & Potters, M. (2000). *Theory of financial risks: from statistical physics to risk management*. Cambridge: Cambridge University Press.

Breymann, W., Ghashghaie, S., & Talkner, P. (2000). A stochastic cascade model for FX dynamics. *International Journal of Theoretical and Applied Finance*, (3), 357–360.

Calvet, L. E. & Fisher, A. J. (2002). Multifractality in asset returns: theory and evidence. *Review of Economics and Statistics, 84*(3), 381–406.

Calvet, L. E. & Fisher, A. J. (2004). How to forecast long-run volatility: regime switching and the estimation of multifractal processes. *Journal of Financial Econometrics, 2*(1), 49–83.

Calvet, L. E. & Fisher, Adlai J. (2008). *Multifractal volatility theory, forecasting, and pricing*. Burlington, MA: Academic Press. ISBN 9780080559964.

Cont, R. (2001). Empirical properties of asset returns: stylized facts and statistical issues. *Quantitative Finance, 1*, 223–236.

Davis, M. (1987). Production of conditional simulations via the LU triangular decomposition of the covariance matrix. *Mathematical Geology, 19*(2), 91–98.

Davis, R. A., & Mikosch, T. (2009). Extreme value theory for GARCH processes. In T. G. Andersen, R. A. Davis, J.-P. Kreiss, & T. Mikosch (Eds.), *Handbook of financial time series* (pp. 187–200). Springer: New York.

Dembo, A., Mallows, C. L., & Shepp, L. A. (1989). Embedding nonnegative definite Toeplitz matrices in nonnegative definite circulant matrices, with application to covariance estimation. *IEEE Transactions on Information Theory, 35*(6), 1206–1212.

Dietrich, C. R. & Newsam, G. N. (1997). Fast and exact simulation of stationary gaussian processes through circulant embedding of the covariance matrix. *SIAM Journal on Scientific and Statistical Computing, 18*(4), 1088–1107.

Engle, R. (2001). GARCH 101: The use of ARCH/GARCH models in applied econometrics. *The Journal of Economic Perspectives, 15*(4), 157–168.

Engle, R. F. (1982). Autoregressive conditional heteroscedasticity with estimates of the variance of United Kingdom inflation. *Econometrica: Journal of the Econometric Society, 50*(4), 987–1007.

Fama, E. F. (1970). Efficient capital markets: a review of theory and empirical work. *The Journal of Finance, 25*(2), 383–417.

Fama, E. F. (1991). Efficient capital markets: II. *Journal of Finance, 46*(5), 1575–1617.

Filimonov, V. (2010). Multifractal Models of Financial Time Series. HSE — Working Paper Series, P1/2010/06, (pp.$\tilde{4}$5).

Filimonov, V. & Sornette, D. (2011). Self-excited multifractal dynamics. *Europhysics Letters, 94*(4), 46003.

Ghashghaie, S., Breymann, W., Peinke, J., Talkner, P., & Dodge, Y. (1996). Turbulent cascades in foreign exchange markets. *Nature, 381*, 767–770.

Gu, G.-F. & Zhou, W.-X. (2010). Detrending moving average algorithm for multifractals. *Physical Review E, 82*(1), 011136+.

Kantelhardt, J. W., Zschiegner, S. A., Koscielny-Bunde, E., Havlin, S., Bunde, A., & Stanley, H. E. (2002). Multifractal detrended fluctuation analysis of nonstationary time series. *Physica A: Statistical Mechanics and its Applications, 316*(1-4), 87–114.

Kolmogorov, A. N. (1941). The local structure of turbulence in incompressible viscous fluid for very large Reynolds numbers. *Dokladi Akademii Nauk SSSR, XXXI*, 299–303.

Kolmogorov, A. N. (1962). A refinement of previous hypotheses concerning the local structure of turbulence in a viscous incompressible fluid at high Reynolds number. *Journal of Fluid Mechanics, 13*(1), 82–85.

Liu, R., Matteo, T., & Lux, T. (2008). Multifractality and Long-Range Dependence of Asset Returns: The Scaling Behaviour of the Markov-Switching Multifractal Model with Lognormal Volatility Components. Kiel Working Papers, (pp. 1–15).

Lux, T. (2009). Stochastic Behavioral Asset-Pricing Models and the Stylized Facts. In *Handbook of Financial Markets: Dynamics and Evolution* (pp. 161–215). Amsterdam: North Holland.

Mandelbrot, B. B. (1975). *Les objets fractals: forme, hasard et dimension*. Paris: Flammarion.

Mandelbrot, B. B. (1982). *The fractal geometry of nature*. San Francisco: W. H. Freeman and Company.

Mandelbrot, B. B. (1985). Self-affine fractals and fractal dimension. *Physica Scripta, 32*(4), 257–260.

Mandelbrot, B. B., Fisher, A. J., & Calvet, L. E. (1997). A Multifractal Model of Asset Returns. Cowles Foundation Discussion Paper #1164.

Mandelbrot, B. B. & Van Ness, J. W. (1968). Fractional Brownian Motions, fractional noises and applications. *SIAM Review, 10*(4), 422–437.

Müller, U. A. (2000). *Specially Weighted Moving Averages with Repeated Application of the EMA Operator* . Technical report.

Muzy, J.-F., Delour, J., & Bacry, E. (2000). Modelling fluctuations of financial time series: from cascade process to stochastic volatility model. *The European Physical Journal B - Condensed Matter and Complex Systems, 17*(3), 537–548.

Pochart, B. & Bouchaud, J.-P. (2002). The skewed multifractal random walk with applications to option smiles. *Quantitative Finance, 2*(4), 303–314.

Richardson, L. F. (1961). The problem of contiguity: an appendix of statistics of deadly quarrels. *General Systems Yearbook, 6*, 139–187.
Saichev, A. & Filimonov, V. (2007). On the spectrum of multifractal diffusion process. *Journal of Experimental and Theoretical Physics, 105*(5), 1085–1093.
Saichev, A. & Filimonov, V. (2008). Numerical simulation of the realizations and spectra of a quasi-multifractal diffusion process . *JETP Letters, 87*(9), 506–510.
Saichev, A. & Sornette, D. (2006). Generic multifractality in exponentials of long memory processes. *Physical Review E, 74*(1), 011111+.
Siven, J. V., & Lins, J. T. (2009). Gain/loss asymmetry in time series of individual stock prices and its relationship to the leverage effect. arXiv:0911.4679v2.
Sornette, D. (2003). *Why stock markets crash: critical events in complex financial systems.* Princeton: Princeton University Press.
Zakoian, J.-M. & Francq, C. (2010). *GARCH models: structure, statistical inference and financial applications*. Oxford: Wiley-Blackwell.

Adaptive Stress Testing: Amplifying Network Intelligence by Integrating Outlier Information (Draft 16)

Alan Laubsch

The future is already here. It's just not evenly distributed yet.
(William Gibson)

Abstract This essay examines lessons from systemic breakdowns, and presents a framework for *Adaptive Stress Testing* to proactively manage systemic risks. The framework is inspired by evolutionary ecosystems, including ecology, economics, technology, psychology, and sociology. Adaptive Stress Testing harnesses network intelligence to integrate early warning signals. We pre-diagnose systemic fragilities by tapping into the marketplace of ideas, and then identify key metrics to monitor market-based early warning signals. We apply the *Technology Adoption Lifecycle* model to develop a theory of *social diffusion of disruptive information* in financial markets. We start by taking a macro view of risk in its hidden potential form, and then focus on phase transition signals as risk becomes visible. This process allows us to better understand key systemic risks, and to more effectively sense and respond to emerging risks.

Keywords Behavioral economics • Black Swan • Complexity • Disruptive innovation • Dragon King • Early warning • Eco-centric risk management • Financial cartography • Financial network analysis • Foreshocks • Groupthink • HeavyTails™ • Integral theory • Network theory • Network visualization • Outliers • Phase transitions • Polarity management • Risk culture • Social adoption • Social market hypothesis • Stress indices • Stress testing • Stress testing • StressGrades™ • Subprime crisis • Systemic risk • VaR backtesting

A. Laubsch (✉)
Financial Network Analytics (FNA), London, U.K.
e-mail: alaubsch@gmail.com

A.K. Bera et al. (eds.), *Financial Econometrics and Empirical Market Microstructure*, DOI 10.1007/978-3-319-09946-0_11

1 Introduction

The sudden onset and severity of crises catch most by surprise. Unpredictable events emerge and quickly escalate out of control. The 2008 US subprime crisis was but the latest of such systemic breakdowns. But not all were equally surprised. Structural fragilities built up over years, and early warning signals escalated from 2006 to 2007. A prescient few foresaw the inevitable bust, mitigated their risk, alerted regulators, and even issued public warnings (largely ignored, unfortunately). Indeed, one might argue that the biggest surprise was the extent of risk myopia despite an abundance of information. Why did some perceive risks that most were blind to? And what can the rest of us learn from them? This essay proposes a methodology to amplify social intelligence in the risk management community.

2 Amplification Mechanisms Drive Systemic Risk

A stress event is a systemic breakdown, which is a form of phase transition. We observe phase transitions in all complex systems. **Phase transitions are triggered after a *critical point* is crossed** at which point self-amplification causes a transformation into a state with radically different properties (e.g., solid, liquid, and gas).

The continual tension between **amplifying and dampening mechanisms** powers complex systems. Financial cycles are driven by the inter-linkage of asset prices, leverage, and risk aversion. Furthermore, the social process of **imitation is a major amplifier**. Imitation is an efficient form of social learning and adaptation, and is prevalent especially during times of uncertainty (Keynes 1930).

Stability increases asset prices and leverage, and lowers risk aversion, which sows the seeds of future instability (Minsky 1992). As bubbles expand and leverage grows, markets become more tightly coupled and vulnerable to collapse. Eventually, a surprise triggers increased risk aversion and a self-amplifying deleveraging spiral (e.g., bank run). Key dampening mechanisms include countercyclical (and symmetric) central bank intervention (Cooper 2008) and contrarian investment strategies.

History is riddled with unpredictable exogenous shocks, or ***Black Swans*** (Taleb 2007). Dramatic examples include extinction events from meteorite impacts or flood basalt eruptions, terrorist attacks like September 11, or technological breakdowns such as the 2011 Fukushima meltdown.[1] And yet, according to Didier Sornette, the majority of financial crises have **endogenous origins** and can be "pre-diagnosed, quantified and predicted to a degree" (Sornette et al. 2009). He calls these ***Dragon***

[1]Also noteworthy is that some Black Swans may be Dragon Kings to those with special insight: astronomers might forecast an asteroid impact, security analysts might uncover a high likelihood of a terrorist attack, while safety engineers might have insight about escalating risks of an industrial breakdown.

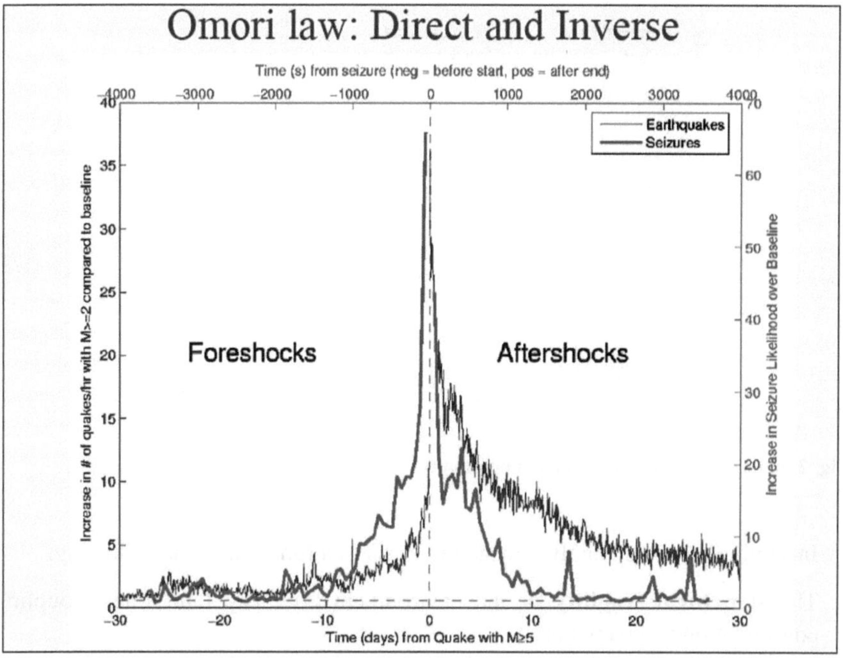

Fig. 1 Amplifying foreshocks. *Source*: Osorio, Sornette et al. (2010)

Kings (Sornette 2009). Systemic collapses can originate from the predicable amplification of small perturbations in a tightly coupled system. Such endogenous crises are our focus in this paper, as these are risks we can and must manage.

Endogenous crises are characterized by escalating ***Foreshocks*** that culminate in a phase transition. We see such patterns throughout nature. Figure 1 from Didier Sornette compares earthquakes and brain seizures, which both exhibit the same pattern of ***amplifying Foreshocks*** and ***mean reverting Aftershocks***.

The presence of Foreshocks implies that we need not be surprised by endogenous crises, and that there is a window of opportunity to mitigate risk. A phase transition progresses as follows:

1. A period of stability is interrupted by an **outlier**, which may be small in absolute terms but unusual from a relative perspective.
2. This initial outlier sets off **amplification mechanisms** which results in a super-exponential rate of change. This initial period of exponential growth is barely perceptible and typically dismissed as noise initially. The window of opportunity for control shuts quickly as the exponential curve gets steep.

Risk managers are continually on the lookout for emerging risks, and recognize that the ability to control risk declines exponentially as risk escalates. This fleeting window of opportunity for exerting control is illustrated (Fig. 2) based on an illustration by reputation risk consultant.

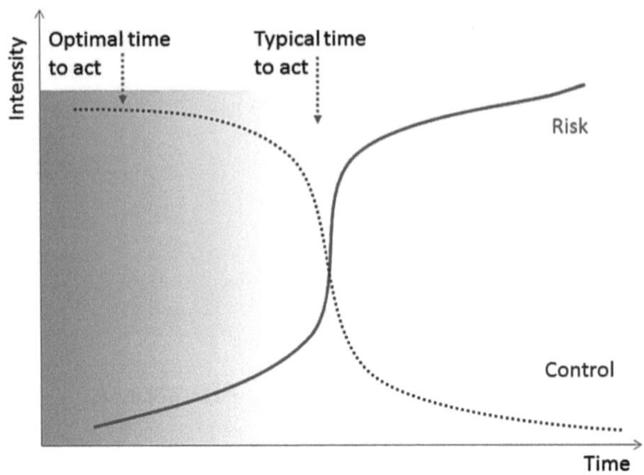

Fig. 2 Window of risk management opportunity

In summary, we can analyze systemic crisis as a function of the following:

1. The **structural fragility** of the environment (i.e., key fault lines, coupling, amplification mechanisms)
2. A **precipitating event** (tremor) which pushes the system beyond a **tipping point**, which sets of a self-amplifying cascade that result in a rapid phase transition (e.g., toppling dominoes)

3 Adaptive Stress Testing Framework

This model of systemic crises leads us to propose our Adaptive Stress Testing framework, which is driven by the integration of top-down and bottom-up perspectives:

I. *Macro*: Identify **potential risks** (e.g., hidden fault lines). Build a ***Stress Library*** by harnessing the intelligence of visionary thought leaders (*Innovators*) who perceive risks in their potential form, while they are still dormant and hidden. An example of an Innovator is Robert Shiller, who warned of the U.S. technology stock bubble in 1999 and the U.S. housing bubble in 2005.
II. *Micro*: Monitor **visible risks** in financial markets (e.g., tremors). Build ***Stress Indices*** for each macro scenario using key market factors, and monitor **early warning** signals (e.g., outliers).

Macro and micro perspectives are interdependent and inform each other. At the macro level, we expand our horizons with potential risks perceived by *Innovators*. Given that innovators are ahead of their time, however, trading based on their views is often a losing proposition. We therefore monitor confirmation that a theme has been adopted by the marketplace, and **transitioned from potential to visible risk**. In summary, (a) tap the social marketplace to identify a wide range of scenarios, and (b) then hone in on emerging scenarios before *critical points* are crossed.

4 Social Diffusion of Disruptive Information

As we have described, social imitation is an important amplifier in markets. As in fashion, new themes constantly emerge and some cross *critical points* to broad adoption due to social imitation. The *Technology Adoption Lifecycle* is a sociological model about how **disruptive innovation** diffuses in the marketplace. See Geoffrey Moore's seminal "Crossing the Chasm" (1999) for a full description of this **punctuated equilibrium** model. Disruptive innovation does not diffuse gradually. Rather, the market remains in stasis as pressure builds up until the conditions are right for a jump to *Early Adopters*.

Malcolm Gladwell's *The Tipping Point: How Little Things Can Make a Big Difference* (2000) describes *Early Adopters* as *Connectors* (social network hubs), *Mavens* (information specialists), and *Sales People* (persuaders). *Early Adopters* play a crucial role in the dissemination of disruptive innovation. It is only after *Early Adopters* buy into a theme that a *tipping point* is crossed, which sets off social amplification (imitation) that results in adoption by the Early and Late Majority. How do entrepreneurs know when they when their disruptive innovation is *crossing the chasm*? It feels like being "Inside The Tornado" (the title of Moore's follow-up book): swept up by super-exponential change and turbulence (Moore 2004).

Figure 3 illustrates the epidemiological jump process of social diffusion of disruptive innovation.

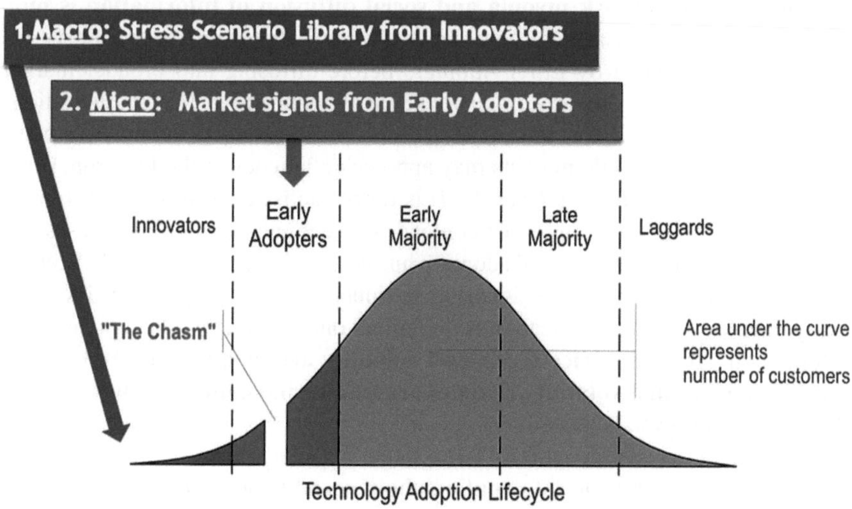

Fig. 3 Social diffusion of disruptive innovation. Graph adapted from Wikipedia

5 Building an Adaptive Stress Library

This insight that **market intelligence is not evenly distributed** drives the design of our Adaptive Stress Library to harness intelligence from *Innovators* and *Early Adopters*.

Innovators foresee potential risks that are imperceptive to most. As Frederic Bastiat recognized in 1850, this ability to foresee is what differentiates visionary and ordinary economists:

> In the department of economy, an act, a habit, an institution, a law, gives birth not only to an effect, but to a series of effects. Of these effects, the first only is immediate; it manifests itself simultaneously with its cause - it is seen. The others unfold in succession - they are not seen: it is well for us, if they are foreseen. Between a good and a bad economist this constitutes the whole difference - the one takes account of the visible effect; the other takes account both of the effects which are seen, and also of those which it is necessary to foresee. Bastiat (1850)

6 Social Markets Hypothesis

This social adoption process suggests an amendment to the Efficient Markets Hypothesis (EMH), which maintains that investors are rational and information is fully reflected in prices. Given that human beings (still) make investment decisions, **extended periods of risk myopia and social diffusion of information is more realistic**. Crucially, disruptive new insights often enter from the periphery and must first be validated by *Early Adopters* before diffusing into broader markets in successive waves. This process can take many months. Far from being rational *homo economicus*, investors are social decision makers, subject to cognitive biases and herd mentality. While markets may approach efficiency in the long run, **in the short run a *Social Markets Hypothesis* is more realistic**. Disruptive information is not evenly distributed, and must contend with entrenched cognitive biases, and jump through a successive social adoption hurdles. It might explain why U.S. equity markets peaked October 2007, ignoring obviously escalating systemic risk for so many months. A major implication is that prices only indirectly reflect all available information. **Absolute price levels and volatility are lagging indicators, while outliers in price and volatility changes are leading indicators**. We will illustrate this theme with several case studies.

As William Gibson recognized: "The future is already here. It's just not evenly distributed yet" (1999). Or, **information about credible potential risks is already here. It just hasn't been widely adopted yet**.[2]

[2] I can't help but think that we see this same effect in the climate change discussion, with climate scientists as Innovators at the periphery of the public network, struggling to cross the chasm of global adoption.

7 Outliers as Early Warning Signals

Outliers play a crucial role in early warning. Outliers are the first visible signal of a regime shift into abnormal markets as *Early Adopters* act on disruptive information. HSBC's February 23, 2007 $10.5 bn subprime loss announcement caused a tripling of AAA subprime spreads in a single day (a 12 standard deviation outlier). Four days later, this disruptive information cascaded into broad equity markets as February 27 saw exceptional downside outliers from China to the U.S. The Dow Jones recording its 6th biggest daily surprise in over 100 years (Finger 2008) on that day. A −3.3 % drop would hardly appear noteworthy, except that it happened just as volatility reached historical lows and therefore represented a 7.8 standard deviation outlier. See Table 1 for the top 10 DJIA surprise since 1900.

The Value-at-Risk (VaR) backtesting[3] graph of the DJIA (Fig. 4) shows how **the February 27th, 2007 outlier signaled the emergence of the subprime crisis in broader markets**. Notice the classic endogenous pattern of escalating systemic risk as pent up invisible risk emerges as visible risk.

Table 1 Top ten DJIA outliers (1900–2008)

Rank	Date	Residual	Return (%)	Volatility (%)
1	26-September-55	−13.3	−6.5	8.1
2	19-October-87	−12.6	−22.6	32.4
3	29-July-27	−10.1	−5.2	8.3
4	13-October-89	−10	−6.9	11.4
5	26-June-50	−8.1	−4.7	9.3
6	27-February-07	−7.8	−3.3	6.8
7	20-January-13	−7	−4.9	11.4
8	30-July-14	−6.7	−6.9	16.9
9	28-July-14	−6.7	−3.5	8.5
10	15-November-91	−6.6	−3.9	9.6

Source: Finger 2008

[3]All VaR backtesting is based on the standard RiskMetrics methodology (exponential weighting with 0.94 decay).

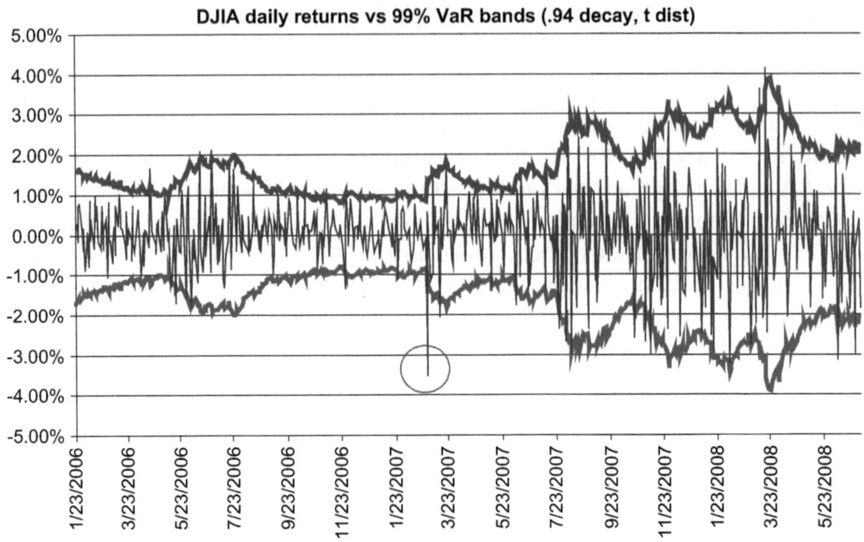

Fig. 4 February 2007 outlier and escalating equity risk. *Source*: Laubsch (2009)

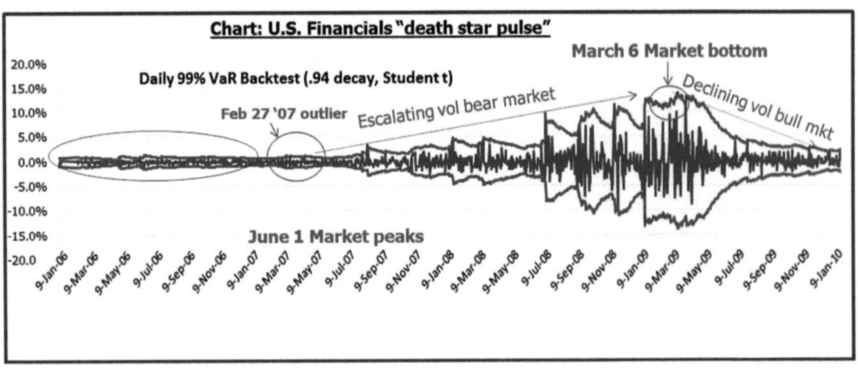

Fig. 5 Financials VaR backtesting chart. *Source*: Laubsch (2010a, b)

The VaR backtesting chart of U.S. Financials (XLF) from 2006 to 2010 (Fig. 5) shows an even more pronounced "death star pulse" of amplifying risk after the February 27 outlier.

7.1 Gold Outlier Case Study: 2012-2013

The gold bubble burst which started in late 2012 is a classic early warning case study. Figure 6 shows that a skew of positive outliers preceded gold's peak in

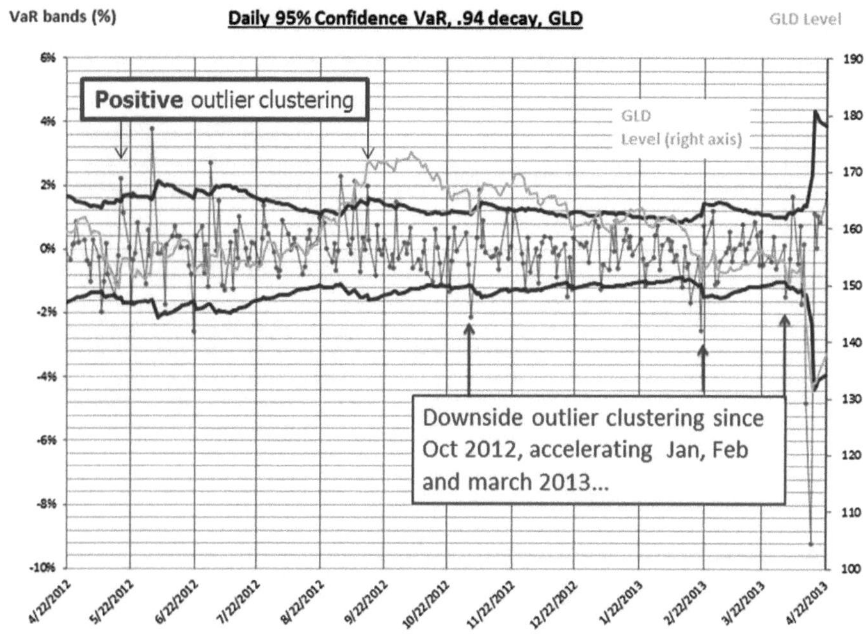

Fig. 6 Gold VaR & outlier graph

October 2012, after which a skew to negative outliers preceded gold's slide and precipitous drop of April 12-15.

7.2 European Divergence Case Study

The European Divergence Scenario illustrates the Adaptive Stress Testing framework well. With the introduction of the Euro, credit spreads converged for all member countries and started an unsustainable cycle of credit growth in high inflation countries like Greece, Italy, Portugal, and Spain. The artificial stability of the Euro currency was a classic Minsky case of stability breeding instability. As hidden imbalances continued to build up in the Euro periphery, Innovators like GaveKal Research analyzed the unsustainability of "PIGS" borrowing levels, and launched a European Divergence Fund in November 2007:

> For ten years, investors have made money on convergence trades (i.e., Italian rates falling to meet German rates).These convergence trades were always based on politics, not economics. However, in the long-run, economics always wins out. And now, as credit conditions tighten around the world, should be the time when this happens.—Louis-Vincent Gave, GaveKal Funds Newsletter, November 11th, 2007

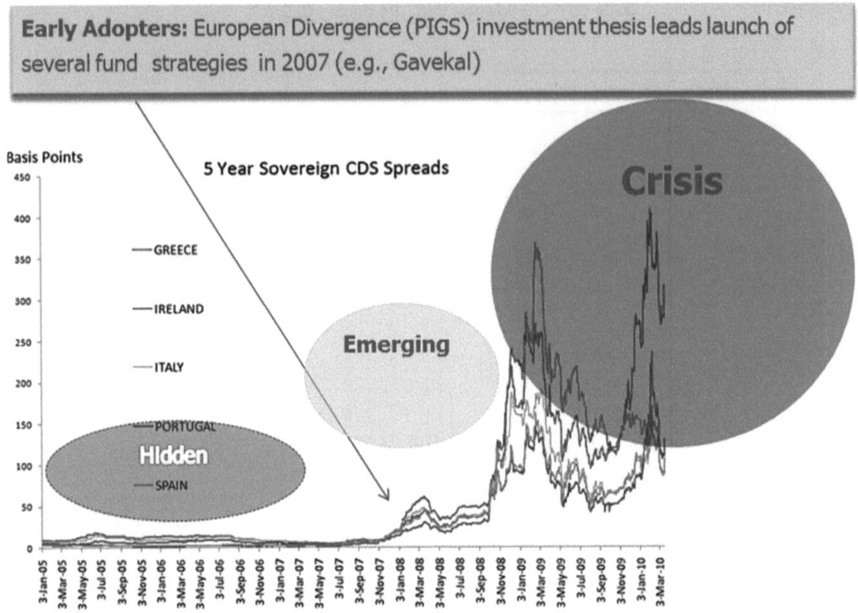

Fig. 7 Five year sovereign CDS spreads for PIGS. *Source*: Laubsch (2010a, b)

Figure 7 shows the escalation of the PIGS sovereign spreads from pre-crisis 2005 when risk was hidden, to 2007 when risk started to emerge and to 2008 when markets went into crisis mode.

As with subprime bonds and equities, outliers were useful early warning signals. Figure 8 shows Greece CDS vs. cumulative VaR outliers. Each wave of escalating spreads is preceded by exceptionally low levels of outliers (e.g., unnaturally low level of variability), and then a rapid phase transition to high volatility marked by escalating outliers.

8 U.S. Subprime Case Study: 2006-2008

Even Nassim Taleb admits that the U.S. subprime crisis was no *Black Swan*. The macro environment in 2006 was increasingly fragile. Classic macroeconomic imbalances built up over years: low rates and easy credit inflated a U.S. housing bubble amidst record levels of financial leverage.[4] Cracks appeared as the housing market started to taper off in mid-2006, and JPMorgan was the first major bank

[4]Deregulation and increased global capital flows were additional systemic warning signals, as noted by Rogoff and Reinhart (2009).

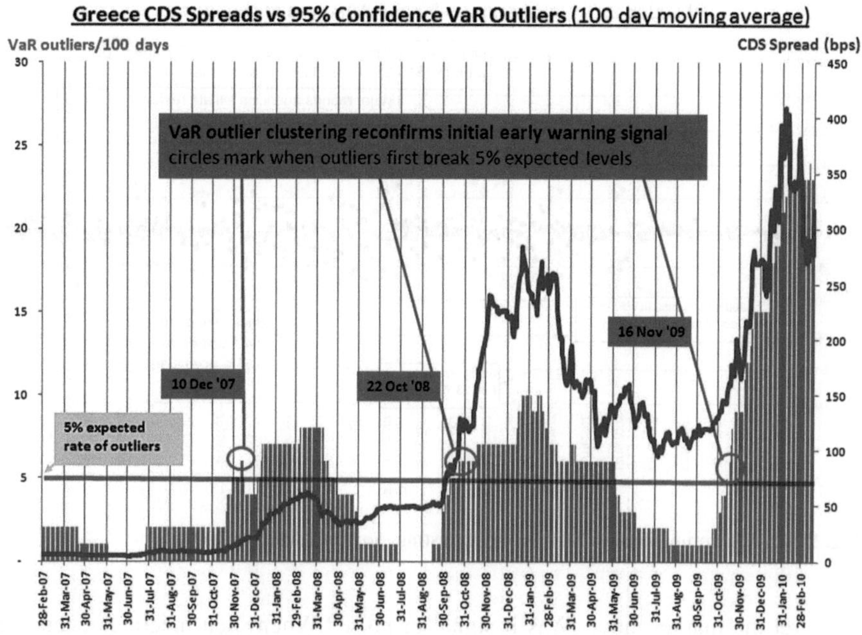

Fig. 8 Greece CDS and VaR outliers. *Source*: Laubsch (2010a, b)

to exit subprime. It had already reduce subprime CDO underwriting as credit spreads compressed to leave little margin for error. CEO Jamie Dimon made the final decision to sell all its subprime holdings after observing a spike in subprime delinquencies in its retail bank (Tully 2008).

8.1 *The First Tremors in December 2006*

And yet, complacency still reigned. Credit markets only registered the first small tremor from December 12–21, 2006, when AAA subprime bond volatility tripled. Most market participants did not notice, because the absolute level of volatility was so low. Five year AAA subprime bonds traded around ten basis points (bps) over Treasury's, and were thus regarded as almost risk free securities.

Daily AAA bond spread volatility around averaged 2 % per day, or about 0.2 basis points. A tripling of volatility only amounted to 0.6 basis points, an increase that was missed by all but the most vigilant institutions which specifically monitored subprime risk. At JPMorgan (where I worked as a risk manager from 1993 to 1998) trading discipline called for meetings whenever VaR limits were breached. At the 95 % daily confidence level, this normally meant such outlier discussions would happen about once month (1 out of 20 trading days). Our objective was to

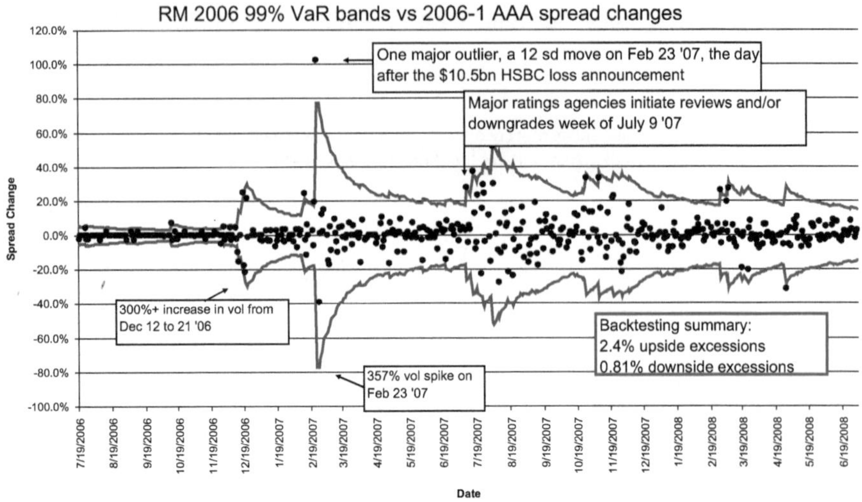

Fig. 9 AAA subprime bond spread VaR backtesting. *Source*: Laubsch (2009)

interpret the market move. Was it signal or noise? When such meetings occurred too often, alarm bells would go off. Risk was not normal and escalating. Goldman Sachs famously decided to "get closer to home" and exit subprime after 10 days of subprime VaR excessions in December 2006. Although absolute losses were relatively low, it quickly became apparent that something was amiss in the subprime world (Nocera 2009).

One crucial insight is that **only investors who closely monitored subprime P&L vs. VaR could observe the December 2006 tremors**. Firms that mixed subprime bonds with regular bonds would have missed these signals. Hence **the importance of defining a *Stress Index* with specific driving factor**s.

Figure 9 is a chart of AAA subprime bond yield changes plotted against 95 % confidence VaR bands. Observe that the biggest outlier was February 23, which was more surprising than any outliers observed during the actual crisis.

8.2 The Second Wave: HSBC's February 2007 Loss

The second major jump in subprime volatility occurred on February 23, 2007, the day after HSBC announced a $10.5 bn loss in their US subprime holdings. It looked like a classic exogenous Black Swan shock, as AAA spreads instantaneously tripled from 11 to 31 bp (an unprecedented 12 sd daily outlier). Spreads soon stabilized below 20 bp again, bolstered by "relative value" trades: traders bet that increasing subprime delinquencies would hurt the first loss equity or mezzanine tranches of CDOs, but thought that the AAA rated securities were immune. And they decided to make the trades "carry neutral" and some even went so far as to characterize this as a "hedge." If BBB's were trading at 200 bp and AAA's were 20 bp, shorting $1 bn BBB meant buying $10 bn AAA. This was history's most penny wise and pound foolish trade, and would later explain Howie Hubler's record setting $9 bn loss at Morgan Stanley.[5]

Evidence of a housing bubble burst continued to mount, as prices fell and subprime delinquencies spiked. Subprime lender NEW's default in March 2007 was no surprise to students of financial statements: 5 months earlier forensic accountant CFRA published a report that they had obfuscated rising delinquencies in their June 2006 earnings release. In May 2007, two Bear Stearns subprime bond hedge funds imploded, and the following month Merrill Lynch failed to sell their AAA rated CDO collateral (bonds they had sold to Bear and held in their own inventory). But even this failure could not shake the market's confidence, as credit spreads and VIX continued to hover at historical lows.

Astonishingly, it took until the week of July 9th, 2007 for S&P, Moody's and Fitch to announce their first subprime CDO downgrade. Only then did risk awareness enter mainstream consciousness, and waves of successive selling followed. Within a few weeks spreads rose to 150 bp before tightening once again to 50 bp, and then widening to 250 bp, and eventually spiking to 400 bp by early 2008.

See Fig. 10 of absolute spread levels, which reveals classic fractal amplification patterns different time scales: from daily ripples to weekly waves to monthly tsunamis.

When looking at the subprime spread chart above, ask yourself when risk was highest. According to VaR, risk peaked in 2008. But when considering hidden risk, the most dangerous time was 2006-2007 as the chase for ever narrowing spreads led banks to unprecedented leverage that would threaten the entire global economic system.

9 Risk Myopia and Disruptive Information

How could markets have been so blind for so long? Credit markets appeared to have outsourced credit risk assessment the ratings agencies, who were asleep at the wheel, not to mention conflicted by lucrative subprime bond underwriting fees. Robert

[5]CFO.com, "Missing Pieces" by Avital Louria Hahn, March 2008, reported: "Morgan Stanley's fixed-income traders built a $2 billion short position on the sector. As protection, they bought $14, billion worth of triple-A mortgage-backed securities. ... Morgan Stanley's hedge collapsed, triggering a $9.6 billion fourth-quarter write-down-nearly triple the $3.7 billion that Colm Kelleher, Morgan Stanley's newly appointed CFO, had forecast a month earlier.

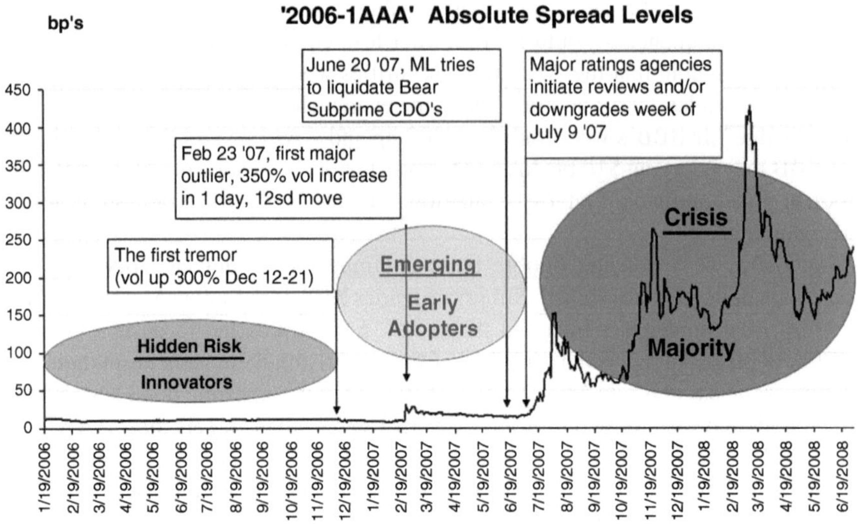

Fig. 10 AAA CDO spreads. *Source*: Alan Laubsch (2009)

Schiller (having warned of housing bubble since 2005) attributed this collective denial to groupthink:

> Suppose you imagine yourself and a group of experts who seem to have converged on an enlightened opinion which has arguments to support it, and it has prominent influential people saying that. It can be difficult for someone to stand up in that room and air what seem to be half-baked or half-formed doubts about it. It can be kind of damaging to your reputation. (Grove 2008)

Pervasive risk myopia is confirmed by behavioral economics. Cognitive biases limit our ability to rationally assess risk and be open to new information. Well known **risk distorting biases** include overconfidence, underestimation of small probability events, confirmation bias, and missing new information when over-focusing (Kahneman 2011). Nate Silver discusses these biases in *The Signal And The Noise* (2012), and observes that they apply to financial professionals and economists just as much as anyone else.

10 Innovation Comes from the Periphery

It therefore makes sense that innovative insights often emerge from the periphery, free from the pressures of groupthink. Michael Lewis explores this theme in his bestselling "The Big Short: Inside The Doomsday Machine" (2010). The visionaries who recognize and acted on the inevitable collapse of subprime bonds before everyone else were a diverse collection of renegades, often at the periphery of the financial community. This included outsiders like Dr. Michael Burry, a medical doctor with Asperger's turned fund manager, and the two inexperienced founders of Cornwall Capital, who not too long ago had started their fund with $100,000 in a Berkeley apartment.[6]

Liberated from the confines of groupthink, outsiders are more receptive to new and disruptive information. Robert Shiller fits this archetypal *Innovator* at the periphery, inspired by diverse interests:

> I think I'm a polymath. I'm interested in everything. When I was a senior in college at the University of Michigan, I was dazzled by the choice set that we had. Young people, you can do whatever you want, and I was disappointed that I had to choose one, realistically. You like to be a renaissance man and do everything. I took long walks trying to decide whether I wanted to be a physicist or a medical doctor or a sociologist, whatever-a scientist, an astronomer. (Grove 2008)

Another characteristic of visionary Innovators is the ability to **integrate information from a variety of sources and to connect the dots.** Nouriel Roubini sources information from taxi drivers to Finance Ministers when traveling. After being named as one of Time Magazines' 100 Most Influential People of the World, he explained:

> In many ways **I simply connected the dot in these different strands of thinking and warnings**... Kenneth Rogoff ... warned early on about ... the US current account deficits and of the global imbalances; Raghu Rajan presented ... analyses of the agency problems and incentive distortions deriving from compensation schemes in financial institutions; ... Stephen Roach, David Rosenberg ... warned about the shopped-out, saving-less, bubble-addict and debt-burdened US consumer; ... William White and his colleagues at the BIS were among the first... to analyze how the "Great Moderation" may paradoxically lead to "Financial Instability", asset and credit bubbles and financial crises (Roubini 2009)

11 Cycle of Hidden vs. Visible Risk

Hidden risk suddenly becomes visible risk similar to the way a spring's **potential energy is released as kinetic energy**. Until February 2007, banks were considered rock solid (despite record debt and subprime concentrations, not to mention warnings from luminaries like Roubini and Shiller). Visible risk (volatility) was at historic lows, as tension mounted below the surface. HSBC's February 23 loss announcement was a classic *Black Swan*, and triggered a jump in volatility that spread from subprime to equities on February 27. Even though risk would keep escalating for the next 2 years, both **the February 23 bond outlier and the February 27 equity outlier would remain the largest surprise**. Surprisingly, though, both equity and bond markets dismissed these outliers as markets recovered after each selloff. Subprime bonds started their bear market only after the July ratings downgrade, and equity markets kept bubbling until their October 2007 peak.[7]

---

[6]As measured by one day standard deviation residual, using dynamic RiskMetrics volatility estimation.

[7]Impressively, October 2007 was the bubble peak forecasted by Didier Sornette's LPPL models.

By the time the market hit bottom on March 6, 2009, volatility had been at sustained record levels and only gradually started to decline. The market's assessment of risk and return was, in effect, exactly the backward. As NBIM's founding CEO Knut Kjaer notes: "The biggest pitfall in investments is herd behavior. Large gains in performance can be achieved by investors with ability to consistently act contrarian" (Kjaer 2011).

12 The Destabilizing Effect of Stability

Hyman Minsky's "Financial Instability Hypothesis" (1992) is often summarized as "stability breeds instability." As Lawrence H. Meyer observed in *Lessons from the Asian Crisis* (1999): "a period of stability induces behavioral responses that erode margins of safety, reduce liquidity, raise cash flow commitments relative to income and profits, and raise the price of risky relative to safe assets–all combining to weaken the ability of the economy to withstand even modest adverse shocks." In the case of the *Asian Crisis*, it was pegged currencies which allowed Asian banks and corporations to raise cheap USD financing. Financial imbalances built up, but did not register in the artificial low volatility of pegged currencies.[8] When the first FX tremors started in Thailand in May 1997, it was only a matter of time before devaluation. Within days chain reaction spread to Indonesia, and the rest of South East Asia. Corporations defaulted, bank NPL's skyrocketed, and economic growth plummeted.[9] The Euro offers similar lessons. With the introduction of the common currency, borrowing costs converged for vastly different economies. Greece and Italy which traditionally ran at high inflation and borrowed at high rates suddenly had access to the same rates as Germany. Unfortunately, it's not possible to legislate risk away. Rather, the artificial suppression of volatility allows imbalances to build under the surface, resulting in hidden fragility. When cracks emerge, the system is threatened with sudden collapse.

This destabilizing effect of stability has also been observed by ecologists: "When the range of natural variation in a system is reduced, the system loses resilience" (Holling and Meffe 1995). And conversely, "the very fact of **low stability seems to produce high resilience**" (Parameswaran 2009). This theme is the focus of Nassim Taleb's latest book, "Antifragile: Things That Gain from Disorder" (2012).

The implication for risk management is to be contrarian. While we must attend to emerging visible risk first, we should not be lulled into complacency by periods of calm. Steady trends with low volatility often points to a lack of diversity in opinion and crowded trades. The most severe reversals come as everyone is forced to exit at

[8]Interestingly implied volatility did spike in THB options prior to the devaluation as an early warning signal, as documented by Malz (2011).

[9]This build-up of hidden risk until a dramatic collapse is a common theme with pegged currencies: Argentina experienced a similar.

the same time. Exceptionally low volatility (e.g., from pegged currencies to tapered bond yields) should ring alarm bells and motivate us to seek out hidden risks.

13 Volatility is not Risk

When asking investors whether they would prefer high or low volatility investments, most opt for low. Few can tolerate the turbulence of volatility. **But low volatility only means low *visible* risk**. What if we put it this way: "Would you prefer low volatility with the possibility of large hidden risk? Or high volatility, but at least what you see is what you get?" Despite record low volatility, early 2007 was the most risky time to be invested. And March 2009 presented exceptional opportunity for returns, despite record short term volatility. As Knut Kjaer observes: "The (future) reward for risk may be at the highest when the market sentiment for risk taking is at the lowest."

The implication is that **as volatility declines, our priority should shift to identifying hidden structural risks**. And during periods of high volatility, contrarian investors might weigh the pain of P&L fluctuations against the potential for superior long term returns opportunities.

The contrarian view of risk and opportunity is supported by mean reverting market volatility. Periods of low volatility lull short term investors into a false sense of security, as hidden risks build up until emerging as volcanic outbursts of volatility. The exodus of investors as visible risk reaches elevated levels creates opportunities for longer term investors who can stomach the adventure of a rocky ride. Hence, it should not be surprising that equities—as the most volatile asset class—offer the most superior long term returns on a diversified basis (Siegel 2007).

Figure 11 of annualized daily volatility for the DJIA index over the last century illustrates this pattern. As with most major developed markets, volatility mean reverts to a 15-20 % range.

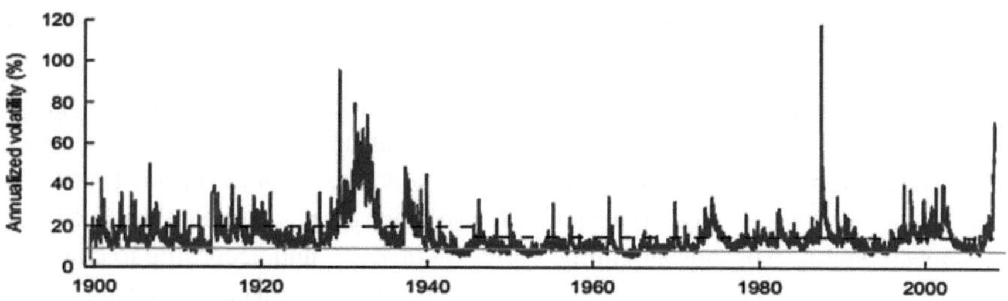

Fig. 11 DJIA index volatility, 1900–2008. *Source*: Finger 2008

14 Macro Micro Polarity Management

As discussed above, the flux between immediate visible risks and longer term fragilities presents a perpetual challenge. It's not a problem that can be solved with better statistical models. Indeed, better data and more precise analytics can lead to overconfidence. This was part of the problem in the subprime crisis ("we were busy looking at sand corns through a microscope when the tsunami hit" recalled a bank risk manager). This is a classic polarity management challenge. Polarities are interdependent opposites which power all complex systems. Barry Johnson's seminal "Polarity Management: Identifying and Managing Unsolvable Problems" (1996) is an excellent primer.

15 Six Macro vs. Micro Risk Management Polarities

As you read the pairs below, consider which requires greater attention at this point in the current market cycle:

1. Potential vs. Visible
2. Long term vs. Short term
3. Top-down vs. Bottom-up
4. Strategic vs. Tactical
5. Qualitative vs. Quantitative
6. Risk vs. Return

While we may have individual preferences, it's important to recognize that **each polarity has both positive and negative attributes.**

- For example, a focus on *Potential* risk allows investors to better prepare for extreme tail risks. And yet over-focus on *Potential* risk can lead to excessive risk aversion and failure to prioritize immediate needs.
- On the other hand, a focus on *visible* risk allows investors to manage risks that matter now, and to be nimble and take advantage of short term opportunities. Yet over focus on *visible* risk can result in myopia and underestimation of structural risks.

Well managed polarities maximize positive attributes and minimize negative ones, sparking a virtuous cycle. Polarities naturally move from the upside to the downside of a polarity, and then to the upside and downside of the opposite polarity, and so on. Poorly managed polarities (e.g., too much focus on one polarity) cause a downward spiral. Appropriate timing varies by process, and according to changing circumstances. The key to managing polarities well is to act on early warning signals that suggest pivoting to the opposite polarity. This cycle illustrated in Fig. 12.

Polarity management is a core life practice. As individuals, we can only see one perspective at a time, so we must keep changing perspectives to better perceive and adapt to our dynamic and multi-faceted world. Indeed, according to developmental psychology, the ability to take different perspectives is central to learning and growth.

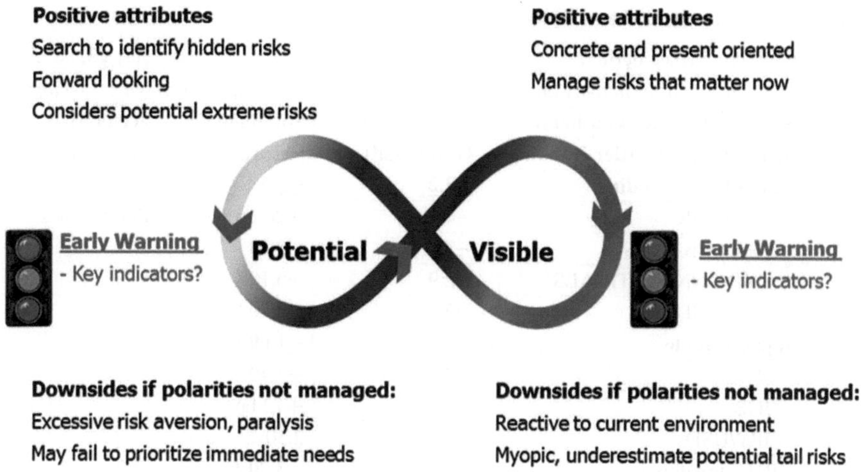

Fig. 12 Potential vs. visible risk focus. *Source*: Laubsch (2010a, b)

From an organizational perspective, the ability to shift perspectives is core to an adaptive risk culture which continually sources new scenarios, while quickly honing in on emerging risks.

As a practical current example, we might consider the monitoring of a potential *China Hard Landing* scenario (Pivot Capital 2011).

1. **Macro** fault lines include credit dependent growth, a real estate bubble, counterparty defaults, rising bank NPLs, reliance on manufacturing and export, slowing economic growth, as well as social instability due to rural poverty and the rising gap between rich and poor, corruption, and repressive government policies.
2. From a **Micro** perspective, we consider key market factors which correlate to an emerging scenario (e.g., Chinese rates & bonds, equities, industrial commodities). Our aim is to build *Stress Indices* with high correlation to specific scenarios. In addition to monitoring emerging risk, *Stress Indices* could be used for hedging or insurance (e.g., purchase 1 year 95 put protection for China Hard Landing).

Just as in our Financial Meltdown scenario, we would plot a time series of our China Hard Landing *Stress Index* (or Indices, as multiple variants are possible) and hone in on outliers and escalating volatility.

16 Outlier Dashboards

Below is a mock dashboard report, which ranks stress themes by surprise (i.e., outlier move as measured in standard deviation). Useful reports will have drill down capability, and the ability to view outlier activity in different dimensions such as theme, country, industry, and asset class (Table 2).

Table 2 Stress outlier dashboard [not real data]

	95 % confidence outliers			July 23, 2013
	Summary: Outlier activity on 4 stress scenarios, 3 asset classes, 2 sovereigns, 5 sectors			
	Stress Scenarios (4 outliers)	Asset Class (3)	Sovereign (2)	Sector (5)
	Stress Theme Outlier Ranking	**Outlier (sd)**	**Return**	Index level
1	**China Hard Landing**	**−3.2**	**−3.5 %**	
	CHINA IR	−4.1	+26 bp	Red to black
	CHINA BANKS	−3.4	−5.20 %	
	CHINA INDUSTRIALS	−2.9	−6.10 %	
	AUD/USD	−2.1	−2.40 %	
2	**Japan Bubble**	**−2.5**	**−1.60 %**	
	N225	−2.1	−3.40 %	
	JGB	−3.5	+12 bp	
	JPY/USD	2.1	2.30 %	
3	**BRIC slowdown**	**−2.1**	**−3.20 %**	
4	**Commodity Bubble**	**−1.9**	**−2.6 %**	

This sample report indicates a −3.2 standard deviation in the *China Hard Landing Scenario*, and the drill down shows exceptional movements in interest rates, banks, industrials, and AUD/USD. Other tabs could be viewed to look at outlier activity in other dimensions (e.g., asset class, country, sector).

17 Introducing StressGrades™

Outlier based early warning is especially useful when considering the many scenarios risk managers shock their positions with. A risk manager at a global bank explained that they run close to 200 daily scenarios against hedge fund counterparties alone, but that the amount of data was overwhelming and therefore largely ignored.

This insight gave rise to the StressGrades methodology to prioritize attention on escalating market based early warning signals. StressGrades are designed to complement the existing stress testing process by (a) drawing attention on escalating visible risk, as well as (b) highlighting abnormally low visible risk themes to search for hidden risk.

18 We Define Three Volatility Based Metrics

1. **PStress** = Market Implied Probability of a Stress Scenario, in bps per annum
2. **DStress** = Distance to stress scenario in standard deviations (e.g., a Z-score)
3. **StressQ** = Quantile (percentile) historical rank of stress scenario (e.g., StressQ = 0.82 implies stress levels have exceeded current levels 18 % of the time)

19 StressGrades Amplify Market Based Risk Signals

As volatility based metrics, StressGrades will not directly uncover hidden structural risk or predict *Black Swans*. StressGrades merely amplify existing market based signals of risk, as a seismograph amplifies geological tremors.[10] And as with earthquakes, the absence of tremors does not imply the absence of risk. Or, to use a medical analogy, a stethoscope allows doctors to listen to what's inside. An experienced doctor can use it to detect imbalances without being fooled into believing it provides a full picture of health. To be effective, StressGrades should be used within the Adaptive Stress Testing framework: (a) prioritize immediately escalating stress themes, and (b) probe for hidden risks in themes with abnormally low volatility.

StressGrades require three major steps:

1. **Design Stress Indices**

   As discussed earlier, Adaptive Stress Testing calls for the construction of market based *Stress Indices* that are correlated with key scenarios. For example, we might consider an oil shock scenario due to conflict with Iran, which we could model at different levels of detail (e.g., oil price, FX prices, country and industry equity sectors, and even down to the specific company level). Note that *Stress Indices* might be constructed using options theory to model non-linearities (e.g., an oil call with a 130 strike + equity put struck at 90). Once we've modeled our Stress Indices, we can start to monitor the tremors for each fault line that indicate escalating risk.
2. **Determine a Stress Point**

   We then determine a critical point which represents a stressed condition. In our examples, we used maximum historical drawdowns over a 10 year period. After calculating the volatility of our ETF time series, we then determine how many standard deviations it would take to achieve such a daily loss to calculate DStress. We implemented the RiskMetrics methodology to estimate daily volatility (i.e.,

---

[10] I am reminded of a statement by a HK hedge fund manager about Goldman Sachs, after we discussed their use of VaR outlier signals to exit subprime. "They're like geologists who make their living right top of all the world's fault lines line, monitoring every tremor."

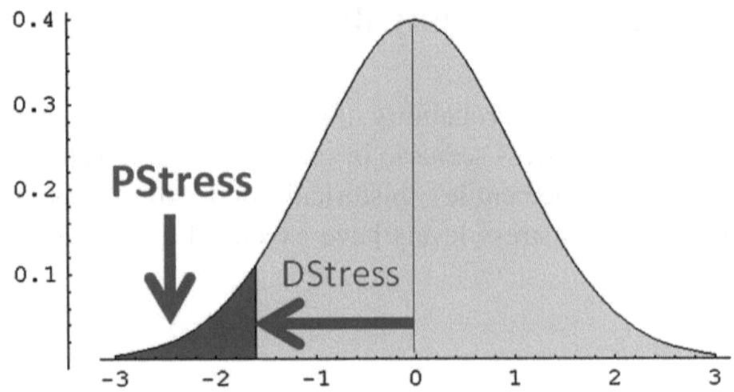

Fig. 13 PStress and DStress assuming normal distribution

exponential weighting with 0.94 decay). Note, however, that in some cases a maximum historical drawdown could be too severe a Stress Point to consider. In practice a 99 % confidence Expected Shortfall over a decade (or a full market cycle) could be reasonable to consider as a Stress Point. Subjective assessments and analysis of similar asset classes should be applied for assets with limited loss history (e.g., subprime bonds in 2006).

3. **Calculate market implied Probability of Stress (PStress)**
   After making a distributional assumption, we can back out the implied Probability of Stress (PStress). For example, assuming Normality, a DStress of −2.33 would imply a PStress of 1 %, while a DStress of 1.65 would imply a PStress of 5 %. In the examples below (Fig. 13), we will use a Normal distribution assumption for simplicity. Clearly, accuracy of PStress could be improved by using a fat tailed distribution such a Student (Zumbach 2007). However, given that our initial objective is to flag outlier changes in market implied stress probability, the use of a Normal distribution is appropriate (future versions will consider Student t and other fat tailed distributions).

To summarize, StressGrades are volatility based metrics which can be used to monitor market implied risk sentiment. To be useful, we need to start with a macro perspective to understand key fault lines, and apply StressGrades both to monitor emerging visible risk as well as identify artificially low levels of visible risk.

20 Backtesting StressGrades

Below we show several are early warning backtesting case studies on ETF's representing major asset classes. We calibrated DStress for each ETF based on the largest daily drawdown dates (e.g., −9.6 % for SPY on December 1, 2008). If StressGrades are predictive, we would expect an escalation in PStress and decline and DStress as we approach the drawdown date (e.g., December 1 for SPY). In other words, we would expect volatility to be high and rising before the peak endogenous stress events.

Again, for simplicity we use the Normal Distribution to calculate PStress in the case studies below.

Given that StressGrades are driven by volatility, we expect StressGrades to fail in predicting Black Swans, but to help in detecting Dragon Kings.

21 S&P 500 (SPY) Case Study

On December 1'08 SPY fell 9.6 % (log return), the biggest daily drop since Black Monday in 1987.

Figure 14 shows a super-exponential increase in (Normal Distribution Implied) PStress leading up to the December 1, 2008 stress event. Note the log scale, so any increase above linear is super-exponential.

The following is noteworthy:

1. On February 27, PStress jumped by 170× from extremely low levels. It was a Black Swan. PStress (i.e., equity market volatility) had no predictive power. However, the extremely low level of volatility/implied stress could be viewed

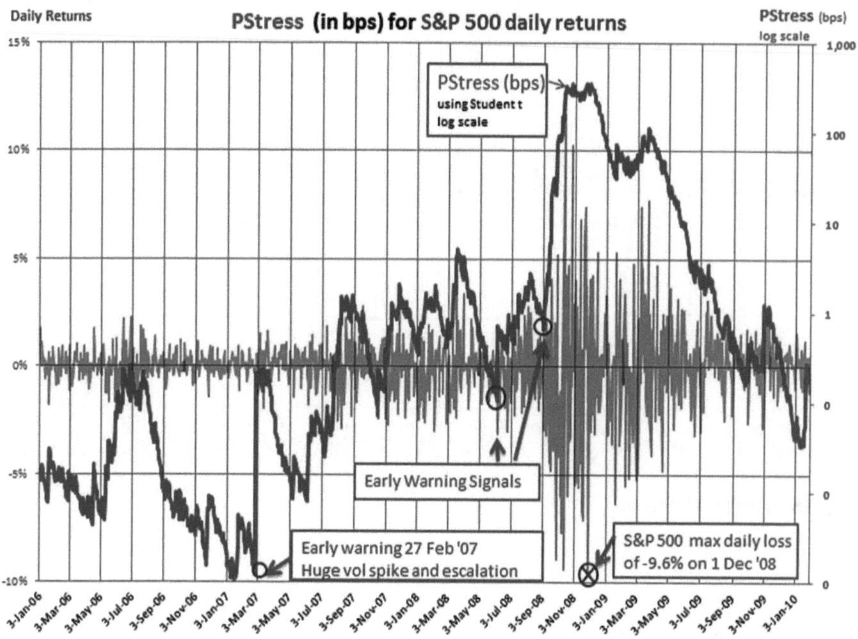

Fig. 14 S&P 500 PStress

as a contrarian signal of high hidden risk and risk myopia/overconfidence as discussed earlier.

2. From that point on PStress spiked over 1300×, implying super-exponential increase of tail risk leading up to the December 1, 2008 drop. PStress picks up well escalating endogenous risk signals.

22 DStress

Figure 15 shows SPY DStress during the same time period. Pre-crisis, a drop of 9.6 % would have represented a distant −24 sd event. After the February 27 outlier DStress jumped to −10 sd, and then further contracted to −2 sd as we approach December 1, 2008. In other words, the actual drop of 9.6 % on 1 December was not much of a surprise by then.

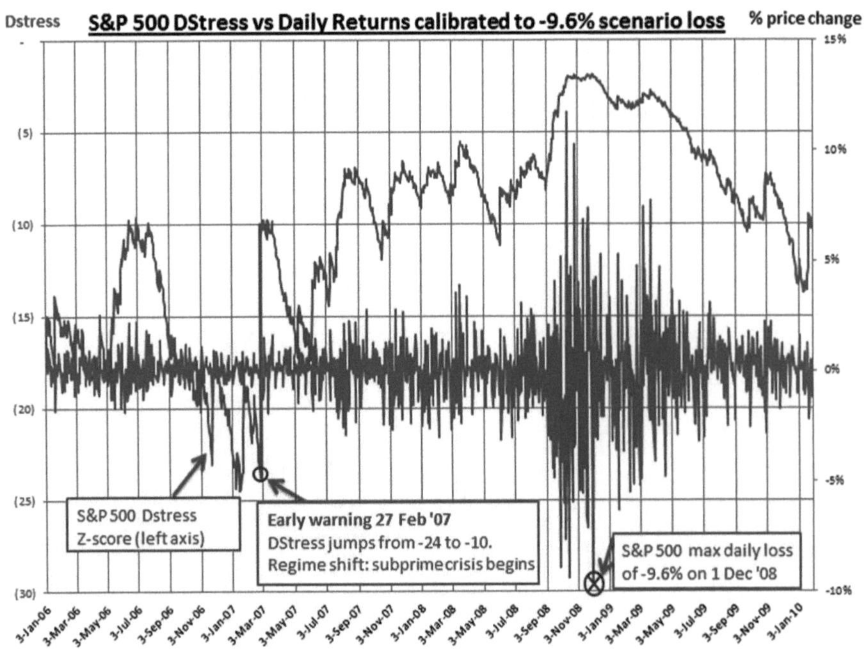

Fig. 15 S&P 500 DStress

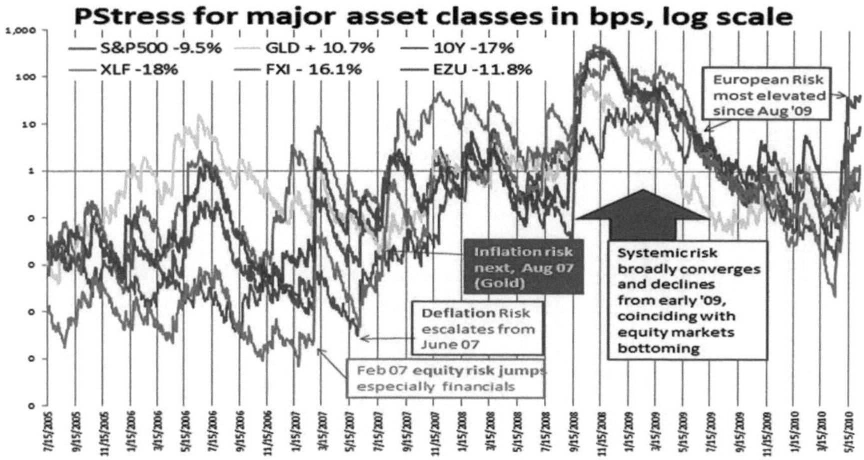

Fig. 16 PStress for major asset classes

23 Cross Asset Class ETF Analysis

StressGrades time series can help us visualize the interrelationship between risk themes. Figure 16 represents major stress themes using ETFs. Observe the sequential cascading of systemic risk starting with the February 27, 2007 equity outlier. In June an outlier drop in 10Y bond yields signaled deflation fear, and in August a jump in GLD signaled escalating inflation fears. Again, note the log scale for PStress .

Especially noteworthy is the increasingly synchronized increase in (Normal Implied) PStress observed across all asset classes after August 1, 2007 as systemic risk increased. Equally noteworthy is the synchronized decline in (Normal Implied) PStress in early 2009, signaling a systemic recovery.

24 StressQ

StressQ is a snapshot of where volatility levels are currently compared to the last year. They can give clues about visible risk, and where we might search for hidden risk. Figure 17 shows StressQ for the major asset classes as of July 14, 2012. We can quickly see that volatility for commodities (DBC, USO, UNG) is at elevated levels, while bonds (esp LQD & TIP) are at very low levels. DBC's StressGrade of 93 means that volatility has only exceeded this level 7 % of the time over the last year. On the other extreme for LQD volatility is lower than 96 % of the time. Most other assets are at average to low volatility levels (58–32), implying broadly moderating volatilities.

A related analysis is to contrast StressQ comparisons with DStress, which considers a longer time horizon anchored to the worst case loss experience by each ETF over the last 10 years. Extremely high DStress for Credit/Interest Rate ETF's

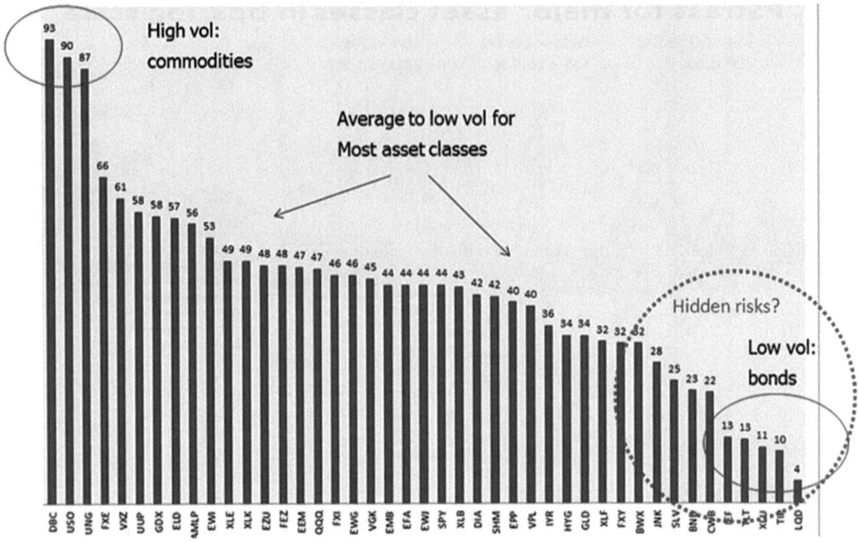

Fig. 17 StressQ for major asset classes as of July 15, 2012

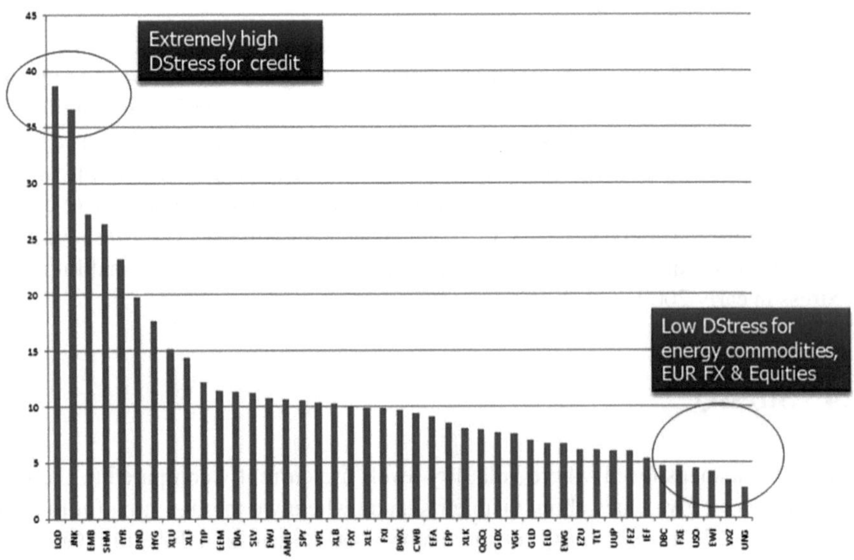

Fig. 18 DStress for major asset classes

shows that market perception of these assets is close to risk free levels, a hint of dangerous overconfidence and complacency. Energy commodities, EUR FX and European stocks on the other hand are at quite elevated levels, less than five standard deviation away from the largest historical moves (Fig. 18).

In summary, StressGrades amplify market based risk signals and are a helpful guide for prioritizing attention to both high and low volatility scenarios.

Micro perspective outlier analysis prioritizes attention the relevant emerging risks. And when markets are calm, we can shift back to explore structural vulnerabilities from a macro perspective.

25 False Positives

When considering market based early warning signals, the million dollar question how often we get false positives (type I errors). Despite Paul Samuelson's famous quip that the "stock market has predicted nine out of the last five recessions" (Samuelson 1966), research suggests that there is predictive value in significant price changes. Jeremy Siegel rebutted Samuelson with a study showing that "...38 of the 41 measured recessions since 1802 have been preceded by and 8 % decline in the stock returns index. There have been twelve "false alarms" using this criterion....Despite these faulty signals there is a significant gain to stock investors from being able to predict turning points in the business cycle over all time periods." (Siegel 1991)

The broad persistence of momentum in stock markets globally (Fama and French 2012) is further evidence of social (as opposed to instant) diffusion of information. Didier Sornette and his Financial Crisis Observatory have a growing track record of bubble forecasting in various asset classes.[11] However, further careful and extensive backtesting should be conducted to confirm that VaR outliers have predictive value. Such backtests might measure conditional returns after VaR outliers in different market regimes. These backtests should help provide insight about what proportion of price change can be attributed to random noise versus signal. Given that noise is likely to be a Gaussian distribution, we should expect that the market's actual fat-tailed distribution to be least be partly attributed to the social diffusion of information. Note that according to RiskMetrics backtesting research, the Gaussian distribution fits markets well until about 95 % confidence (e.g., 1.65 standard deviations). After that, the accuracy of the Gaussian drops significantly, and a Student t distribution with 5 degrees of freedom is significantly more accurate in volatility forecasting from 1 day to 1 year (Zumbach 2007).

[11] See www.er.ertz.ch for updated information.

From a practical perspective, the monitoring and discussion of outlier signals should be part of investment discipline. At a minimum, VaR outliers should trigger a formal discussion, as is the practice at active traders such as J.P. Morgan and Goldman Sachs. Given that outlier signals are often ambiguous, expertise and judgment is required to connect the dots. Therefore, outlier risk management should not be formulaic, but rather based on a discipline of rigorous social intelligence.

26 Adaptive Stress Testing Visualizations

Visualization is a crucial component to Adaptive Stress Testing. Visualization draws attention on emerging risks, and can help build intuition on how different risks are interconnected.

We can use heatmaps to prioritize attention to escalating high probability scenarios (red) and then escalating lower probability scenarios (Fig. 19).

Risk managers will focus first on the imminent threats, or escalating high PStress scenarios. Escalating low PStress scenarios are emerging scenarios, and still offer the potential for exerting control through proactive risk management. Black Swans could be lurking underneath stable low PStress scenarios, which calls for harnessing social intelligence to probe deeper into hidden fault lines.

We can also use network graphs to visualize emerging risk themes. Network graphs are a great way to build intuition about how interrelationships are changing over time, and we will show some examples of such graphs follows.

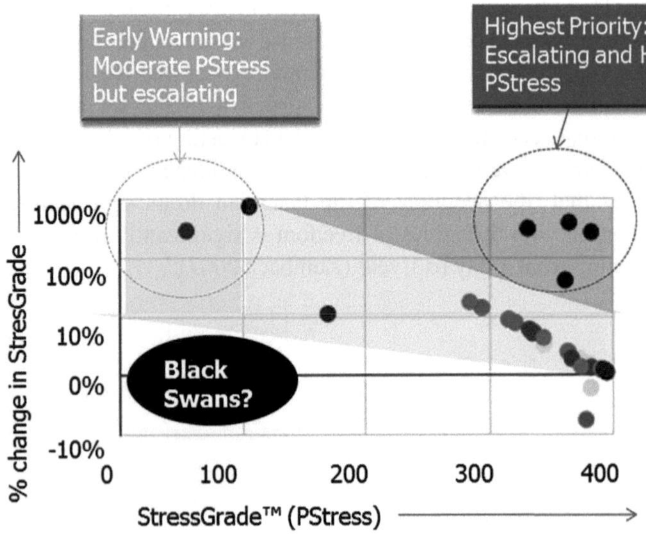

Fig. 19 StressGrades heatmap

27 Network Stress Testing in Practice

Network approaches are very effective for designing comprehensive stress scenarios. For example, consider the **partial correlations** stress testing methodology developed by FNA, which shows how stress scenarios are likely to cascade through a network. Partial correlations measure pairwise correlations between two random variables, taking out the effect of other variables.

28 Japan Case Study

We'll use a recent Japan stress test case study to illustrate how this would work. As a consequence of massive quantitative easing, Japanese equities and bonds entered an exceptional bull market from 2012 to early 2013. By April 4, 2013 Japan was entering bubble territory with JGB yields reaching a historical low at 35 bp and with equities up 80 % over the last year. JGB downside (price) outliers then started escalating as Gold crashed on April 12 and 15. Then on May 23, after the Fed announced potential tapering of QE and after a lower than expected China PPP announcement, Japanese equities dropped by 7.32 % in 1 day. This was a classic early warning signal—after a long ebullient run tied to low interest rates, risk finally returned. This analysis was noted in our PRMIA Emerging Stress Scenarios community on NextThought.com, and we contemplated the potential repercussions of a Japan meltdown on global markets.

The visualization (Fig. 20) shows how a 10 % daily drop in the Japanese Stock Index (EWJ) would likely affect other asset classes, using partial correlation analysis. As opposed to using current data, we applied a historical stressed period during the GFC (a month period starting 4 May, 2008).

Statistically significant partial correlations are shown as links between the nodes (ETFs). Link widths denote the strength of dependence. Node sizes scale with the predicted 1-day return on the day of the stress event. Node color denotes positive (green) or negative (red) returns.

This analysis shows three layers of relatively weak connection between Japan and other asset classes. Japan only shows a moderate primary link to EAFE (logical, given its 21 % weighting in the index). So it was not surprising that as Japan stocks continued to slide by over 10 %, EAFA dropped by just under 5 %, while broader markets hardly reacted.

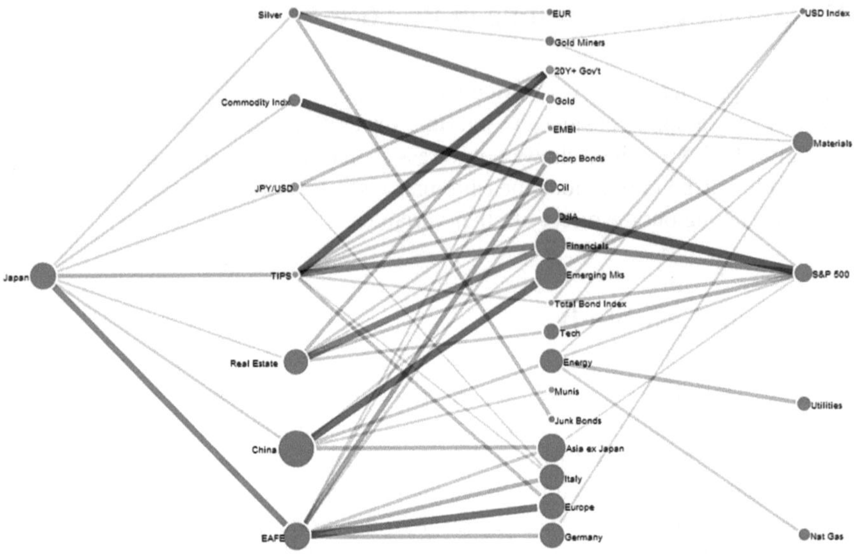

Fig. 20 Japan equities partial correlations based stress scenario. *Source*: Samantha Cook, FNA

29 AIG Case Study Using FNA HeavyTails™ Outlier Monitor

FNA's recently launched HeavyTails application is network implementation of the Adaptive Stress Testing framework. The initial focus of HeavyTails is to monitor global market outliers.[12]

AIG's collapse during the U.S. subprime crisis is a classic early warning case study. As policy makers wrestled with the implosion of Lehman Brothers, they were blindsided by AIG, which would require a record $182 bn bailout. Policy makers realized AIG was on the precipice only "days before its imminent collapse" recounts Phil Angelides, Chair, Financial Crisis Inquiry Commission.

When analyzing the unusual price action of AIG stock, it becomes evident that major market participants suspected AIG's precarious state for many months. Outlier based early warning signals would have been invaluable to policy makers in prioritizing their focus.

Figure 21 shows the *Minimum Spanning Tree* (Mantegna 1999) correlation network of major financial institutions on September 15, 2008, the day Lehman defaulted and AIG's shares collapsed. Outliers are highlighted in red, and ranked by size. Lehman was the biggest surprise, as a 3.92 standard deviation (sd) loss, followed by AIG, a 3.68 sd surprise.

[12] A detailed white paper of HeavyTails is available upon request, and visit HeavyTails.com for more information.

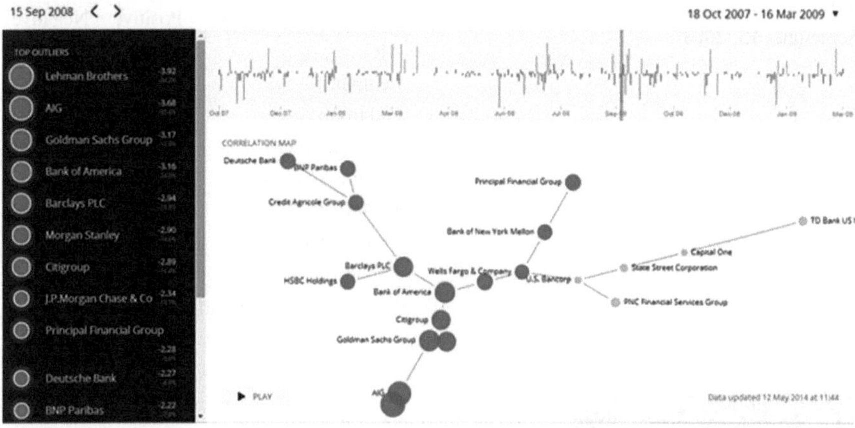

Fig. 21 September 15, 2008: Banking HeavyTails Network Graph. *Source*: FNA HeavyTails

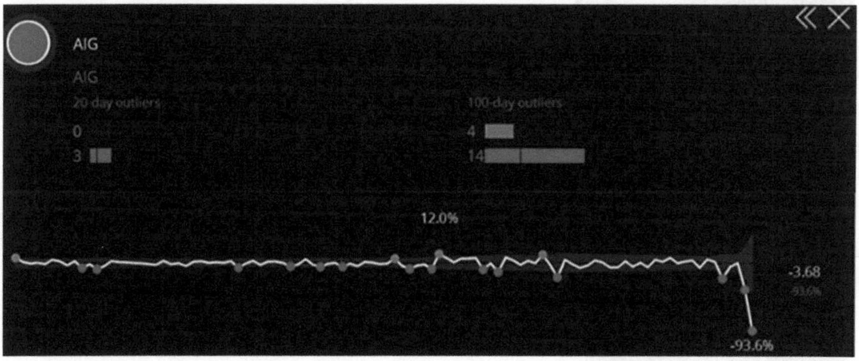

Fig. 22 AIG 95 % VaR graph, September 15, 2008. *Source*: FNA HeavyTails

AIG exhibited exceptional outlier activity leading up to the crisis. Figure 22 shows that there were 14 negative 95 % Confidence VaR outliers in the previous 100 trading days, almost triple the expected 5 % level. The probability of seeing 14 or more negative 5 % outliers in 100 days (assuming IID) is 0.0004632734.

AIG's negative outliers were also exceptional when compared with other financial institutions. Table 3 ranks major institutions by number of negative VaR outliers as of market close on September 15, 2008.

AIG's unusual level of negative outliers commenced almost a year before its eventual collapse. In Fig. 23 we can see that AIG's stock experienced a −3.1 sd outlier (12.5 % decline) on February 11, 2008 as it announced "material accounting weakness" in its credit derivative portfolio. Again, outliers provided early warning, as AIG had been running at 9 % downside outliers vs. 3 % upside in 100 days as seen in Fig. 24. Notice AIG was the only major financial institution that experienced an outlier on that day.

Table 3 VaR outliers as of September 15, 2008

VaR outliers/100 days	Positive	Negative
AIG	3	14
Lehman	2	11
Citigroup	3	9
Bank of America	6	9
Goldman Sachs	2	6
JPMorgan	3	6
Morgan Stanley	3	5
Barclays	6	4

Source: FNA HeavyTails

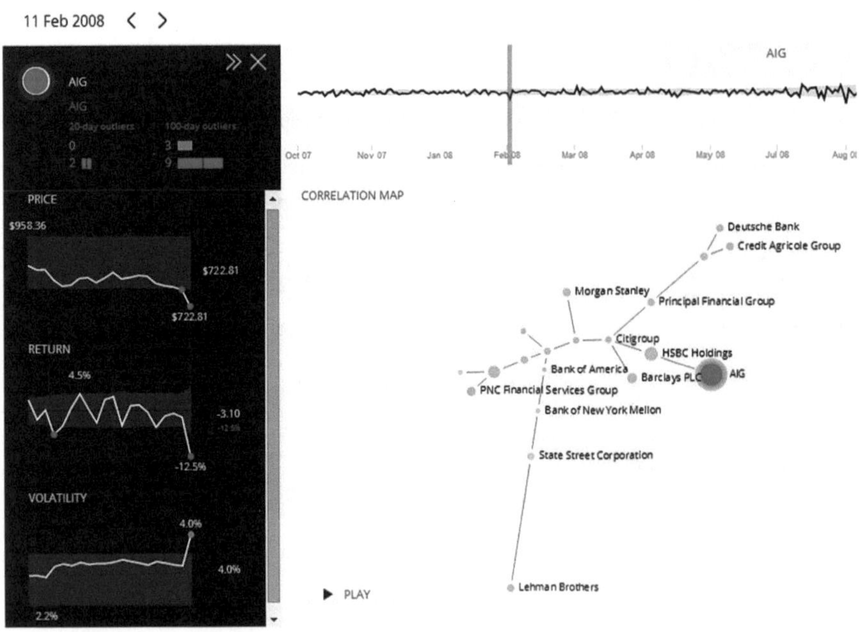

Fig. 23 AIG negative outlier February 11, 2008. *Source*: FNA HeavyTails

30 AQAL Risk Management

We'll apply Integral Theory philosopher Ken Wilber's **All Quadrant All Levels (AQAL)** Framework to put the major components of Adaptive Stress Testing in a larger context (Wilber 2001). Integral refers to "balanced, comprehensive, interconnected, and whole" (Wilber 2006).

Wilber proposes consider at least four interdependent perspectives for an integral understanding of reality. These perspectives consist of two pairs of polarities:

1. Objective Exterior vs. Subjective Interior
2. Individual vs. Collective

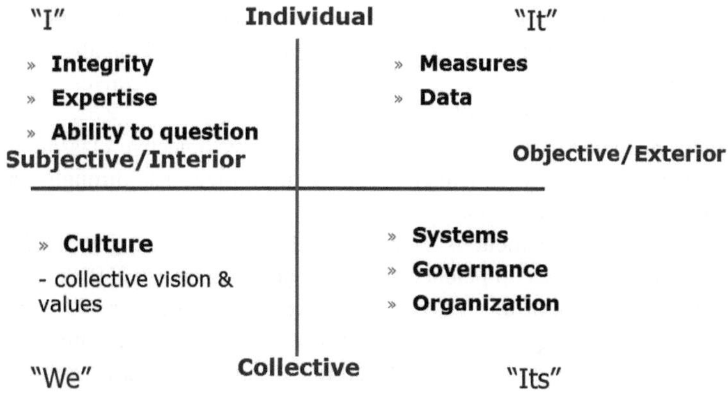

Fig. 24 AQAL framework. *Source*: Laubsch (2010b)

This results in four interdependent perspectives shown in Fig. 24. The quadrants can be summarized as "I" (Interior Individual), "We" (Interior Collective), "It" (Exterior Individual) and "Its" (Exterior Collective).

Starting in the upper right, metrics are important (It), but we also need to take the systems perspective (Its) and ensure sound processes and governance. This is the science half of risk management. Without it, we're flying blind. The left quadrants are subjective, and are the art of risk management. In the upper left (I), we need individuals with integrity, expertise, and the ability to question and be contrarian. But even the best risk managers can't help unless there is a risk culture (We). Culture is defined by a group's collective vision and values. It is the primary evolutionary driver of organizations. As people and systems come and go, culture determines the evolution and persistence of organizational learning. It's an organization's evolving DNA.

Each of these interdependent quadrants plays an important role for *Adaptive Risk Management*. It starts with the Innovators who are free from groupthink (I).[13] In our *Adaptive Stress Library* we aggregate credible scenarios by Innovators (We), and then look for metrics that indicate early adoption of a theme (It), and put systems and governance structures in place to act on such intelligence (Its).

31 Levels of Development

Both individual and collective learning progresses through sequential stages of development (or Levels, in AQAL terminology). In simplified terms, you could represent three sequential stages of development.

[13]Innovator would consider all four quadrants (and more perspectives) in their risk assessment.

1. ***Intuitive*** risk management is dominated by a subjective view of risk. Risk management (or lack thereof) is typically dominated by principals, who go with gut instinct and are unwilling to consider multiple of perspectives. The focus is on the parts, and not the whole.
2. ***Predict and control*** risk management is driven by objective classification and measurement of risks. Managers seek to minimize risk through traditional hierarchy, rules, and processes.
3. ***Integral*** risk management is characterized by networked intelligence. Risk management is a core competence and risk culture is pervasive. Risk is viewed as both danger and opportunity, and hence a higher tolerance for taking conscious risks that don't endanger the organization as a whole. Individuals are empowered, risks are continually communicated, and the organization learns from survivable failures.

These simplified stages represent an organizational center of gravity, where of certain perspectives are predominant.

To summarize, risk management depends on healthy organizational development, which is driven by the integration of ever more perspectives (e.g., from Subjective to Objective to Integral). Perspectives from each stage are important.[14] For example, the qualitative opinion of a trader may give essential color, while a researcher might provide useful objective analysis. To excel at managing risk, however, organizations must efficiently process multiple streams of intelligence. Therefore, Integral risk management builds on all previous stages (e.g., with checklists[15] and metrics that were implemented at the "Predict & Control" stage), but goes beyond measurement to include qualitative dimensions such cultivating a broad range of ideas, promoting a pervasive risk culture, and embracing risk as danger and opportunity.

32 Adaptive Learning

Risk management is a core discipline in a rapidly changing world. From finance to ecology, we face unprecedented systemic risks from increasingly coupled global systems. Non-linearities render long term predictions futile, and require consideration of many possible paths. Indeed we've seen a paradigm shift from "*Command and Control*" to "*Sense and Respond*" (Haeckel 2004). As in an ocean sailing race, organizations must navigate changing conditions using *dynamic steering* (Robertson 2010) with continuous feedback. "*Managing Uncertainty*" has replaced "*Change*

[14]A famous example is legendary investor George Soros who developed gut instincts about risk. He was known to presciently exit positions by listening to his body's stress signals.

[15]Dr. Atul Gawande's "Checklist Manifesto" (2009) provides great insights about importance of well designed checklists for managing risk, with case studies from medicine, aviation, investments, and construction.

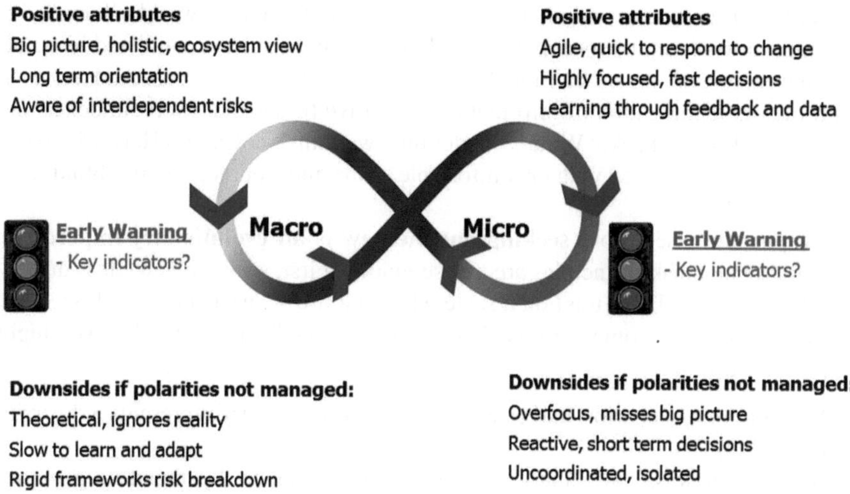

Fig. 25 Macro and micro polarity management. *Source*: Laubsch (2010a, b)

Management" in leadership seminars (change management makes no sense if the direction of change is not clear).

Figure 25 shows how we can spark a positive organizational learning spiral by managing macro and micro polarities.

This general macro to micro framework applies universally in risk management. Duhigg (2012) describes a tragic case study of the consequences of macro & micro level breakdowns in the 1987 London King's Cross Fire case study. Outside experts (fire brigade) who pre-diagnosed the fire hazard were ignored for years by an overconfident organization with disaster myopia (no previous loss of life from fire). Actionable early warning signals (a passenger reporting smoke) were not transmitted in a siloed organization. A brief window of opportunity to extinguish the fire and/or evacuate was lost, and 81 people were killed when the fire erupted in a giant explosion.

33 Seek Out the New: Harnessing Network Intelligence

Adaptive Stress Testing builds **a creative tension between contrarian views of *Innovators* and the "wisdom of crowds"** (Surowiecki 2004). *Innovators* are contrarians who perceive hidden risks that are not yet accepted by the market. Given that there are potential risks that never materialize, we incorporate the agile sensory intelligence of *Early Adopters* who are attuned to emerging market themes. *Adaptive Stress Testing* helps mitigate systemic risk by proactive stress testing while risks are still in potential form, as well as by counter cyclical investment (e.g., leaving crowded theatres early).

As hyper-connected individuals, we can access the entire world's knowledge with unlimited computational power. No longer bound by slow centralized forecasting and management, we can rapidly adapt to changing conditions. And yet even with the democratization of information, cognitive biases like overconfidence and groupthink hold us back. "We are blinder than we think" notes Tim Harford (2011). Even worse, when faced with uncomfortable truths, most opt for "willful blindness" (Heffernan 2011).

Despite the discomfort, **seeking out the new is an evolutionary imperative**. Harford's Palchinski Principles are a wise guide: "First, seek out new ideas and try new things; second, do [this] on a scale where failure is survivable; third, seek out feedback and learn from your mistakes as you go along" (Harford 2011). We might summarize this as:

1. Seek out new scenarios, and ensure that your stress library represents a diversity of thinking.
2. Seek out new signals emerging from the marketplace, by monitoring outlier activity and super-exponential rates of change.

Polarity management helps manage this process. We first analyze contrarian views at the periphery, and then hone in on signals emerging by *Early Adopters* who are connected to the broader marketplace.

34 Summary

Adaptive stress testing is a blend of art and science which continually integrates qualitative macro and quantitative micro perspectives. The first challenge in stress testing is to conceive of a wide range of credible potential threats before they materialize. Let's tap into the marketplace of ideas for scenarios, and harness the ability of visionaries to perceive risk in potential form (think Albert Einstein). After constructing *Stress Indices* to reflect scenarios, we monitor outliers, which are precursors to regime shifts. The StressGrades methodology amplifies market-based risk signals, which highlights cascading risk (e.g., super-exponential increases in PStress). StressGrades are also useful in identifying assets with exceptionally low volatility, which should be stressed for hidden risks (e.g., high DStress & low StessQ). Let's never forget that volatility only represents visible risk and that risk managers must be contrarian and uncover risks that are invisible to most. Low volatility is a temporary respite which allows us to search for hidden risks and rebalance to build more resilient portfolios and institutions. By being intelligently contrarian, we can mitigate systemic risks and transform future crises into opportunity.

Conclusions: Spark Network Intelligence

Evolutionary adaptation is a learning process: we sense changes in the environment and respond with learning experiments. Failures are not only inevitable, but essential to learning. As Tim Harford elegantly observes in *Adapt: Why Success Always Starts With Failure (2011)*: "the art of success is to fail productively." But to be able to learn from failure, we must be able to survive and keep playing. The obvious priority for risk managers is to ensure that their organization can withstand credible stresses. And yet paradoxically, many risk strategies that are designed to reduce individual risk (e.g., portfolio insurance, stop-loss limits, and liquidity hoarding in crisis situations) increase coupling, and often even precipitate crises. *In A Demon Of Our Own Design (2008)* Richard Bookstaber shows that many crises were precipitated by flawed safety mechanisms. When faced with complexity, tightly coupled systems eventually break down.

To manage systemic risks, we must look beyond individual nodes and understand the non-linear processes driving ecosystems. In "Rethinking Capitalism" Nick Hanauer and Eric Liu implore us to transcend "Machinebrain" linear thinking:

> In the Gardenbrain story, markets are not perfectly efficient, but they are effective if managed well. Humans are not perfectly rational, calculating and selfish; they are emotional, approximating and reciprocal. And outcomes are not just as they should be; rather, they reflect the kinds of compounding and feedback loops—virtuous circles or death spirals—that distort all complex systems. (Hanauer and Liu 2012)

Industrial capitalism has fuelled economic growth and expanded wealth worldwide. But it also comes with new liabilities (externalities), many of which are in hidden form. We face serious disruptive threats across all our global ecosystems.[16] As Otto Scharmer writes in "Leading from the Emerging Future: From Ego-System to Eco-System Economies" (2013), individually oriented approaches are unsustainable:

> What's dying is an old civilization and a mindset of maximize "me"—maximum material consumption, bigger is better, and special-interest-group driven decision-making that has led us into a state of organized irresponsibility, collectively creating results that nobody wants.

Throughout history, humans have faced a basic choice when meeting challenges: conflict or cooperation. Conflict, while unavoidable at times, is negative sum. Cooperation yields far better results, and indeed is the

(continued)

[16]Major potential risk fault lines include rising economic inequality and environmental degradation due to pollution, overuse of resources, and loss of biodiversity.

foundation for sustainable growth and innovation (Johnson 2010). As inventor Dean Kamen puts it: "if you have an idea and I have an idea and we exchange them, then we both have two ideas. It's nonzero (Diamandis and Kotler 2012)."

We see the benefits of cooperation throughout natural systems. Evolutionary leaps occur when individual "holons" (Koestler 1967) cooperate, for example as in the emergence of multi-celled organism, or hive insects like ants and bees. Thriving ecosystems are characterized by "cooperative relationships, self-regulating feedback cycles, and dense interconnectedness" (Benyus 2002).

The specter of disruptive global risks calls for ***mass collaboration platforms*** to better share information and coordinate responses. Nate Silver (2012) makes a case for predictive markets for economic data. Dan Tapscott shows many practical examples of effective mass collaboration platforms in *Macro Wikinomics* (2012), such as the mobile and Google Maps based platform that helped coordinate the Haiti earthquake relief efforts. Why not build sharing platforms for financial risk management, and specifically around stress testing? Indeed, network visualization platforms such as FNA might serve as the shared Google Maps of financial cartography, to help us better understand and communicate about the dynamic financial landscape.'

A new ***sharing economy*** has emerged. Social networks have connected us in online communities, and every like, tweet, and update has the potential to increase collective intelligence. Each of us has the potential to contribute in uniquely. Evolution, after all, is not an abstract force. We each embody evolutionary intelligence, and are all co-creators in a world where a "flap of a butterfly's wings in Brazil [could] set off a tornado in Texas" (Lorenz 1972). Imagine a neuron within a vast network of neurons, each sensing and responding to an ever changing world. Let's spark an evolutionary leap in intelligence by participating in collaboration platforms to share information about the risks that affect us all. It's not technology that's holding us back. The challenge is mindset, and a transition from an ego-centered to an eco-centric perspective of risk.

Acknowledgments I'd like to express gratitude to the many minds who have inspired this work. Firstly, to Sergey Ivliev of Prognoz for organizing the unique gathering of Perm Winterschool, and encouraging this paper.

Thank you to my colleagues at FNA. Kimmo Soramaki opened my eyes to financial cartography. Sam Cook generated our case study network graphs. And Eugene Nevdov provided valuable feedback.

Deep gratitude to the RiskMetrics family. Ethan Berman nurtured the open and creative culture that brought the best out in us. Our credo: "Change the world. Have fun. Make money. In that order." Allan Malz's crisis early warning research was seminal. I've referenced Chris Finger's research throughout, and am proud that we have finally realized our idea of a global outlier based systemic risk monitor with FNA HeavyTails. It was great to work with Pete Benson on riskcommons.org and to produce the first generation of StressGrades analytics. Gilles Zumbach's RiskMetrics 2006 time series research were invaluable. It was an honor to work with Knut Kjaer on next generation risk management, which evolved into the Adaptive Stress Testing framework. It was always a joy to brainstorm with my RiskMetrics labs partner Ron Papanek. Marty Nemeth was also a great sounding board, overflowing with ideas. Alvin Lee was my first mentor at JPMorgan and has always supported new ideas and a path of growth and adventure. And it was great to work with Ken Parker, Tom Stockdale, and the NextThought.com team to produce our online Adaptive Stress Testing course.

Thank you to PRMIA for much support. Lori Ramos-Marilla offered constant encouragement and enabled the opportunity to present the work at several conferences. Alex Voicu has been a creative force in enabling this research. He established a bridge to the global risk community by organizing many excellent workshops and producing the Adaptive Stress Testing online course at PRMIA University.

I deeply appreciate the insightful conversations with Anne Lalsing of Citibank, who inspired the StressGrades methodology and has provided so much thoughtful feedback.

Thank you to my Winhall Consulting partner David Shimko for encouraging early warning research, an area he had pioneered many years ago at JPMorgan.

I am grateful to philosopher Ken Wilber who inspired Integral Risk Management, and to the Boulder Integral community (especially Jeff Salzman and Nomali Perera).

Thank you to the editors at Springer for their detailed attention and patience.

And finally, I hope that Didier Sornette's foundational Dragon King research will empower the global community to be more proactive in managing systemic risks before irreversible tipping points are crossed.

References

Bastiat, F. (1850). That which is seen, and that which is not seen. *Library of Economics and Liberty*.

Benyus, J. M. (2002). *Biomimicry: Innovation inspired by nature*. New York: William Morrow.

Bookstaber, R. (2008). *A demon of our own design: Markets, hedge funds, and the perils of financial innovation* (pp. 161–164). Hoboken: Wiley.

Cooper, G. (2008). *The origin of financial crises: Central banks, credit bubbles, and the efficient market fallacy*. New York: Vintage.

Diamandis, P. H., & Kotler, S. (2012). *Abundance: The future is better than you think*. New York: Free Press.

Duhigg, C. (2012). *The power of habit: Why we do what we do in life and business*. New York: Random House.

Fama, E. F., & French, K. R. (2012). Size, value, and momentum in international stock returns. *Journal of Financial Economics, 105*(2012), 457–472.

Finger, C. C. (2008). Doomed to repeat it? *RiskMetrics Group (now MSCI) Research*.

Gawande, A. (2009). *The checklist manifesto: How to get things right*. New York: Metropolitan Books.

Gladwell, M. (2000). *The tipping point: How little things can make a big difference* (pp. 21–69). Boston: Little Brown.

Grove, L. (2008). *The world according to Robert Shiller*. Portfolio.com.

Haeckel, S. H. (2004). Peripheral vision: Sensing and acting on weak signals: Making meaning out of apparent noise: The need for a new managerial framework. *Long Range Planning, 37*(2), 181–189.

Hahn, A. L. (2008). *Missing pieces*. CFO.com.

Harford, T. (2011). Adapt: Why success always starts with failure. *Picador.*

Hanauer, N., & Liu, E. (2012, Winter). Rethinking capitalism. *RSA Journal*.

Heffernan, M. (2011). *Willful blindness: Why we ignore the obvious at our peril*. New York: Walker & Company.
Holling, C. S., & Meffe, G. K. (1995). Command and control and the pathology of natural resource management. *Wiley Online, Conservation Biology, 10*(2), 328–337.
Johnson, B. (1996). *Polarity management: Identifying and managing unsolvable problems*. Amherst: HRD Press.
Johnson, S. (2010). *Where good ideas come from: The natural history of innovation*. London: Penguin Group.
Kahneman, D. (2011). *Thinking fast and slow*. New York: Farrar, Straus and Giroux.
Keynes, J. M. (1930). *A treatise on money*. London: Macmillan.
Kjaer, K. N. (2011). New ideas on future portfolio design. *Trient Asset Management Research*.
Koestler, A. (1967). *The ghost in the machine* (p. 48). New York: Macmillan.
Laubsch, A. (2009, March). Was the credit crisis a Black Swan - An unforecastable extreme event? *Infinance Magazine*.
Laubsch, A. (2010a). *Equities as collateral in U.S. securities lending transactions*. A Study Implemented by The RMA Executive Committee on Securities Lending.
Laubsch, A. (2010b, March 10). *Integrated risk management - Early overview*. Working Paper. RiskMetrics Group (now MSCI).
Lorenz, E. (1972). *Predictability: Does the flap of a butterfly's wings in Brazil set off a Tornado in Texas*? Address at the 139th Annual Meeting of the American Association for the Advancement of Science.
Malz, A. M. (2011). *Financial risk management: Models, history, and institutions* (pp. 588–559). Hoboken: Wiley Finance.
Mantegna, R. N. (1999). Hierarchical structure in financial markets. *European Physical Journal B, 11*, 193–197.
Meyer, L. H. (1999). *Lessons from the Asian crisis: A central banker's perspective*. Levy Economics Institute Working Paper No. 276.
Minsky, H. P. (1992). *The financial instability hypothesis* (pp. 6-8). Working Paper No. 74.
Moore, G. A. (2004). *Inside the Tornado: Strategies for developing, leveraging, and surviving hypergrowth markets*. New York: HarperBusiness.
Moore, G. A., & McKenna, R. (1999). *Crossing the chasm: Marketing and selling high-tech products to mainstream customers*. New York: HarperBusiness.
Osorio, I., Frei, M. G., Sornette, D., Milton, J., & Lai, Y.-C. (2010). Epileptic seizures, quakes of the brain? *Physical Review E, 82*(2), 021919.
Parameswaran, A. (2009). *Minsky's financial instability hypothesis and Holling's conception of resilience and stability*. Macroresilience.com.
Pivot Capital Management. (2011). *China's investment boom: The great leap into the unknown*. Pivot Capital Management.
Robertson, B. (2012, January 30). *Riding a bicycle by committee*. BIGTHINK.COM.
Rogoff, K., & Reinhart, C. (2009). *This time is different: Eight centuries of financial folly*. Princeton: Princeton University Press.
Roubini, N. (2009). *The thinkers who predicted early on many aspects of this financial crisis*. EconoMonitor.com.
Samuelson, P. (1966). Science and stocks. *Newsweek*.
Scharmer, O., & Kaufer, K. (2013). *Leading from the emerging future: From ego-system to eco-system economies*. San Francisco: Berrett-Koehler Publishers.
Siegel, J. (1991, January 1). *The behavior of stock returns around N.B.E.R. turning points: An overview*. Rodney L. White Center for Financial Research, The Wharton School, University of Pennsylvania.
Siegel, J. (2007). *Stocks for the long run: The definitive guide to financial market returns and long-term investment strategies* (4th ed.). New York: McGraw-Hill.
Silver, N. (2012). *The signal and the noise*. New York: Penguin Group.
Sornette, D. (2009). Dragon-kings, black swans and the prediction of crises. *International Journal of Terraspace Science and Engineering, 2*(1), 1–18.

Sornette, D., & Woodard, R. (2009). Financial Bubbles, real estate bubbles derivative bubbles, and the financial and economic crisis. In *Proceedings of APFA7 (Applications of Physics in Financial Analysis), Conference series entitled Applications of Physics in Financial Analysis focuses on the analysis of large-scale Economic data, organized by Misako Takayasu and Tsutomu Watanabe.*

Surowiecki, J. (2004). *The wisdom of crowds: Why the many are smarter than the few and how collective wisdom shapes business, economics, society and nations: Why the many ... business, economies, societies and nations (1st ed.).* Little, Brown.

Taleb, N. N. (2007). *The Black Swan*. Random House.

Taleb, N. N. (2012). *Antifragile: Things that gain from disorder*. New York: House Random.

Tapscott, D. (2010). *Macrowikinomics: Rebooting business and the world*. New York: Portfolio Hardcover.

Tully, S. (2008). Jamie Dimon's Swat Team. *Fortune Magazine*.

Wilber, K. (2001). *A theory of everything*. Shambhala.

Wilber, K. (2006). *Introduction to the Integral Approach (and the AQAL Map)*. Kenwilber.com.

Zumbach, G. (2007). *The riskmetrics 2006 methodology* (pp. 1-12). RiskMetrics Group.

On Some Approaches to Managing Market Risk Using VaR Limits: A Note

Alexey Lobanov

Abstract Market risk has been traditionally considered in a single-period setting, with fixed positions in a static portfolio and losses caused by price volatility over a specified time horizon. In the real world, however, trading losses are generally a product of both position changes and adverse market movements. Market risk limits have been widely used in the industry for controlling both ex-ante and ex-post losses from traders' actions, but the interplay of risk limits with risk measurement has been scarcely studied in the literature. This note aims to provide insights into the broad concepts of using limits in market risk management, as well as some approaches to setting and managing market risk limits in a dynamic setting.

Keywords Market risk • Positions limits • Traders' actions • Trading desk • VaR limits

JEL Classification G21, G32

1 Introduction

A common view on market risk presumes that losses are caused by adverse price movements, while positions are fixed over a holding period. For example, in RiskMetrics™ a value-at-risk (VaR) measure is calculated for a static portfolio over a 1-day holding period under the assumption that changes in the portfolio structure and/or composition can be neglected and hence daily P&L is entirely driven by market movements (J. P. Morgan/Reuters 1996).

The views and opinions expressed herein are those of the author and do not necessarily reflect the official position of the Bank of Russia.

A. Lobanov (✉)
Banking Regulation Department, Bank of Russia, Moscow, Russia
e-mail: alobanov@akado.ru

A.K. Bera et al. (eds.), *Financial Econometrics and Empirical Market Microstructure*, DOI 10.1007/978-3-319-09946-0_12

In the real world, however, trading losses are a complex product of both price movements and changes in the portfolio structure.[1] Traders' actions require revising risk estimated after each material change in the portfolio structure and thus "contaminate" the P&L of a trading desk used in VaR backtesting (Basel Committee on Banking Supervision 1996, 2006). This problem can be mitigated for asset managers and, less effectively, for bank trading desks by checking the results of such "dirty" backtesting, based on theoretical P&L driven only by price movements against "clear" or "cleaned" backtesting[2] (Deutsch 2009). In high-frequency trading, the information value of conventional risk measures, such as VaR, tends to expire, as the holding period may barely exceed a time interval between consequent price movements. For shorter holding periods, traders' actions come to the forefront as a distinct risk factor, which needs to be understood and managed. Controlling traders' activity by enforcing position and risk limits should, therefore, be viewed as a primary risk management tool that complements hedging strategies and economic capital.

2 Trading Strategy as a "Shadow" Part of Market Risk

When considering the contribution of a trading strategy to the overall risk of the position, we need to examine what determines a trader's attitude to risk. Admittedly, traders have a greater participation in the upside than in the downside of their trades (Allen 2003). This means that a typical trader's compensation resembles a payoff on a longcall option with bonuses linked to profit which are potentially unlimited, while the financial share a trader bears in losses is capped at his salary and any deferred payments. Some of the highest disclosed traders' compensations are a good illustration of the upside potential, for example; Driss Ben-Brahmin (Goldman Sachs) reportedly earned about £30 in 2006 (BBC 2004), Brian Hunter (Amaranth) received over $100 m in 2005 (Petzel 2006), and Adam Levinson (Fortress) was remunerated with £156 m in 2008 (Antonowicz 2008).

Since vega of a long option position is positive, traders have strong incentives for risk-loving behavior because it increases their expected payoff. As long as a profitable trading strategy keeps producing alpha (and, consequently, bonuses for its owner), traders are reluctant to share the details about their strategies and risks they are about to take with their peers and risk managers.

[1]This viewpoint is consistent with Sharpe's (1992) decomposition of a mutual fund's return into two components: the "style" (i.e. asset-class factors, such as large-cap stocks, growth stocks etc.) and "selection" (i.e. an uncorrelated residual).

[2]The "cleaned" P&L is calculated in the same way as the "dirty" P&L, but without taking into account position changes during the VaR horizon. Paid and received fees and commissions are omitted from the calculation.

This information asymmetry is inherent to any financial firm and means that the risk of individual trading strategies and their undesirable interactions may not be properly detected and controlled by the firm's risk managers. This, in turn, enables rogue traders to take and pile up hidden risks that, if realized, may increase the trader's propensity for operational risk, as illustrated by the collapse of Baring's.

Ironically, losses incurred as a result of rogue risk-taking do not necessarily mean an end of the trader's career. Some evidence indicates that the trader's market value may be negatively correlated with failures. According to Allen (2001), "*Even firing does not have that large an effect – the tendency is for firms to hire traders who have had spectacular blowups elsewhere, figuring they've learned a lesson (at someone else's expense). Nick Leeson going to jail was an aberration (possibly due to different attitudes in Singapore than in the West).*" The years which have passed since the demise of Baring's have shown that Mr. Leeson's case was a precedent rather than an exception. Since Baring's collapse, the rogue trader "hall of fame" has expanded to include Yasuo Hamanaka from Sumitomo (8 years in jail following a \$2.6 bn loss on copper trades in 1996), John Rusnak from the Allied Irish Bank (7.5 years in jail after \$691 m loss on FX options), Brian Hunter from Amaranth (hedge fund liquidated after a loss of \$6.69 bn on natural gas futures), Jérôme Kerviel from Societé Géneral (5 years in jail following a loss of €4.9 bn), and Kweku Adoboli from UBS (7 years in jail after a loss of \$2.3 bn on stock index futures).[3] In hindsight, risk management in these institutions should have been held responsible for failing to prevent these losses.

The trader's risk appetite can be curbed by means of more symmetrical compensation schemes (e.g. 'golden cuffs' and cash bonus clawbacks), internal controls (e.g. regular audits, phone conversation recording), or even pre-committing traders to specific loss limits by incentivizing them to share their forecasts with risk managers[4] (Miller 2001). It is more common, however, to limit the risk traders may take from the top-down rather than bottom-up, by having a trader stick to externally set limits. A typical market risk limit structure in a financial firm includes various position limits, P&L (stop-loss) limits, limits on specific risk parameters (e.g. rate buckets, "the Greeks," markets, and liquidity), total risk limits (VaR/CVaR limits), and limits based on stress testing. Now we will consider more closely the interplay of risk limits, with VaR as a risk measure, and exposure (position) limits.

3 The Role of VaR Limits in Risk Budgeting

In market risk management, limit setting is driven by economic capital allocation, and is normally conducted from the top-down. Economic capital is viewed as an internal solvency constraint on a firm's value-maximization function (Beeck et al.

---

[3]Source: Wikipedia.

[4]Expressed by the formula, "Lack of Identification of Risk + Unexpected Loss = Disciplinary Action/Dismissal by Business" (Miller 2001).

1999; Schroeck 2002). More specifically, economic capital serves as a probabilistic loss bound over a target time horizon and is typically measured by value at risk (VaR). Within the trading area, economic capital is allocated assuming that market risk is consistently measurable at all levels.[5] Breaching a limit should entail a book closure and a re-allocation of limits set for other trading desks. The frequency of capital allocation and, consequently, a limit re-setting, varies from quarterly to annually (e.g. Johanning 1998; U.S. Bankruptcy Court 2010).

VaR has been used in the industry for setting and managing risk limits since the 1990s. G-10 regulators have required that banks using the internal models approach for calculating capital for market risk (Basel Committee on Banking Supervision 1996, 2006) also employ their VaR models for setting trading limits.[6] Thus, integrating a VaR model used for calculating the regulatory capital into the limit setting process has been considered a necessary requirement of the model use-test in a bank.

Designing a market risk limit system requires a number of problems to be solved along both the spatial and temporal dimensions, as shown in Fig. 1. Most research has been focused on the coherent treatment of risk diversification at all levels of corporate organizational structure (e.g. Kimball 1998; Kuritzkes et al. 2003). Scarce literature exists, however, on the interrelated problems of consistent time scaling of risk limits, adjusting limits for traders' P&L, and accounting for model risk in limit setting and management. Progress made in these areas is further reviewed in this note.

4 Do Risk Limits Add Value?

According to the Wall Street adage, one of the best ways to make money is not to lose it. The importance of binding and enforceable internal limits has long been recognized in financial firms, yet not always observed in practice. Five years after the collapse of the Lehman Brothers, it appears that lax exposure limits could have been one of the major causes of the firm's failure (U.S. Bankruptcy Court 2010).

Let us consider an argument about two investment funds presented by Lo (2001). Fund A has a portfolio with an expected return of 10 % p.a. and an annual volatility of 75 %. Fund B replicates the portfolio of Fund A, but enforces a stop-loss limit every time its annual returns fall down to −20 %. Assuming that the portfolio returns of Fund A follow a log-normal distribution, it can be shown that Fund B would

[5] A typical hierarchy within a trading function includes trading books run by trading desks, which, in turn, are operated by individual traders. The Basel Committee on Banking Supervision has recently attempted to give a regulatory definition of a trading desk (Basel Committee 2013).

[6] "The risk measurement system should be used in conjunction with internal trading and exposure limits. In this regard, *trading limits should be related to the bank's risk measurement model* in a manner that is *consistent over time* and that is *well understood by both traders and senior management*." (Basel Committee on Banking Supervision 2006, §718(Lxxiv)-f).

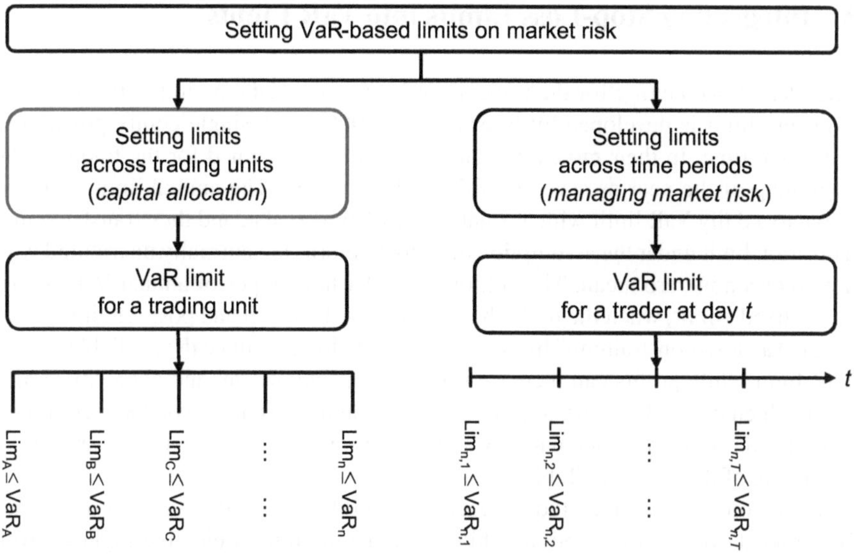

Fig. 1 Divisional and temporal dimensions of setting VaR limits for market risk (Straßberger 2002)

exhibit not only a lower volatility of 67 %, but also a higher expected return of 21 %, more than double that of Fund A. Thus, truncating the loss tail of the returns distribution reduces its variance and skews it towards the positive side. For a normal distribution with mean μ and variance σ^2, a one-sided truncation at level z (i.e. enforcing a stop-loss limit) ensures a higher expected value and a lower variance:

$$\mathrm{E}\left(X \middle| X > z\right) = \mu + \sigma \times \lambda(a) > m,$$

$$\sigma^2\left(X \middle| X > z\right) = \sigma^2 \times (1 - \delta(\alpha)) < \sigma^2(X),$$

where $\alpha = (z - \mu)/\sigma, \lambda(\alpha) = \phi(\alpha)/(1 - \Phi(\alpha)), \delta(\alpha) = \lambda(\alpha)\,(\lambda(\alpha) - \alpha),$

$\phi(\cdot)$—the probability density function of the standard normal distribution,
$\Phi(\cdot)$—the cumulative probability function of the standard normal distribution.

Besides stop-loss limits, truncated distributions appear in other risk management applications including margin trading (loss tail truncation at a margin call level), private equity funds (profit tail truncation at an exit level), and hedging (e.g. two-sided truncation at strike prices in bull/bear spreads).

Focusing on stop-loss limits in our further analysis, let us consider how these ex-post limits can be embedded into ex-ante exposure limits.

5 Integrating Stop-Loss Limits into VaR Limits

The idea of accounting for the traders' P&L over a specific period in the available exposure limit is developed by Beeck et al. (1999) for a single equity position in a discrete time. In their model, an annual risk limit is defined for the trader at the beginning of the year as a yearly VaR of the position. The annual limit is scaled down to a daily VaR limit with a square-root-of-time scalar, and then translated into a position limit under the assumption that stock returns are normally distributed with a zero or a non-zero mean. The authors consider three types of annual VaR limits: (1) a "fixed" limit, when the trader has the same risk budget and position limit every day, (2) a "loss-constraining" limit, when realized losses reduce the available annual VaR limit while profits can increase it back to its initial size, and (3) a "dynamic" limit, which differs from the stop-loss limit in that there is no cap on the recognition of realized profits in the annual VaR and the limit may increase above its size at the beginning of the year (see Table 1).

Based on a simulation study, Beeck et al. (1999) show that the stop-loss limit is the most conservative option, while the dynamic limit yields the highest profit potential at the expense of the largest P&L volatility. The results of the study also indicate that enforcing the risk limits makes the actual confidence level of the yearly VaR much lower than the one presumed in the VaR model.

Lobanov and Kainova (2005) extend the approach of Beeck et al. (1999) to include historical simulation for calculating VaR limits. They also propose an approach to adjusting position limits for model risk based on the results of the regulatory back-testing (Basel Committee on Banking Supervision 2006) and an alternative procedure of live-testing.

The methodology developed by Beeck et al. (1999) can be used for managing a linear position with a single risk factor to ensure a single exposure limit for a given risk limit. Extending this approach to a portfolio with multiple risk factors leads to non-unique solutions for the position limit, i.e. the composition of the portfolio.

Other drawbacks of this approach make its use problematic even for a single-factor position. Straßberger (2002) observes that an annual VaR implies that positions are fixed for a 1-year horizon, which is unrealistic for proprietary trading. Scaling an annual limit down to the daily VaR using the square-root-of-time rule leads to an underestimation of the daily position limit and to severe underutilization of economic capital.[7] Alternatively, if scaling is done with a square-root-of-time remaining to the year-end, this leads to an uneven distribution of limits over the year.

Finally, the approach by Beeck et al. (1999) appears to be overly conservative, as the annual VaR will almost never be exceeded by losses if trading is halted after

[7] E.g. a scaled daily risk limit is only 6.25 % of the yearly risk limit with $T = 256$ trading days.

Table 1 Deriving position limits for a single-factor position

Limit type	Annual risk limit (*YL*)	Daily risk limit (*DL*)	One-day position limit (*V*)
Fixed	$YL = YL_0 = const$	$DL_t = \frac{YL}{\sqrt{T}} = const$ (for $\mu = 0$) $DL_t = YL\frac{\overline{\mu} - k_\alpha \overline{\sigma}}{\overline{\mu}T - k_\alpha \overline{\sigma}\sqrt{T}}$ (for $\mu \neq 0$)	$V_t = \frac{DL}{k_\alpha \sigma_t}$ (for $\mu = 0$) $V_t = -\frac{DL_t}{\mu_t - k_\alpha \sigma_t}$ (for $\mu \neq 0$)
Loss-constraining	$YL_t = \begin{cases} YL_0, & \sum_{s=1}^{t} \Delta V_{t-s+1} \geq 0, \\ YL_0 + \sum_{s=1}^{t} \Delta V_{t-s+1}, & \sum_{s=1}^{t} \Delta V_{t-s+1} < 0 \end{cases}$	$DL_t = \frac{YL_t}{\sqrt{T}}$ (for $\mu = 0$) $DL_t = YL_t\frac{\overline{\mu} - k_\alpha \overline{\sigma}}{\overline{\mu}\, T - k_\alpha \overline{\sigma}\sqrt{T}}$ (for $\mu \neq 0$)	$V_t = \frac{DL_t}{k_\alpha \sigma_t}$ (for $\mu = 0$) $V_t = -\frac{DL_t}{\mu_t - k_\alpha \sigma_t}$ (for $\mu \neq 0$)
Dynamic	$YL_t = YL_0 + \sum_{s=1}^{t} \Delta V_{t-s+1}$	$DL_t = \frac{YL_t}{\sqrt{T}}$ (for $\mu = 0$) $DL_t = YL_t\frac{\overline{\mu} - k_\alpha \overline{\sigma}}{\overline{\mu}\, T - k_\alpha \overline{\sigma}\sqrt{T}}$ (for $\mu \neq 0$)	$V_t = \frac{DL_t}{k_\alpha \sigma_t}$ (for $\mu = 0$) $V_t = -\frac{DL_t}{\mu_t - k_\alpha \sigma_t}$ (for $\mu \neq 0$)

Notation: *YL* Annual VaR limit, *DL* Daily VaR limit, *V* Daily position limit, $\overline{\mu}$ Average expected return, $\overline{\sigma}$ Average standard deviation of returns, k_α Quantile of the standardized normal distribution the annual risk limit is depleted. In reality, however, the daily position limit is not always fully utilized by the trader, and the trading book is not necessarily closed till the year-end after the cumulative loss has surpassed the allocated annual risk limit.

6 Setting VaR Limits Based on Portfolio Insurance and Quantile Hedging

These drawbacks are addressed in a dynamic model proposed by Straßberger (2002). In this model, the market risk of a stock portfolio[8] is managed through VaR limits in a continuous time. The underlying idea is a combination of portfolio insurance with synthetic put options (Rubinstein and Leland 1981) and "quantile hedging" (Föllmer and Leukert 1999). As in the model by Beeck et al. (1999), the annual risk limit is defined as a maximum cumulative loss over a year, and is dynamically adjusted for the trader's daily P&L. However, the annual risk limit is translated not into a daily VaR limit, but directly into a daily position limit using the daily VaR parameters.[9] The daily position limit is adjusted using a risk-aversion scalar (a_t) and, by construction, is equal to or smaller than the annual risk limit. This scalar is a function of the position delta and the standard Black–Scholes parameters of a synthetic put option used to hedge the portfolio. Using the notation from Table 1, the algorithm for deriving the daily position limit is shown in Table 2.

The portfolio insurance is implemented as follows. The stock position is delta-hedged by a long European-style synthetic put option replicated with a short position in the stock and a long position in a risk-free asset:

$$\textit{Long stock} + \textit{Long synthetic put} = \textit{Net long position in stock} \\ + \textit{Long position in risk-free asset}.$$

The strike price of the put option (i.e. the insurance bound) is set to achieve the confidence level implied in the VaR model. Delta of the put option is continuously

Table 2 Deriving the position limit for a single-factor position

Limit type	One-year risk limit	One-day risk limit	One-day position limit
Dynamic	$YL_t = YL_0 + \sum_{s=1}^{t} \Delta V_{t-s+1}$	$DL_t = a_t\, YL_t$	$V_t^{\max} = \frac{DL_t}{\mu_t\, T - k_\alpha \sigma_t \sqrt{T}}$ (for $\mu \neq 0$)

[8]According to Straßberger (2002), this approach can be extended to a stock portfolio. In our following discussion, we consider a single position in a stock.

[9]These are the mean and the standard deviation of portfolio returns, as VaR is calculated using the variance–covariance approach.

estimated, and the position is rebalanced accordingly.[10] For a European put option, delta is derived from the Black–Scholes model under an assumption of $T^* = 0.5T$:

$$\delta_t = \mathrm{N}\ (d_1)\ -1;$$

$$d_1 = \frac{\ln\left(S_t/K\right) + \left(r + \sigma^2/2\right) T^*}{\sigma\sqrt{T^*}};$$

where $K \in [0; V_0 - YL_t]$ is the strike price of the put option.

The risk-aversion scalar is defined as a ratio of the resulting net long position in stock to the maximum available daily position limit, which is assumed to be fully utilized by the trader:

$$a_t = \frac{(1+\delta_t)\ V_t}{(1+\delta_t)\ V_t + M_t} = \frac{V_t\,\mathrm{N}\,(d_1)}{V_t\,\mathrm{N}\,(d_1) + K\,(1-\mathrm{N}\,(d_2))};$$

where $M_t = K(1 - \mathrm{N}(d_2))$ is the size of the position in a risk-free asset under the assumption of a zero risk-free rate;

$$d_2 = d_1 - \sigma\sqrt{T^*}.$$

Straßberger (2002) further improves the model by using a synthetic European knock-out barrier option, as it ensures a minimum hedging cost (Föllmer and Leukert 1999). The stock position is insured through a synthetic down-and-in put option with a barrier price U equal e.g. to portfolio initial value. If $V_t > U$, the option disappears, and its zero delta makes the annual risk limit fully available for the trader.

In both the models, the algorithm for managing the risk limits over time is the same as summarized in Table 3.

It can be shown that if the strike price of the put option is set exactly at $K = V_0 - YL_t$, we obtain the same dynamic VaR limit as in Beeck et al. (1999). In the more conservative case of $K < V_0 - YL_t$, the probability of keeping the position value above the annual risk limit can be set equal to the VaR confidence level.

Table 3 Continuous management of market risk limits

Scenario	Parameters	Risk limits
$t = 0$, *or* $\mathrm{P\&L}_t = 0$	$V_0 > K$; $\delta_t \approx 0$, $M_t \approx 0$, $a_0 \approx 1$	$YL_t = YL_0$; $DL_t = YL_t$
$\mathrm{P\&L}_t > 0$	$V_t > V_0$, $V_t \gg K$; $\delta_t \approx 0$, $M_t \approx 0$, $a_0 \approx 1$	$YL_t > YL_0$; $DL_t = YL_t$
$\mathrm{P\&L}_t < 0$	$V_t < V_0$; $\delta_t < 0$, $M_t > 0$, $a_0 < 1$	$YL_t < YL_0$; $DL_t < YL_t$

[10]Delta for a portfolio with long and short positions is calculated for changes in daily position limits and not in the stock prices.

Lokareck-Junge et al. (2000) use a Monte-Carlo simulation to estimate the option strike price K sufficient to insure a stock position at a specified confidence level. They show that, for instance, for a confidence level of 95 % the strike price of the put option can be set at 50 % of the difference between the initial position value V_0 and the available annual risk limit YL_t, which is approximately equal to 51 % of the initial value of the position. Obviously, the hedging cost decreases with the strike price of the put option.

This theoretically appealing marriage of VaR limits and quantile hedging has many advantages over the management of risk limits in a discrete time. Firstly, it ensures that the annual risk limit is consistent with the definition of VaR while making a higher daily risk limit available for the trader. In this approach, the risk limits are consistently managed over time and the risk aversion of the firm management is explicitly and promptly reflected in the risk limit. Setting a barrier price allows for more flexibility in achieving the desired confidence level.

While conceptually attractive, the high-frequency management of risk limits is problematic in practice due to high transaction costs. Resetting the limits for complex portfolios becomes prohibitively computer-intensive. In principle, a re-allocation of all risk limits across the firm from the top down is required after any material adjustment of the annual risk limit for any single portfolio. Besides, human issues are likely to emerge, as traders will find it difficult to operate within constantly changing limits. Finally, this approach is relatively complex for understanding by senior management compared to more conventional techniques for setting VaR limits.

Conclusion and Issues for Research

Along with market movements, the actions of traders or trading algorithms are a distinct risk factor that can be effectively controlled both ex-post, by means of stop-loss limits, and ex-ante, by enforcing exposure or risk limits. Using internal VaR models for setting and managing market risk requires a number of complex problems to be solved, from the allocation of economic capital across trading desks, which would correctly account for risk diversification, to deriving daily exposure limits from longer-term risk budgets and making the exposure limits sensitive to the traders' P&L.

The overview of some of the theoretical approaches to solving these problems given in this note calls for empirical evidence. For instance, it would be interesting to investigate the distribution of a trader's P&L as a single asset and decompose it into the "market" and "trader" components.

Another issue that definitely merits more research is modeling the interaction of traders' P&L at the desk and firm levels. While the Basel Committee on Banking Supervision (2013) suggests using empirical correlation between market returns in both the revised approaches (i.e. the internal models and

(continued)

the standardized ones), the observed correlations of revenue returns between various product segments may be spectacularly low, as found by Perold (2005) (Table 4):

Table 4 Correlations between trading revenues of businesses within major product segments in a major New York investment bank

	Interest-rate	Equity	Foreign exchange	Commodity
Interest-rate	1			
Equity	0.135	1		
Foreign exchange	0.053	−0.111	1	
Commodity	0.057	−0.007	−0.002	1

Source: Perold (2005).

The imperfect correlations between traders imply that risk limits may be systematically underutilized from a firm-wide perspective. This warrants research into ways of enhancing the utilization of exposure limits given risk constraints.

In light of the fundamental review of the trading book (Basel Committee on Banking Supervision 2013), it is worth studying how the expected shortfall as well as other risk measures (e.g. lower partial moment) can be used for setting trading limits, though employing the same internal model for purposes other than calculating regulatory capital charge is no longer proposed by the Basel Committee.

Ultimately, designing a manageable and incentive-compatible limit system remains one of the major challenges in market risk management. The recent cases of huge trading losses in some of the largest banks indicate that more research is needed into the roots and factors of vulnerabilities in financial institutions that can hardly be prevented, if not magnified, by technological advances and regulatory changes.

References

Allen, S. (2001). *Institutional background in financial risk management, Course material*. New York: New York University.

Allen, S. (2003). *Financial risk management: A practitioner's guide to managing market and credit risk*. Hoboken, NJ: Wiley.

Antonowicz, A. (2008, 13 August). Wall Street Trader gets largest bonus ever – £156 million. *Mirror News*. Available at http://www.mirror.co.uk/news/top-stories/2008/08/13/wall-street-trader-gets-largest-bonus-ever-156-million-115875-20695469/.

Basel Committee on Banking Supervision. (1996, January). *Amendment to the capital accord to incorporate market risks*.

Basel Committee on Banking Supervision. (2006, June). *International convergence of capital measurement and capital standards: A revised framework*, Comprehensive version.

Basel Committee on Banking Supervision. (2013, October). *Fundamental review of the trading book*. Consultative Document.

BBC. (2004, 18 January). City trader's £30m record bonus. *BBC News*. Available at http://news.bbc.co.uk/2/hi/business/3406965.stm.

Beeck, H., Johanning, L., & Rudolph, B. (1999). *Value-at-Risk-Limitstrukturen zur Steuerung und Begrenzung von Marktrisiken im Aktienbereich*. OR Spektrum.

Deutsch, H.-P. (2009). *Derivatives and internal models* (4th ed.). Palgrave: Macmillan.

Föllmer, H., & Leukert, P. (1999). Quantile hedging. *Finance and Stochastics, 3*(3), 251–273.

Johanning, L. (1998). *Value-at-Risk zur Marktriskosteuerung und Eigenkapitalallokation*. Bad Soden/Ts: Uhlenbruch Verlag.

Kimball, R. C. (1998, July/August). Economic profit and performance measurement in banking. *New England Economic Review*, 35–53.

Kuritzkes, A., Schuermann, T., & Weiner, S. M. (2003). Risk measurement, risk management and capital adequacy of financial conglomerates. In R. Herring & R. Litan (Eds.), *Brookings Wharton papers in financial services 2003* (pp. 141–194). Washington: Brookings Institution Press.

Lo, A. (2001, June), *Risk management for hedge funds: Introduction and overview*. Working paper.

Lobanov, A., & Kainova, E. (2005). Sravnitel'nyĭ analiz metodov rascheta VaR-limitov c uchëtom model'nogo riska na primere rossiĭskogo rynka aktsiĭ [Methods for setting VaR limits with an adjustment for model risk on the Russian Stock Market: A comparative study]. *Upravlenie finansovymi riskami* (Vol. 1, pp. 44–55).

Lokareck-Junge, H., Straßberger, M., & Vollbehr, H. (2000). Die Ermittlung von Value-at-Risk-Handelslimiten zur Kontrolle und Steuerung von Marktrisiken bei kontinuierlicher Überprüfung. In K. Inderfurth, et al. (Hrsg.), *Operations research proceedings 1999* (S. 317–322). Berlin.

Miller, S. (2001). *Identifying specific vulnerabilities, presentation to GARP 2nd global risk management convention*. New York: JPMorgan, Marriott WTC.

J.P. Morgan/Reuters. (1996). *RiskMetrics™ technical document* (4th ed.). New York/London: J.P. Morgan/Reuters.

Perold, A. (2005). Capital allocation in financial firms. *Journal of Applied Corporate Finance, 17*(3), 110–118.

Petzel, T. (2006, September/October). Hedge funds: Lessons learned from Amaranth. *GARP Risk Review* (32), 4–5.

Rubinstein, M., & Leland, H. E. (1981). Replicating options with portfolios in stock and cash. *Financial Analyst Journal, 37*(4), 63–72.

Schroeck, G. (2002). *Risk management and value creation in financial institutions*. Hoboken, NJ: John Wiley & Sons, Inc.

Sharpe, W. F. (1992). Asset allocation: Management style and performance measurement. *Journal of Portfolio Management, 18*, 7–19.

Straßberger, M. (2002). *Risikokapitalallocation und Marktpreisrisikosteuerung mit Value-at-Risk-Limiten*. Lohmar: Josef Eul Verlag.

U.S. Bankruptcy Court. (2010). In re Lehman Brothers Holding et. al.: Report of Anton R. Valukas. *Examiner, 8*.

Simulating the Synchronizing Behavior of High-Frequency Trading in Multiple Markets

Benjamin Myers and Austin Gerig

Abstract Nearly one-half of all trades in financial markets are executed by high-speed autonomous computer programs—a type of trading often called high-frequency trading (HFT). Although evidence suggests that HFT increases the efficiency of markets, it is unclear how or why it produces this outcome. Here we create a simple model to study the impact of HFT on investors who trade similar securities in different markets. We show that HFT can improve liquidity by allowing more transactions to take place without adversely affecting pricing or volatility. In the model, HFT synchronizes the prices of the securities, which allows buyers and sellers to find one another across markets and increases the likelihood of competitive orders being filled.

1 Introduction

Financial markets have changed considerably over the last 20 years. During this time, most exchanges have switched from floor-based to fully electronic trading where orders can be sent to the market and executed with little or no human involvement (MacKenzie 2012). As a result, automated trading has flourished. One particular type of automated trading, known as high-frequency trading (hereafter HFT), has especially grown in size and importance. HFT exploits short-term price fluctuations and seeks a small profit per transaction many times throughout the day, without taking on significant overnight positions. Although difficult to determine

B. Myers
Department of Physics, University of Oxford, Oxford, UK
e-mail: myers.benjamin.s@gmail.com

A. Gerig (✉)
CABDyN Complexity Centre, Saïd Business School, University of Oxford, Oxford, UK
e-mail: austin.gerig@sbs.ox.ac.uk

A.K. Bera et al. (eds.), *Financial Econometrics and Empirical Market Microstructure*, DOI 10.1007/978-3-319-09946-0_13

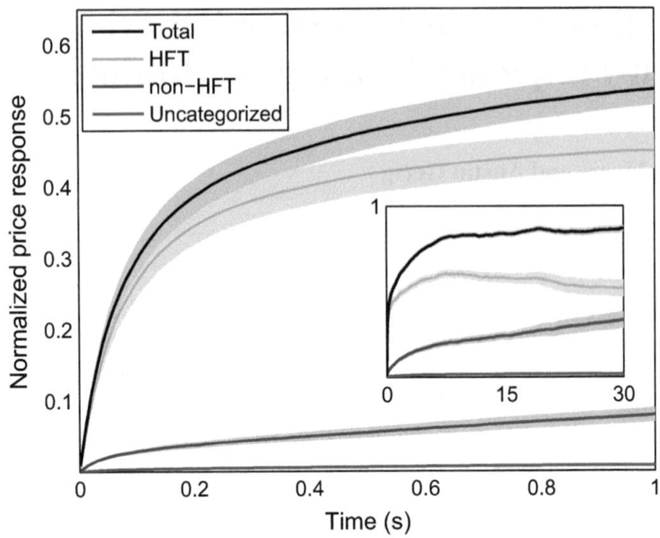

Fig. 1 Normalized price response of stock i due to stock $j \neq i$, for 40 US stocks traded on NASDAQ, decomposed into the amount due to HFT activity (*green*), non-HFT activity (*blue*) and uncategorized activity (*red*). Standard errors of the sample means are indicated in the shaded color. Taken from (Gerig 2012) (Color online)

its true size, most studies estimate that about one-half of all transactions on major exchanges are due to HFT.[1]

This study focuses on one particular effect linked to HFT—the synchronizing of price responses across multiple related securities (Gerig 2012; Gerig and Michayluk 2010). Figure 1 (taken from a recent article) shows this effect. Here, to analyze price synchronization in more detail, we simulate two markets where an identical security is traded and compare investor welfare when the prices in these markets are and are not aligned by the actions of HFT.

In our simulation, investors are modeled in a zero-intelligence framework (Gode and Sunder 1993; Farmer et al. 2005). This treatment strips out the idiosyncrasies of individuals' behavior and assumes only local interactions are of significance—investors are only interested in meeting their own specific price expectations and they do not use complex strategies. To consider the effect of HFT, we simulate two zero-intelligence markets where an identical security is traded and allow HFT to

[1]Several research firms provide estimates of HFT activity for subscribers; examples are the TABB Group, the Aite Group, and Celent. Publicly, this information is available in articles such as "The fast and the furious", Feb. 25, 2012, *The Economist* and "Superfast traders feel the heat as bourses act", Mar. 6, 2012, *Financial Times*.

connect orders between the two markets when their prices cross. We show that HFT activity (as defined in the model) increases the probability that a typical investor entering the market will transact. Furthermore HFT activity reduces volatility so that prices are closer to their fundamental value.

2 Model

The model simulates the continuous double auction, a market structure common to most modern exchanges. Traders submit bids and offers to buy and sell respectively at the best price they are willing to transact at. If prices cross—a bid meets or exceeds a previous offer, or the converse—a transaction takes place at the earlier listed price. If an incoming order is unable to transact with any existing orders, it is placed in the limit order book. This consists of two lists, the bid book and the ask book which contain the previously unfilled orders on the buy and sell side respectively.

At each time step in our model, a random order of unit size is generated. Orders have equal probability of being a buy or sell and are given a price drawn from a uniform probability distribution with limits 1 and 200. These hard limits on order prices should not be thought of as boundaries that would exist in real markets, but instead are assumed for simplicity. In real markets, we would expect participants to place limit orders according to some humped shaped distribution around the equilibrium clearing price (which would change through time). For simplicity, we assume this humped distribution is a uniform distribution with limits and that the clearing price is constant through time. Using a dynamic clearing price and/or a different distribution with open limits (such as a Gaussian), although perhaps more realistic, would not change the main results of the paper.

Orders fill the limit order book until a transaction takes place. When a transaction occurs, all unfilled orders in the limit order book are cleared and the process of generating new orders is started again. Figure 2a illustrates this diagrammatically.

The HFT interaction is modeled by running two identical exchanges simultaneously. If transactions are unable to take place on either market in a given time step, but would occur if the two markets were combined, HFT is permitted to transact between the two markets. Figure 2b illustrates this diagrammatically. We make an idealized assumption that HFT is perfectly competitive so that their profit is zero. Therefore, the HFT transactions in the two markets take place at the same price, which we set to the midpoint between the bid price and offer price of the two orders in the two markets.

For example, let the bid price on market 1 be denoted by b_1 and the ask price on market 2 be denoted by a_2, where $b_1 \geq a_2$. When HFT transacts with these orders, the transactions take place at price $(b_1 + a_2)/2$ in both markets.

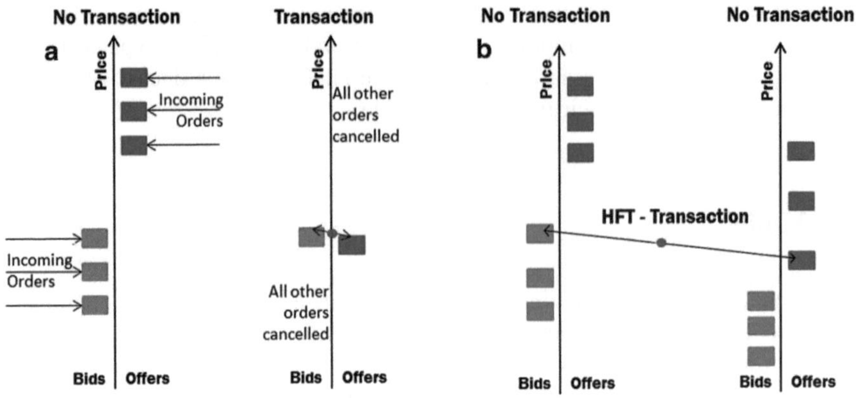

Fig. 2 A diagram of the order book in both scenarios modeled. (**a**) exhibits the market without HFT, at time steps with and without a transaction occurring. Note that transactions can only occur if a bid exceeds an offer. (**b**) shows the connecting effect HFT. The order book is cleared entirely after each transaction

3 Results

When buying or selling a security, investors typically are interested in the following three questions: How likely am I to transact? What price am I likely to receive? How much does this transaction price vary? We therefore estimate the following three observables in the model both with and without HFT: (1) the probability that a submitted order will result in a transaction, (2) the average transaction price of filled orders, and (3) the volatility of the transaction price of filled orders (the standard deviation of the transaction price). We run the simulation 100 times for 10000 time steps both with and without HFT, and we record average values of the relevant observables. The results are shown in Figs. 3 and 4 and in Table 1, which we discuss in more detail below.

The model reproduces several empirical findings that have otherwise been difficult to explain (Hendershott et al. 2011; Hasbrouck and Saar 2013; Brogaard et al. 2013):

1. Transaction prices are more accurate when HFT is present, i.e., they are closer to the equilibrium value.
2. Volatility is reduced when HFT is present.
3. Liquidity is increased when HFT is present.

The equilibrium price, defined as the intersection of the expected aggregate supply and demand curve in the simulation, is just the mean of the uniform distribution of prices, i.e., 100.5. As seen in Table 1, the average transaction price both with and without HFT converges to the equilibrium value within the 2 standard error range that defines a 95 % confidence interval. However, the variance around the equilibrium value is reduced when HFT is added. The reduction in variance is

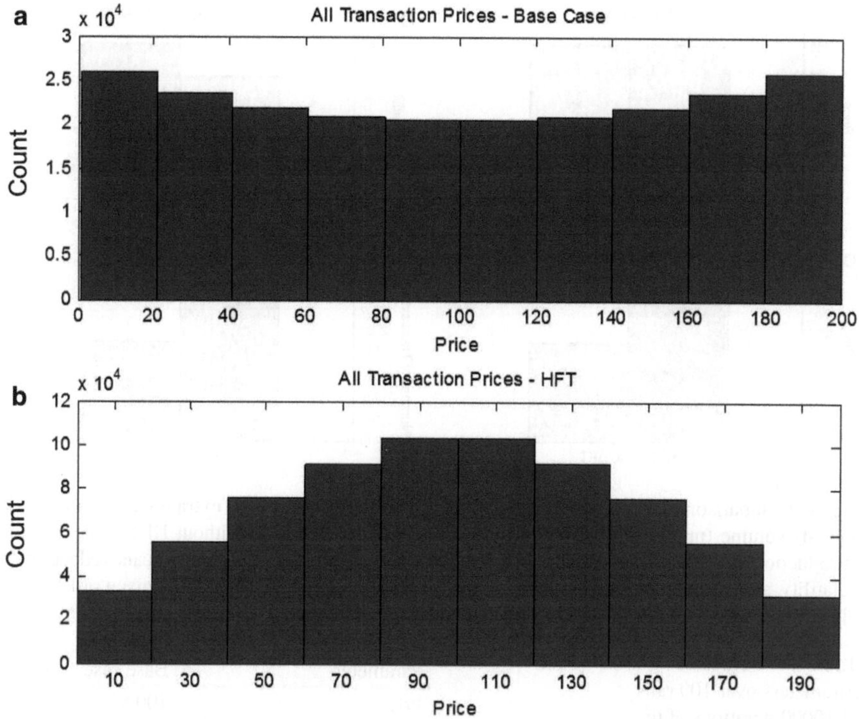

Fig. 3 Histogram of transaction prices in the simulation. (**a**) shows the market without HFT; (**b**) shows the market with HFT, including transactions over both exchanges. Note that without HFT the average transaction price is not observed as often as prices at the extremes

shown in Table 1 and can be seen in the comparison of the histogram of transaction prices in Fig. 3a,b. As seen in the figure, HFT causes more transactions to occur near the equilibrium price. This result matches previous empirical studies that have shown algorithmic trading in general and HFT specifically increases the accuracy of prices in markets (Brogaard et al. 2013; Hendershott et al. 2011).

Empirical studies have also found that HFT reduces intraday volatility (Hasbrouck and Saar 2013). Our simulation reproduces this result as well (see Fig. 4b). Because the equilibrium price is constant in the model, any variance in transaction price can be interpreted as excess volatility. Because HFT reduces the variance of execution prices, it also reduces the volatility of the market.

The final metric we consider is liquidity. An asset is liquid if "it is more certainly realizable at short notice without loss" (Keynes 1930). Liquidity can be defined quantitatively in a number of ways. However, our model accounts for the requirement of short notice, as orders are canceled if they do not result in a transaction, and when they do transact, the price must satisfy the reservation price initially generated. As a result, our measure of liquidity is the number of transactions that take place per simulation, or the probability that an order transacts. As shown

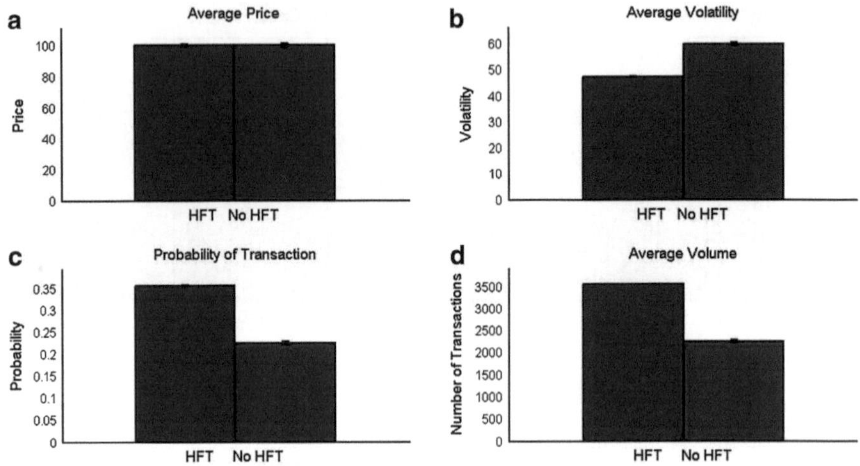

Fig. 4 Comparison plots of (**a**) average transaction price, (**b**) volatility, (**c**) transaction probability, and (**d**) volume (number of trades) in the simulation both with and without HFT. Note that the introduction of HFT has no discernible effect on price, but statistically significant reduction of volatility, along with an increase in the number of trades and the likelihood of a given order being filled. Error bars denote 95 % confidence intervals

Table 1 Average of parameters over 100 runs of 10000 iterations of the simulation. Standard deviations are shown in parentheses below the average. For the HFT case, the average is taken over both markets

Parameter	Base case	HFT
Price	100.5	100.5
	(1.38)	(0.71)
Volatility	60.1	47.4
	(0.6)	(0.4)
Volume	2255	3580
	(33)	(6)
Probability of Transaction	0.226	0.358

in Table 1 and Fig. 4c,d, orders are more likely to be filled when HFT activity is present in our model. Again, this result matches empirical findings (Hasbrouck and Saar 2013).

Conclusions

In this chapter, we analyzed the effects of high-frequency trading in a simulated environment. With the premise that HFT activity connects orders across markets, we found that prices are closer to their equilibrium value, volatility is reduced, and liquidity is increased when HFT is present. These results suggest that connecting order flow across similar securities is important for investor welfare, and to the extent that HFT performs this function, it serves an important purpose in modern financial markets.

Acknowledgements This chapter is a modified version of Benjamin Myers MPhys thesis originally entitled "Agent Based Simulations of High-Frequency Trading in Financial Markets." This work was supported by the European Commission FP7 FET-Open Project FOC-II (no. 255987)..

References

Brogaard, J., Hendershott, T., & Riordan, R. (2013). High Frequency Trading and Price Discovery. Working paper, http://ssrn.com/abstract=1928510.

Farmer, J. D., Patelli, P., & Zovko, I. I. (2005). The predictive power of zero intelligence in financial markets. *Proceedings of the National Academy of Sciences USA, 102*(6), 2254–2259.

Gerig, A. (2012). High-Frequency Trading Synchronizes Prices in Financial Markets. Working paper, http://ssrn.com/abstract=2173247.

Gerig, A., & Michayluk, D. (2010). Automated Liquidity Provision and the Demise of Traditional Market Making. Working paper, http://ssrn.com/abstract=1639954.

Gode, D. K., & Sunder, S. (1993). Allocative efficiency of markets with zero-intelligence traders: market as a partial substitute for individual rationality. *Journal of Political Economy, 101*(1), 119–137.

Hasbrouck, J., & Saar, G. (2013). Low-latency trading. *Journal of Financial Markets, 16*(4), 646–679.

Hendershott, T., Jones, C. M., & Menkveld, A. J. (2011). Does algorithmic trading improve liquidity. *Journal of Finance, 66*(1), 1–33.

Keynes, J. M. (1930). *A treatise on money*, Vol 2: *The applied theory of money*. New York: Harcourt.

MacKenzie, D. (2012). Mechanizing the Merc: The Chicago Mercantile Exchange and the Rise of High-Frequency Trading. Working paper, University of Edinburgh.

Raising Issues About Impact of High Frequency Trading on Market Liquidity

Vladimir Naumenko

Abstract The aim of this paper is to consider some problems with evaluation of the impact of high frequency trading on market liquidity. The first part is devoted to difficulties of disentangling the impact of high frequency on market liquidity from other relevant factors. The remainder of the paper is intended to discuss some issues affecting the evaluation of the influence of high frequency trading on particular aspects of market liquidity.

Keywords Depth • High frequency trading • Market liquidity • Resiliency • Tightness

Over the last years, high-frequency trading (HFT) has become the object of ever-increasing attention on the part of market participants and academics as well as regulators. Plenty of academic studies were devoted to evaluating the impact of HFT on different aspects of market quality such as liquidity, volatility, and informational efficiency. It is not surprising that they have different, and sometimes diametrically opposed views about HFT's impact on modern financial markets. Despite the fact that the opinions of experts in the field of financial markets split, the general background of statements in mass media concerning HFT can be characterized as negative. In addition, a number of initiatives were proposed by regulators all over the world in response to changes in the nature of trading in financial instruments, largely due to the prevalence of HFT (IOSCO 2011). These proposals also require serious consideration, since their implementation in some cases can lead to far-reaching consequences for the market quality, while not always clear and definite in advance.

The importance of thorough examination of the matter is confirmed by evidences that a significant increase in number of trades and quotes attributable to HFT has occurred over the past 5–10 years in many trading venues. Despite the stabilization or even a slight decrease in the proportion of HFT in developed capital markets, there is no doubt that in the near future the impact of this phenomenon on the

V. Naumenko (✉)
JSC "PROGNOZ", Risk Lab, Perm, Russia

National Research University Higher School of Economics, FERM Lab, Moscow, Russia
e-mail: naumenko.v@prognoz.ru

A.K. Bera et al. (eds.), *Financial Econometrics and Empirical Market Microstructure*, DOI 10.1007/978-3-319-09946-0_14

market quality will continue to remain significant. In any case, despite the decline in market share of HFT, there is no serious rationale to believe that HFT for any reason will voluntary quit the markets in the coming years, albeit under the above mentioned negative pressure from the mass media and scrutiny from the regulators. Moreover, in the less developed trading venues, especially those based in developing countries, a trend of inflow of HFT traders to markets remains (WFE 2013). In these circumstances, it is of great importance to study the behavior of trading algorithms and other things necessary to elaborate procedures in order to encourage positive behavior and eliminate or at least mitigate possible negative effects.

1 How to Disentangle the Impact of HFT on Market Liquidity from Other Factors?

There are a number of challenges for evaluation of the influence of HFT on market quality. First, it is very difficult to disentangle the impact of HFT on the market quality from other technological and regulatory innovations which led to substantial changes in the market structure of many trading venues, e.g., decimalization in U.S. equity markets. Then, it is also difficult to identify HFT traders even if researchers have access to agent-resolved data, which is extremely rare (Kirilenko et al. 2011; Hagströmer and Nordén 2013; Benos and Sagade 2012). One explanation suggests that many traders pursue a variety of strategies which both provide and take liquidity depending on market conditions. Then, it is possible that a single trading account represents an omnibus account which may be used by several different agents, e.g., in the case of sponsored market access, provided by financial intermediaries to their clients. Next, much research in this area refers to HFT as a certain homogenous entity, while there is a multitude of trading strategies that have a different impact on the market quality. Sometimes this is due to the above-described difficulties in identifying the HFT traders.

Nevertheless, the positive changes in the market quality are sometimes attributed to HFT as a whole. In this case, as rightly noticed by Tse et al. (2012), it is possible that the positive effects of some HFT strategies (e.g., market-making and statistical arbitrage) outweigh the negative effects produced by other strategies, thereby masking the negative side effects of HFT on market quality. For instance, academic studies examining the phenomenon of HFT consider market-making strategies (Hendershott et al. 2011; Kirilenko et al. 2011), unwittingly spreading their effects on HFT as a whole. It would be better to consider the impact of peculiar trading strategies equipped with HFT technology on the market quality. One can then try to analyze the combined effect of the strategies considered on market quality. Ideally, one should conduct a comparative analysis of market quality with and without HFT (Hendershott et al. 2011). Unfortunately, it seems pretty far from feasibility.

It is reasonable to pose the question how to determine the impact of HFT on market quality when at the times of appearance and increasing use of HFT there were many other significant changes in the market structures which could not but affect the nature of trading in financial instruments. Even if one finds correlation between the increase in HFT and improvement (deterioration) in some market quality, correlation is not necessarily causation. "The challenge is to measure the incremental effect of HFT beyond other changes in equity markets" (Jones 2013). Ideally, one would like to track changes in the market structure which led to an increase in the proportion of HFT in the market, for example, autoquote dissemination on the NYSE in first half of 2003 (Hendershott et al. 2011). Then, one needs to compare the state of the market before and after the changes have occurred. It is advisable in this case to establish a causal link between the increase in the share of HFT and changes in the market structure of some trading venue. What is really important is what metrics will be used to reflect a certain quality of the market. Moreover, the task can be complicated by the fact that the metrics often do not reflect all aspects of some dimension of market quality, especially dealing with market liquidity. In other words, one needs to be more careful in the conclusions and not to make hasty statements in the spirit of "post hoc ergo propter hoc".

"HFT is not a strategy but a technology" (WFE 2013) that facilitates the implementation of many traditional trading strategies whose effects on the market quality vary considerably. In other words, the nature of the trading strategies is likely to determine the effects of HFT on market quality rather than computerization of these strategies in itself. HFT liquidity providers, in fact, have replaced many of the traditional market makers which became less effective in highly automated order-driven markets. It is obvious that the algorithms are better at monitoring market conditions and adjusting orders than humans, for example, specialists on the NYSE. According to Biais et al. (2010), the machines are more effective because they obviously have no problems with limited attention or concentration required for simultaneously implementing multiple tasks. Undoubtedly, algorithms are much better at detecting and eliminating arbitrage opportunities, reducing, in fact, their lifetime to a few milliseconds (Sorkenmaier and Wagener 2011), to be precise, up to the time delay of the signal (latency) on a given trading platform. Ideally, when assessing the impact of HFT on market quality, one should not consider HFT as a whole but focus on individual trading strategies, using HFT technology. At the same time it would be great to establish whether the use of this high-speed technology exacerbates the problem generated by the strategy, e.g., market manipulation. However, there is a problem herein with the fact that myriad of strategies can be used by market participants, including those who process information on the number of financial instruments and simultaneously trade on multiple trading venues. Even having access to the source code and scripts underlying certain trading strategies it can be difficult to determine their behavior in real markets.

As with any technology, HFT can bring more good than harm, or vice versa. Ideally, appropriate application of technology can enhance market quality which significantly reduces liquidity premium and subsequently the firm's cost of capital. Therefore, the question of the prohibition looks at least weird and can be even viewed as an attempt to stop scientific and technological progress. However, one should understand how to behave with particular classes of trading strategies using HFT technology to gain a competitive advantage. In other words, it is necessary to highlight key points, i.e., a close attention must be paid to particular trading strategies, while the technology for their application should be considered from the point of view of the possibility of worsening the market quality. However, it is clear that disruptive behavior exists and can be reinforced with HFT technology. Thus, it is important to thoroughly consider these scenarios, stay aware of their consequences and be ready to eliminate or mitigate their negative effects.

HFT strategies can be divided into "good" and "bad" according to their relation to the short-term mispricing (Tse et al. 2012). Those strategies that profit from detecting short-term mispricing and correcting it should be referred to the "good" strategies that improve the market quality. Strategies that profit from the creation of short-term mispricing and its subsequent removal may be considered as "bad" strategies, which have a negative impact on market quality.

2 How to Evaluate the Impact of HFT on Particular Aspects of Market Liquidity?

In general, market liquidity can be defined as ability to trade when you want to trade (Harris 2002). To be more specific, a liquid market can be described as a market where participants can rapidly execute large-volume transactions with a small impact on prices (BIS 1999). Even the last definition is not precise enough, since it's not clear what the following expressions mean: "rapidly execute", "large-volume transactions" and "small impact". In order to somehow evaluate such elusive characteristic of market quality, Kyle's approach is usually applied in market microstructure research (Kyle 1985). Its key idea is to consider separately three different aspects of market liquidity: tightness, depth, and resiliency. Tightness is the cost of opening and closing a position over a short period of time. It is well characterized by the bid-ask spread. Depth denotes the volume of incoming order required to change the price a given amount or the total amount of orders in limit order book. Resiliency refers to the speed with which market recovers from a random, uninformative shock. Next, we consider the impact of HFT on each of these aspects of market liquidity.

Tightness. High-frequency traders (HFTs) have largely replaced traditional market makers because they are able to post more competitive quotes, thereby providing tighter bid-ask spread. In market microstructure research the bid-ask spread is usually decomposed into three following components: order-processing costs, asymmetric information costs and inventory-carrying costs (Huang and Stoll 1997). Technological and regulatory changes gave to these "new" market makers (Menkveld 2013) an advantage over the traditional ones. Firstly, they can intermediate trades at lower costs. This is partly from the automation of trading process that has brought lower costs comparing with manual execution of trades. Therefore, they have smaller order-processing component of their bid-ask spread. However, they obviously have to incur additional costs of developing, testing and maintaining algorithms as well as make investments in hardware and software to implement them. Perhaps, there are some economies of scale in this case. Recently, HFTs most likely had to incur serious start-up costs which were likely reimbursed afterwards by profits from market-making due to speed advantage over traditional market makers. It is possible, that this suggestion is supported by researchers' findings on increases in realized spreads and other measures of liquidity supplier revenues (Hendershott et al. 2011).

Secondly, due to automation of access to markets they can react more quickly to any new information about multiple financial instruments. Furthermore, computerized trading strategies have no problems with concentration or tiredness when monitoring market conditions. So, they can make timely response to a relevant event, if any. Thus, they reduce their exposure to the risk of being picked off by informed traders that, in turn, reflects in smaller asymmetric information component.

Thirdly, HFTs are also more efficient in inventory management due to holding relatively small positions, keeping them for a very short period of time, not carrying inventories overnight and having more diversified portfolio because of trading more financial instruments. Thus, they reduce their exposure to the inventory risk that, in turn, reflects in smaller inventory-carrying component.

Therefore, all of these effects lead to narrower bid-ask spreads. Many studies (Angel et al. 2010; Hasbrouck and Saar 2013; Hendershott et al. 2011; Kirilenko et al. 2011) support the narrowing of bid-ask spreads up to the size of the minimum price increment (tick). Obviously, it witnesses improving of tightness. It turns out that market participants, arranging small-volume transactions (i.e., not exceeding the quoted depth), have benefited as their transaction costs have dramatically reduced.

Depth. And what has happened with profitability of market participants making large-volume transactions, i.e., exceeding the quoted depth? It seems that this question cannot be answered definitely. On the one hand, they also have benefited from lower bid-ask spreads, thereby reducing the value of their implicit transaction costs as for the volume not exceeding the quoted depth. On the other hand, some market participants indicate a decrease in the depth of the market, linking this phenomenon primarily with a decrease in tick size (e.g., decimalization in U.S. markets). Under these new conditions it is much easier to rearrange limit orders to advance in the queue. This process resembles leapfrog, as Larry Harris describes it (Harris 2002). In this case, implementing front-running strategies becomes cheaper, since one more tick (i.e., one cent or penny nowadays) does not significantly increase the costs of execution of trading strategies. It turns out that it makes little sense to put a large amount of limit orders as faster market participants can easily stand ahead in order to benefit from this situation (see "quote matching" for more information on this type of front-running). As a result, many market participants reflect their trading intentions less intensively in their limit orders, thereby increasing the amount of hidden liquidity, hanging over the market.

For the sake of justice it should be noted that before the proliferation of HFT submission of large-volume limit orders often affected adversely on the financial performance of their initiators. Thus, they also could become a victim of one of the front-running strategies. The difference with the past is the following: under the conditions of small tick sizes in markets pursuing such "parasitic" (Harris 2002) strategies has become much less expensive, i.e., front-running has become more feasible. Moreover, some market participants have a priori advantage in speed of access to the market, which affects the distribution of the balance of power between the market participants. In other words, without the same technology withstanding front-running and protecting the value of the embedded option in limit order from extraction is much more difficult. Submitting a limit order, especially with large volume, this market participant provides to other traders, in fact, a free option (Copeland and Galai 1983). In the event that it becomes "in-the-money", HFT traders having technological advantages in speed of access to the market will be able to extract its value faster than the participant will be able to cancel this limit order. In such circumstances, limit order submission is a luxury. Thus, the displayed depth of the market likely has most likely deteriorated, but first of all it depends on the tick size, which determines the profitability of the front-running strategy.

Tightness improvement together with depth deterioration has an ambiguous effect on the implicit transaction costs of the market participants. On the one hand, it decreases the value of the bid-ask spread. On the other hand, it increases the costs of market impact. The final result will depend on the volume distribution between the components of implicit transaction costs. In order to have a positive result the additional gain referred to the volume below the quoted depth have to exceed the additional loss from walking the book. So, the market participant faces trade-off and needs to choose the volume to balance the bid-ask spread with costs of market impact.

Market impact depends on the structure of the limit order book, particularly on the distribution of volumes among price levels and the presence of gaps in the book. By the way, it is possible that the rest of the amount will be executed at prices that would have been inside the market when compared to the previous bid-ask spread, or worse just one penny (by reducing the tick size). In general, one needs to evaluate the entire magnitude of the implicit transaction costs for different levels of volume. It might be supposed herein that upon reaching a certain level transaction costs will increase, while remaining at a lower level in the new environment, and then they will exceed the previous total implicit transaction costs after passing that level.

It should be considered whether such volumes were traded in the past, i.e., before changes in market structure induced by HFT. Possibly, it wasn't so feasible. Then, there is no question herein. Before the era of automated trading block traders used the services of intermediaries from upstairs market, i.e., the services of so-called block broker/dealers, or acted in the market through a single or multiple floor traders, who "quietly work the order". It seems very plausible that the emergence of algorithmic trading (especially after the seminal work of Almgren and Chriss

2000) led to the switch of block traders from over-the-counter markets to organized markets where securities were primarily listed. In this case, the performance of block traders would be increased as it became possible to make transactions more quickly and at lower cost in comparison with those arranged at the discretion of the floor traders. In this case, there were eliminated any conflicts of interest associated with the use by floor brokers information about the positions of block traders to pursue their own interests to the detriment of the interests of their clients which is the breach of fiduciary duty. Can it be said that some of the increase in the total volume of trade associated with the appearance of algorithmic trading relates to institutional investors trading in large volumes (block traders)? Some researchers share this opinion, e.g., (Jorion 2007). However, the task of identifying block traders is extremely difficult if one has no access to agent-resolved data, because it requires integrating the small orders ("child" orders) in single "meta-order" ("parent" order). Although it is possible that in the data there are certain patterns, reflecting a presence of order-splitting strategy.

However, this has decreased the apparent depth of the limit order book, as shown above. Perhaps for some market participants it has become more difficult to sell their volumes, despite lowering bid-ask spreads. It is still necessary to check what outweighs: gain from narrowing bid-ask spread or losses from the reduction in the depth of the best available orders. Nevertheless, it is conceivable that most of the volumes which became relatively more expensive to trade were hardly traded before. It is possible that some of the volumes were executed for the next several price levels of limit orders, and now it takes more price levels, i.e., one has to walk deeper the book. However, the difference between two adjacent levels of ticks has most likely decreased. How to unravel this tangle: the decimalization led to an increase in algorithmic trading (it has become easier to rearrange the best limit orders, thus reducing the value of time priority as an order precedence rule). It is unlikely that anyone at once traded volumes of more than 5 % of the daily volume, a famous ad-hoc rule used by traders to determine market liquidity. However, it needs empirical checking. At the same time there are dark pools which are rather effective substitutes of upstairs markets. However, trading at dark pools, as a rule, is not conducted continuously. One might pose the question whether large volumes participate in the price discovery given that they are directed to the dark pools, where the price is usually taken from other trading venues. In other words, does this practice lead to the fact that not all the information is reflected in the price? Do we need then these dark pools? But in the upstairs markets the usual practice was almost the same. It would be necessary to compare the two regimes, which is practically impossible because of the lack of necessary data. Is it worth so worrying about reducing the depth of the market after all? We compare the current state of markets with what it was before, or with what we would like to have? Moreover, all the consequences of this "ideal" world we can hardly evaluate. And then there is the hidden liquidity, which is simply not reflected in the limit order book (partly also due to the front-runners), but always ready to join in the action (fundamental buyers and sellers in the terminology of Kirilenko et al. 2011).

However, there are studies that claim that the depth of the market has increased (Angel et al. 2010; Hasbrouck and Saar 2013). Can it be contributed to quote stuffing, at least partly? Submission and immediate cancellation of orders located deep inside the book really does not increase the depth of the market, unless the random order with large volume is executed during the lifetime of this fleeting order. However, not every methodology for measuring the liquidity of the market will be able to properly take into account (more precisely, to exclude from consideration) this effect. Most likely, the market depth metrics, taken on various time frames, would demonstrate an increase in depth even after aggregating.

Resiliency. Even if the depth has decreased, then the slower execution of the order (implying a splitting of the total order into smaller parts) may reduce market impact costs subject to a decrease in time of market resiliency. There is evidence that the algorithmic liquidity providers closely monitor the situation with the anomalous expansion of the bid-ask spread. In this case HFTs promote liquidity replenishment due to speed advantage over other market participants (Brogaard 2010). Furthermore, this effect can even be used to detect HFTs, i.e., detecting who submit limit orders (which tighten the bid-ask spread) immediately after execution of market or marketable limit orders which led to widening of the bid-ask spread.

However, under certain conditions (e.g., during the Flash Crash large amounts were systematically dumped to one side of the market) liquidity providers become liquidity takers (Kirilenko et al. 2011). It can be brought about by pursuing their intentions to keep the level of inventory in the area of the preset target which leads to a further increase in the bid-ask spreads and a sharp price movement in an unfavorable direction. Put forward the assumption that up to a certain level of the bid-ask spread new liquidity providers contribute to the resiliency of the market, and above this level there comes a realization of systemic risk. So, liquidity providers face another trade-off. However, under normal market conditions the time of resiliency likely decreases considerably, thus compensating to a certain extent the decrease of the market depth.

Thus, under these new conditions order-splitting has become even more meaningful. For those who want to immediately sell significant volumes there are different dark pools. Rather, in terms of fragmented liquidity there will be some combination of rational order splitting and the use of dark pools. The answer to the question what and when to use will depend on the current market conditions, the rules of engagement into dark pools and pricing rules. Moreover, in order to improve the efficiency of this strategy one needs to split the block not only in time but also in space (across different trading venues) using smart order routing technology.

Thus, market microstructure theory identifies several positive and negative effects on market liquidity which could be produced by HFTs. Ultimately, determining net effect is an empirical question given that methodological choices are reasonable enough to reflect multi-faceted nature of market liquidity.

References

Almgren, R., & Chriss, N. (2000). Optimal execution of portfolio transactions. *Journal of Risk, 3*(2), 5–39.

Angel, J., Harris, L., & Spatt, C. (2010). *Equity trading in the 21st century*. Working paper. Carnegie Mellon.

Committee on the Global Financial System. (1999). *Market liquidity: Research findings and selected policy implications*. Bank for International Settlements – Monetary and Economic Department.

Benos, E., & Sagade, S. (2012). *High-frequency trading behaviour and its impact on market quality: evidence from the UK equity market*. Working paper. Bank of England.

Biais, B., Hombert, J., & Weill, P. (2010). *Trading and liquidity with limited cognition*. Working paper. Toulouse University, IDEI.

Copeland, T.. & Galai, D. (1983). Information effects on the bid-ask spread. *The Journal of Finance, 38*, 1457–1469.

Hagströmer, B., & Nordén, L. (2013). The diversity of high frequency traders. *Journal of Financial Markets, 16*, 741–770.

Harris, L. (2002). *Trading and exchanges: Market microstructure for practitioners* (p. 656). New York: Oxford University Press.

Hasbrouck, J., & Saar, G. (2013). Low-latency trading. *Journal of Financial Markets, 16*, 646–679.

Hendershott, T., Jones, C., & Menkveld, A. (2011). Does algorithmic trading improve liquidity? *The Journal of Finance, 66*, 1–33.

Huang, R., & Stoll, H. (1997). The components of the bid-ask spread: A general approach. *Review of Financial Studies, 10*, 995–1034.

Jones, C. (2013). *What do we know about high-frequency trading*? Working paper. Columbia Business School.

Jorion, P. (2007). *Value-at-risk: The new benchmark for managing financial risk* (3rd ed., p. 602). New York: McGraw-Hill.

Kirilenko, A., Kyle, A., Samadi, M., & Tuzun, T. (2011). *The flash crash: The impact of high frequency trading on an electronic market*. Working paper. University of Maryland and CFTC.

Kyle, A. (1985). Continuous auctions and insider trading. *Econometrica, 53*, 1315–1336.

Menkveld, A. (2013). High frequency trading and the new-market makers. *Journal of Financial Markets, 16*, 712–740.

Sorkenmaier, A., & Wagener, M. (2011). *Do we need a European "National Market System"? Competition, arbitrage, and suboptimal executions*. Working paper. Karlsruhe Institute of Technology.

Technical Committee of the International Organization of Securities Commissions. (2011). *Regulatory issues raised by the impact of technological changes on market integrity and efficiency* (60 pp.).

Tse, J., Lin, X., & Vincent, D. (2012). *High frequency trading – The good, the bad, and the regulation*. Credit Suisse. AES Analysis. Market Commentary, 5 December 2012. https://edge.credit-suisse.com/edge/Public/Bulletin/Servefile.aspx?FileID=23284&m=1815212669

World Federation of Exchanges. (2013). *Understanding High Frequency Trading (HFT)* (5 pp.).

References

[illegible]

Application of Copula Models for Modeling One-Dimensional Time Series

Vadim Onishchenko and Henry Penikas

Abstract This paper proposes method of detecting a structural break/shift in time series such as AR(1) with a nonlinear dependence structure of lagged value and the estimation of the break point, based on nonparametric estimations of the dependence's copulas and comparison with some existing tests. However, we assumed the time series to be stationary and homoscedastic. This paper compares the efficiency of the standard test, considering only linear autoregressive dependence nature. A suggested technique is given, some modifications of the evaluation scheme is offered and a more flexible method of detecting structural break is proposed, usefulness of our methodology is demonstrated through some applications to a few macroeconomic and financial time series.

The paper is organized as follows: the first section contains a selective literature review. The second section describes the generation's procedure of time series, used in further calculations. The problem of detection of the structural break with respect to the nonlinear time series is formulated in the third section. The fourth section contains results of evaluations using simulated data. In Sect. 5 we provide examples of our suggested technique. The final section contains "Conclusions".

Keywords Copula • Nonlinear time series • Nonparametric estimation of copula • Structural break

1 Literature Review

Modern papers about economics and finance are increasingly using dependence modeling with copulas. This approach has advantages of taking account of nonlinear dependence structure. Moreover, unlike many other common dependence measures (for example, Pearson's correlation coefficient), the copula model is applicable

V. Onishchenko (✉)
Economics Department, National Research University Higher School of Economics, Moscow, Russia

H. Penikas
Department of Applied Economics, International Laboratory of Decision Choice and Analysis, National Research University Higher School of Economics, Moscow, Russia

A.K. Bera et al. (eds.), *Financial Econometrics and Empirical Market Microstructure*, DOI 10.1007/978-3-319-09946-0_15

when fat tails of distributions are observed, which lead to large values of fourth moments that conflicts with the Gaussian character of distributions. Moreover copulas provide a certain flexibility, allowing for model joint distribution of real values separately from marginals.

Formally, a copula is defined as a function on the n-dimensional unit cube $[0; 1]^n$ with values in segment [0; 1], satisfying further conditions:

$$\text{–for any i} = 1 \ldots \text{n is correctly } C(u_1, \ldots \ u_{i-1} \ , \ 0, \ u_{i+1} \ \ldots \ u_n) = 0 \tag{1}$$

$$\text{–for any i} = 1 \ldots \text{n is correctly } C(1, \ldots 1, \ u_i \ , \ 1 \ldots 1) = u_i \tag{2}$$

– C is n-increasing function, which means that it's integral over any parallelepiped contained in n-dimensional unit cube is non-negative: if $B = \prod_{i=1}^{n} [x_i; y_i] \subset [0; 1]^n$ then

$$\int_B dC\left(u_{1,\ldots u_n}\right) \geq 0. \tag{3}$$

According to Sklar's theorem (1959), any multivariate distribution function $H(x_1, \ldots x_n) = P(X_1 \le x_1, \ \ldots \ X_n \le x_n)$ of random vector $(X_1, \ \ldots \ X_n)$ with marginal distribution functions $F_1(x) = P(X_1 \le x), \ \ldots \ F_n(x) = P(X_n \le x)$ could be considered as $H(x_1, \ldots x_n) = C(F_1(x_1), \ \ldots \ F_n(x_n))$, where C—some n-dimensional copula, and if $F_1, \ \ldots \ F_n$ continuous, then copula C is unique.

Common theoretical aspects of applicability copula models to Markov chains were formulated by Darsow et al. (1992) and extended to the case of Markov chains of arbitrary order by Ibragimov (2009). In particular, the sufficient and necessary conditions for the submission of a stationary Markov process of arbitrary order using the copula have been introduced and have satisfied the Chapman–Kolmogorov equation. To do this, an operation ^{k} over the copulas was determined defined by the following: if A and B are two copulas of dimensions m and n respectively such that:

$$\begin{aligned} A(u_1 = 1, \ldots u_{m-k} = 1, \ v_1 \ldots v_k) &= A(u_1 = 1, \ldots u_{n-k} = 1, \ v_1 \ \ldots \ v_k) \\ &= C(v_1 \ \ldots \ v_k), \end{aligned} \tag{4}$$

then the result of operation $A^k B$ will be (m + n – k)—dimensional copula defined by the follow formula:

$$\begin{aligned} A^k B(u_1, \ldots u_{n+m-k}) = \int_0^{u_{m-k+1}} \ldots \int_0^{u_m} \frac{\partial A(u_1, \ldots u_{m-k}, \ v_1 \ldots v_k)}{\partial v_1 \ldots \partial v_k} \Bigg/ \frac{\partial C(v_1 \ldots v_k)}{\partial v_1 \ldots \partial v_k} \\ * \frac{\partial B(v_1 \ldots v_k, \ u_{m+1} \ldots u_{m+n-k})}{\partial v_1 \ldots \partial v_k} \Bigg/ \frac{\partial C(v_1 \ldots v_k)}{\partial v_1 \ldots \partial v_k} \ C(dv_1, \ldots dv_k) \end{aligned} \tag{5}$$

Then stochastic process $\{X_t\}_{t \in T}$ is the Markov process of order k if and only if for any $t_1 \leq t_2 \leq \ldots t_n$, n > k, $t_i \in T$ (i = 1 … n) it's true that

$$C_{t_1 \ldots t_n} = C_{t_1 \ldots t_{k+1}}{}^k C_{t_2 \ldots t_{k+2}}{}^k \ldots {}^k C_{t_{n-k} \ldots t_n} \tag{6}$$

in particular when stationary consequences of random variables is $C_{t_1 \ldots t_{k+1}} = C_{t_{1+\tau} \ldots t_{k+1+\tau}}\ \forall\, \tau$.

If the marginal distributions doesn't change, then the first order Markov process $\{X_t\}_{t=1}^{N}$ follows:

$$\mathrm{G}\,(y_t; y_{t-1}) = \mathrm{C}\,(\mathrm{F}\,(y_t);\ \mathrm{F}\,(y_{t-1}))\,. \tag{7}$$

A general problem of determining the structural shift in using the copula was illustrated by Brodsky et al. (2009). It has been formulated for some samples $\{X_1; X_2, \ldots X_N\}$ of independent random m-dimensional vectors $X_i = (x_{i1}, \ldots x_{im})$ with cumulative distribution functions $V_i = V_i\ (x_{i1}, \ldots x_{im})$. Marginal distributions $F_1 \ldots F_m$ are not changed. It is assumed that either $V_1 = V_2 = \ldots = V_N$, or at some point of time there is a change in joint distribution function of the components of vectors X_i, which could be defined using a copula. Formally, the problem can be formulated as follows:

$$V_i\,(x_1, \ldots x_d) = \begin{cases} G_1\,(F_1\,(x_1)\,, \ldots F_m\,(x_m)) & i = 1 \ldots L \\ G_2\,(F_1\,(x_1)\,, \ldots F_m\,(x_m))\quad . & i = L+1 \ldots N \end{cases} \tag{8}$$

Then the null hypothesis could be stated as H_0: $G_1 = G_2$ against the natural alternative. In the case of rejection of the null hypothesis, we must construct the estimation $\widehat{L}$ of the structural break point. For this, the estimations of empirical dependence's copulas between $x_1 \ldots x_k$ are built at any time L = 1 … N − 1 based on all observations before the anticipated structural shift and after:

$$D_L^{before}\,(\mathrm{u}) = \frac{1}{L}\sum\nolimits_{i=1}^{L} I\,(U_{i.L} \leq u) = \sum\nolimits_{i=1}^{L}\prod\nolimits_{j=1}^{m} I\,\left(U_{ij.L} \leq u_j\right)$$

$$D_{N-L}^{after}\,(\mathrm{u}) = \frac{1}{N-L}\sum\nolimits_{i=L+1}^{N} I\,(U_{i.N-L} \leq u) = \sum\nolimits_{i=L+1}^{N}\prod\nolimits_{j=1}^{m} I\,\left(U_{ij.N-L} \leq u_j\right) \tag{9}$$

where $U_{i.L} = (U_{i1.L}, \ldots U_{im.L})$ and for any j = 1, 2 … m

$$U_{ij,L} = \frac{L}{L+1}F_{j,L}\left(x_{ij}\right) = {rank\left(x_{ij}\right)}\Big/{L+1}\,,\ \mathrm{i} = 1, \ldots \mathrm{L}$$

$$U_{ij,L} = \frac{L}{L+1}F_{j,L}\left(x_{ij}\right) = {rank\left(x_{ij}\right)}\Big/{N-L+1}\,,\ \mathrm{i} = \mathrm{L}+1, \ldots \mathrm{N} \tag{10}$$

Then the following Kolmogorov–Smirnov statistic is constructed based on two empirical copulas:

$$\Psi_{L,N-L}(\mathrm{u}) = \left(D_L^{before}(\mathrm{u}) - D_{N-L}^{after}(\mathrm{u}) \right) * \sqrt{L(N-L)} \Big/ N \tag{11}$$

and we find its maximum and point of maximum:

$$T_N = max_{[\beta N] \le L \le [(1-\beta)N]} \{sup_u abs(\Psi_{L,N-L}(\mathrm{u}))\} \tag{12}$$

$$\hat{m}_N = argmax_{[\beta N] \le L \le [(1-\beta)N]} \{sup_u abs(\Psi_{L,N-L}(\mathrm{u}))\} \tag{13}$$

If the statistic value T_N exceeded critical value, then it is interpreted as the presence of a structural break, the estimation of which is value $\widehat{m}_N$. Some properties of this estimator have been identified, namely an exponential decrease in probabilities of type I and type II errors with increasing sample size, assuming independence of sample s (random vectors X_i) and permanence marginal distribution functions. Moreover, critical values have been calculated for Clayton and Gumbel copulas with some identical parameters and a few lengths of samples using the Monte Carlo method with 500 replications.

In the third part of this paper we describe the method of detecting a structural break according to nonlinear time series like AR(1). Moreover, a modified Cramér–von Mises statistic is proposed, with a changed coefficient by calculating $\Psi_{L,N-L}(\mathrm{u})$ in (11). In that formulation, a copula-based structural break detection method has already been used by Penikas (2012) on an example of the U.S. quarterly GDP from 1947 to 2012. A dependence was identified and showed statistical significance of non-linear dependence only with the first lag. Next, a proposed copula test revealed the presence of a structural break in 1980, in what could be interpreted as a consequence of the oil crisis serving as an external shock. The Andrews–Zivot test pointed to significantly earlier dates, reasonable interpretation of which is quite difficult.

A previously proposed parametric estimation method using the empirical copula is not only possible—the use of kernel estimates copula, according to Omelka et al. (2009) is one of the natural ways to expand this work.

Azam (2012) proposed using a Bayesian approach, which extends this method to the case of discrete marginal distributions. This survey is not exhaustive due to the rapid popularity of the use of copula models in many modern problems of economics, statistics and actuarial mathematics, including the analysis of time series.

2 Generation of Time Series

This section describes how to generate a time series such as AR(1) with non-linear structure of dependence, defined by using a two-dimensional copula C(u, v). The series is stationary, but the procedure can be easily generalized to non-stationary series. Thus, let u = F(x_t), v = F(x_{t-1}). According to (5) the joint distribution of the time series is as follows:

$$\mathrm{F}\left(x_1, x_2 \ldots x_T\right) = C\left(x_1, x_2\right)^1 C\left(x_2, x_3\right)^1 \ldots \mathrm{C}(x_{T-2}, x_{T-1})^1 \mathrm{C}\left(x_{T-1}, x_T\right) \quad (14)$$

Let known realization x_t. Joint distribution x_t and x_{t+1} are settings per copula C(F(x_{t+1}), F(x_t)) = C(u, v) where the denoted u = F(x_{t+1}), v = F(x_t). Then conditional distribution x_{t+1} is set as

$$\mathrm{C}\left(x_{t+1}\middle|x_t\right) = \frac{\partial C\left(u,\ v\right)}{\partial v}\bigg|_{v=F(x_t)} = \frac{\partial C\left(F\left(x_{t+1}\right),\ v\right)}{\partial v}\bigg|_{v=F(x_t)} \quad (15)$$

It is worth noting we can take advantage of copulas' remarkable property that allows us to model the joint distribution of C(u_1, ... u_k) separately from the marginal $F_1(x_1)$, ... $F_k(x_k)$, and move from observations x_1 ... x_k to pseudo-observations—probabilities, obtained by marginal distribution functions $u_1 = F_1(x_1)$, ... $u_k = F_k(x_k)$.

Thus, the process of generation is constructed as follows:

1. First observation x_1 generates from marginal distribution F(x).
2. For current observation of the time series, x_t derives a pseudo-observation with marginal distribution function: $v^* =$ F(x_t).
3. Then it obtains a conditional distribution function with a conditional copula: $\mathrm{G(u)} = \mathrm{C(u|v)} = \frac{\partial C(u,\ v)}{\partial v}\Big|_{v=v^*}$
4. Generate number α from uniform distribution on segment [0,1]
5. Then we find solution u^* of equation G(u|v) = α. For this next procedure, sequentially enumerated values 0, 0.1, 0.2 ... 0.9. We find the number β_1 such, that G(β_1) − α < 0 and G(β_1+0.1) − α ≥ 0. Further, sequentially enumerated values β_1, $\beta_1 + 0.01$, $\beta_1 + 0.02$, ... $\beta_1 + 0.09$ find such β_2, that G(β_2) − α < 0 and G($\beta_2 + 0.01$) − α ≥ 0 and so on. Thus, you could easily find a solution of G(u|v) = α up to any number of digits after the decimal point.
6. Knowing the value of u^*, we restore the value of time series with the quantile function F(x): $x_{t+1} = F^{-1}(u^*)$.

It is worth nothing that if we generate the entire time series at once, we could not recalculate every observation and pseudo-observation. If we only work with pseudo-observations, then the first observation is generated from a uniform distribution in [0, 1], and then we use only steps (3)–(5), then we use the quantile function to find the number of observations. Moreover, this procedure could be easily generalized to arbitrary dimension k, generating a pseudo-observation of the conditional distribution function

$$\mathrm{G}\left(\mathrm{u}\right)=\mathrm{C}\left(\mathrm{u}\left|v_1..v_{k-1}\right.\right)=\frac{\partial^{k-1}C\left(u,\ v_1\ldots v_{k-1}\right)}{\partial v_1\ldots\partial v_{k-1}} \tag{16}$$

For generating the time series we can use six types of copulas: Clayton, Frank, Gumbel, independent, Farlie–Gumbel–Morgenstern (FGM) and Plackett. The first three are Archimedean copulas: they satisfy a certain analogy between the axiom of Archimedes (for any positive numbers a and b could find natural number n, in that n*a > b); for any u and v from [0;1] we could find natural number n, in that M_C^n(u) > v where operation M_C^n(t) is defined as follows via Archimedean copula C: for any k M_C^{k+1}(t) = C(t, M_C^k(t)). Functionally, the Archimedean copula have the function generator φ(t) and are of the form C(u_1 … u_k) = $\varphi^{-1}(\varphi(u_1)+\ldots+\varphi(u_k))$. The fourth copula is an independent copula or product copula; it is the simplest example and specifies independent distribution of random variables, i.e. their joint distribution is the precise product of the partial distribution functions. The last two copula can't be assigned to any of the conventional classes. Farlie–Gumbel–Morgenstern (FGM) copula, is often used to model the random variables with small absolute values of the rank correlation. Plackett copula is also often used in applied research. Every copula (except product) is characterized by parameter θ, which takes values from some range of admissible values. Each value of the parameter θ corresponds to a specific value of rank correlation, but not necessarily to each value of the rank correlation that corresponds to a parameter θ of range of admissible values. In the two-dimensional case, Spearman's pho and Kendall's tau are expressed through copula C(u, v) = C(F(x), G(y)) = H(x, y) as follows:

$$\rho_S\left(\mathrm{x},\ \mathrm{y}\right)=12\iint_{[0;1]^2}\left(C\left(u,\ v\right)-u*v\right)du\,dv \tag{17}$$

$$\tau_K\left(\mathrm{x},\ \mathrm{y}\right)=4\iint_{[0;1]^2}C\left(u,\ v\right)dC\left(u,\ v\right)-1. \tag{18}$$

The annex contains the formulas of used copulas for two-dimensional cases, the formulas of conditional copulas (for Archimedean copulas-specified generator functions) and ranges of admissible values for parameters. It also contains some examples of generated time series for the above copulas and scatter plots of dependence between pseudo-observation and lagged pseudo-observation.

3 Problem Statement

Following Brodsky et al. (2009), the problem of structural break detection could be formulated through nonparametric estimation of copulas applied to time series. In contrast to the original paper, this approach was stated for structural break detection in time series; the hypothesis about independence of multidimensional vectors of observations is not proven.

Let us look at a batch of observations of time series $x_1 \ \ldots \ x_N$. Assuming time series of the form AR(m) with nonlinear dependence structure of previous observations, in any time of moment t = m + 1 ... N, it can be assumed that dependence from lagged values is defined through some continuous (m + 1)-dimensional copula $C_t(x_t; x_{t-1}; \ \ldots \ x_{t-m})$. The problem of determining structural break is that hypothesis $H_0: C_2 = \ \ldots \ = C_N$ about the permanence of dependence copula is true against alternative H_1: $\begin{cases} C_{m+1} = \cdots = C_K = C^* \\ C_{K+1} = \cdots = C_N = C^{**} \end{cases}$ where $C^* \neq C^{**}$. In the case of rejecting the null hypothesis, it's required to find a consistent estimation $\widehat{K}$ of the structural break moment.

The proposed technique is based on estimation in every moment of existing dependence before and after the suspected moment, and if the difference between the dependences is large enough, we could identify changing of dependence copula in this moment of time. For estimation, the nonparametric method is used. Nonparametric methods are based either on estimation of empirical copula or kernel estimations (Penikas 2010). We construct an empirical copula at first estimates marginal distribution functions for x_t, L(x_t), ... $L^m(x_t)$ where L = lag operator, or $\{L^s(x_t)\}_{s=0}^{m}$:

$$F_s^{emp}(\mathrm{x}) = \frac{1}{N-m}\sum\nolimits_{i=s+1}^{N-m+s} I\left(x_i \leq x\right), \qquad \mathrm{s} = 0,\ 1\ldots\mathrm{m} \tag{19}$$

Where I(A)—indicator of event A. Then we found estimations of pseudo-observations:

$$x_{si} = F_s^{emp}(x_i), \qquad \mathrm{i} = \mathrm{s}+1,\ \mathrm{s}+2\ldots\mathrm{N}-\mathrm{m}+\mathrm{s}; \mathrm{s} = 0,\ 1,\ldots\mathrm{m}. \tag{20}$$

Omelka et al. (2009) used Monte Carlo modeling to show that it's better to use asymptotically equivalent estimations:

$$x_{si} = \frac{N-m}{N-m+1}F_s^{emp}(x_i), \qquad \mathrm{i} = \mathrm{s}+1,\ \mathrm{s}+2\ldots\mathrm{N}-\mathrm{m}+\mathrm{s}; \mathrm{s} = 0,\ 1,\ldots\mathrm{m}, \tag{21}$$

in which small shifts move pseudo-observations to zero and works better on finite samples. Thus N-m (m + 1)-dimensional observations of current and m lagged values are derived, the dependence between which is assumed to be defined through the corresponding (m + 1)-dimensional copula.

It is worth noting that in this stage of derived empirical marginal distribution functions $F_s^{emp}(x)$, the stationary condition could be checked by comparing any two obtained functions with two-sample test (Kolmogorov–Smirnov, Cramér–von Mises, Anderson–Darling, chi-squared).

Further, for every time of moment L = m + 1 ... N estimates of empirical copulas are based before and after an anticipated moment of break:

$$C_L^{before}(u_0, \ldots u_m) = \frac{1}{L-m}\sum_{i_0=1}^{L-m}\sum_{i_1=2}^{L-m+1}\cdots\sum_{i_m=m+1}^{L} I\left(u_0 \geq x_{0i_0}\right) \\ *I\left(u_1 \geq x_{1i_1}\right) * \cdots * I\left(u_m \geq x_{mi_m}\right) \tag{22}$$

$$C_L^{after}(u_0, \ldots u_m) = \frac{1}{N-L}\sum_{i_0=L-m+1}^{N-m}\sum_{i_1=L-m+2}^{N-m+1}\cdots\sum_{i_m=L+1}^{N} I\left(u_0 \geq x_{0i_0}\right) \\ *I\left(u_1 \geq x_{1i_1}\right) * \cdots * I\left(u_m \geq x_{mi_m}\right) \tag{23}$$

Copula evaluations may also be used depending on nuclear and evaluation. Then C_L^{before} and C_L^{after} will be smooth multidimensional functions, weakly converging to the true distributions.

A measure of difference between copulas C_L^{before} and C_L^{after} could be applied via modified Kolmogorov–Smirnov statistic as suggested by Brodsky et al. (2009) and used by Penikas (2012). At each time moment L = m + 1 ... N following function is constructed:

$$\Psi_L(u_0, \ldots u_m) = \text{abs}\left\{C_L^{before}(u_0, \ldots u_m) - C_L^{after}(u_0, \ldots u_m)\right\} * \text{W}(\text{L}), \tag{24}$$

where W(L)—special correction factor, depending on proximity of the L moment to the middle of the sample of observations.

Then as a measure of value of statistic it's accepted that:

$$T_{KS} = max_{L\in B(N,\beta)}\left\{sup_{(u_0,\ldots u_m)\in[0;1]^{m+1}}\left(\Psi_L(u_0, \ldots u_m)\right)\right\}, \tag{25}$$

and as a estimation of break moment

$$K_{KS} = argmax_{L\in B(N,\beta)}\left(sup_{(u_0,\ldots u_m)\in[0;1]^{m+1}}\left(\Psi_L(u_0, \ldots u_m)\right)\right), \tag{26}$$

where $B(N, \beta)$—set of moments of time m + 1 ... N not including share β of first values and $(1-\beta)$ of last values: (N, β) = m + [β*(N-m-1)] + 1, m + [β*(N-m-1)] + 2, ... N − [β*(N-m-1)] − 2, N − [β*(N-m-1)] − 1. Due to the small quantity of observations in estimation, one of the empirical copulas could have obtained unlikely statistic values. In this approach, the difference between multidimensional functions is determined by the maximal value of difference between copulas over all points of the (m + 1)-dimensional unit cube and over all observations, except for some shares from two ends of the sample.

In (Brodsky et al. 2009) they suggested W(L) = $\sqrt{(L-m)*(N-L)}\Big/_{N-m}$, however the results of the next chapter convince, that more accurate results are obtained, using the square of the coefficient: W(L) = $(L-m)*(N-L)/(N-m)^2$. For convenience, in the denominator you can use the first degree. It will not affect the assessment of the time shift, and will only increase the value of the statistics at the time of L and the critical value of the statistic, which will be discussed later.

Moreover, we can use the difference between copulas, integrated over the unit cube. The obtained modified statistic (modified Cramér–von Mises statistic) will be expressed as follows:

$$T_{CM} = max_{L\in B(N,\beta)}\left\{\iint_{[0;1]^{m+1}} \Psi_L\left(u_0,\ldots u_m\right) \mathrm{d}u_0 \ldots du_m\right\} \tag{27}$$

$$K_{CM} = argmax_{L\in B(N,\beta)}\left(\iint_{[0;1]^{m+1}} \Psi_L\left(u_0,\ldots u_m\right) \mathrm{d}u_0 \ldots du_m\right),\tag{28}$$

For every L finding maximum $\Psi_L(u_0, \ldots u_m)$ and integration is carried out numerically on grid with mesh size Q for each of the axes, i.e. with nodes of type $({}^{i_0}\!/_Q; \ldots {}^{i_m}\!/_Q)$, $i_0, \ldots i_m = 0, 1, \ldots$ Q.

4 The Results of Evaluation on Generated Data

A suggested method of detecting and estimation of structural breaks in a time series was used on a generated time series with a specified pattern of dependence.

For analysis we can use six types of copulas, described in Table 1. Dependence from lagged value was generated at different levels, corresponding to Kendall's rank correlations −0.8, −0.6, −0.4, −0.2, 0, 0.2, 0.4, 0.6, 0.8. Since not all copulas can describe all of the above levels of dependence, we only used 34 copulas.

Table 1 Generated copulas

Copula	Value of rank correlation
Clayton	−0.8, −0.6, −0.4, −0.2, 0.2, 0.4, 0.6, 0.8
Frank	−0.8, −0.6, −0.4, −0.2, 0.2, 0.4, 0.6, 0.8
Gumbel	0, 0.2, 0.4, 0.6, 0.8
Product	0
FGM	−0.2, 0, 0.2
Plackett	−0.8, −0.6, −0.4, −0.2, 0, 0.2, 0.4, 0.6, 0.8

For a fixed length of time series N = 1,000 and for every three parameters of moment of structural break $\theta = 0.3$, $\theta = 0.5$, $\theta = 0.7$ where $\theta = {}^{m}/_{N}$, m-observation, in which structural break occurs, we generated a time series for all possible pairs of copulas from the 34 described above. In total we turned 3*34*34 = 3,468 various time series. In 3*34 = 102 of them, the copula before structural break are the same as after, i.e. there is no structural break. In 3*74 = 222, rank correlation changed without changing of copula. In 3*210 = 630, rank correlation didn't changed but copula changed. In 3*838 = 2,514 changing occurred both in rank correlation and copula. For every time series we calculated the suggested statistics of Kolmogorov–Smirnov and Cramér–von Mises and Andrews–Zivot test (all three modifications). For nonlinear statistics, values were found on a uniform grid of a unit square of dimension of 50 at 50 nodes.

The Andrews–Zivot test, in general, did not lead to accurate results. The null hypothesis of this unit root test (which means rejection structural break existence) was not rejected only in two of 3,468 cases. In other time series tests, statistics have pointed to existence of a structural break in the beginning or end of the sample, and this result didn't depend on whether or at what point there was a structural break. Regardless, it is the specification of a structural shift test (in trend, intercept, or both).

In Fig. 1, histograms of Andrews–Zivot statistic maximum's distribution are introduced (for every moment of break (in 300-th, 500-th and 700-th observation of 1,000) and every specification of the test).

For nonlinear copula-based tests, the same trend of large quantity of false signals about the structural break in both ends of sample are observed only with small enough changes in rank correlation, and more for Kolmogorov–Smirnov statistic. Overall, the Cramér–von Mises statistic gives more acceptable results. With the Kolmogorov–Smirnov statistic, the anomaly of high values in edges occurs more often. Using a grid with larger number of nodes decreases this effect, but doesn't

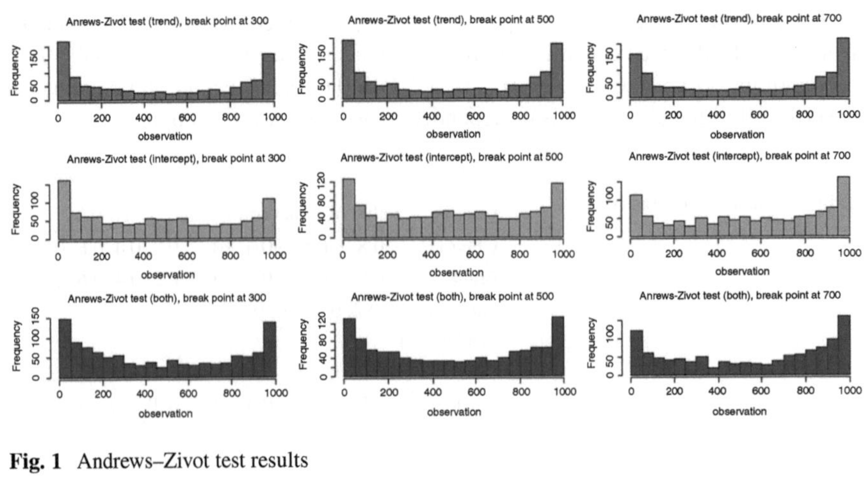

Fig. 1 Andrews–Zivot test results

eliminate it, so the values of statistics have been adjusted for otherwise than suggested by Brodsky et al. (2009). The correction factor equals the squared original factor, which means more weight for values from the middle of sample.

This a reasonable compromise: decreasing the probability of receiving false signals about structural breaks at the edges, thereby decreasing the chance to determine if an actual shift occurred in the same place. Due to the small number of observations, the statistic will show significantly less accurate approximations of real dependence. Moreover, in time series without structural breaks, the large values in the beginning or end of the row also have been observed, so the statistic's maximum in that range couldn't be interpreted as an indicator of structural break for small changes in rank correlation.

As an example, maximum's histograms of Kolmogorov–Smirnov and Cramér–von Mises statistic for structural breaks in the 700th observation and different rank correlation changes are introduced in Figs. 2 and 3. The statistics do not count large values in the first and last 5 % of observations.

Critical values in absolute scale for different pairs of copulas differ slightly, which lets us calculate critical values as corresponding quantiles of the statistic's values sample in all different points of time and all possible combinations of copulas before and after the structural break. Values are calculated for different levels in rank correlation changes, and applied in time series generation: there are only 9 values from 0 to 1.6 increments by 0.2. Additionally, some calculations have been performed by the same scheme for time series of length N = 250 and N = 500 observations. We traced the same character of revealed results, and concluded that with equal change in rank correlation, the statistic value is bigger as a rule if there is a change in copula. Critical values were calculated separately for observations with and without change in copula for significance levels of 90, 95 and 99 %. Obtained critical values were larger than found by Brodsky et al. (2009) approximately two to three times. This explains the considerable difference in the problem statement,

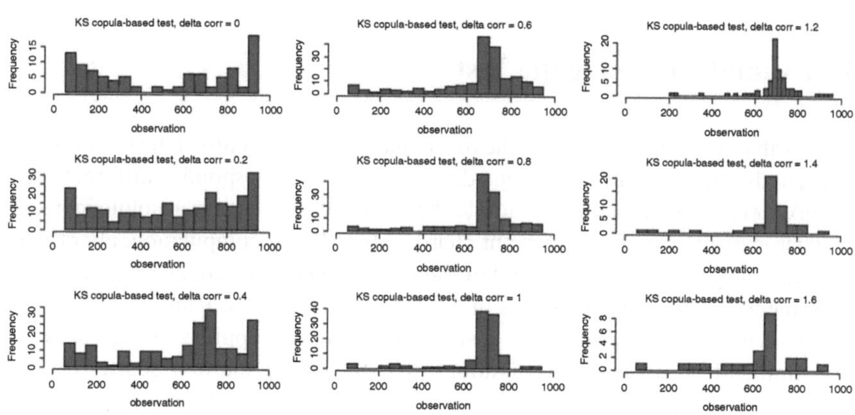

Fig. 2 Kolmogorov–Smirnov test results

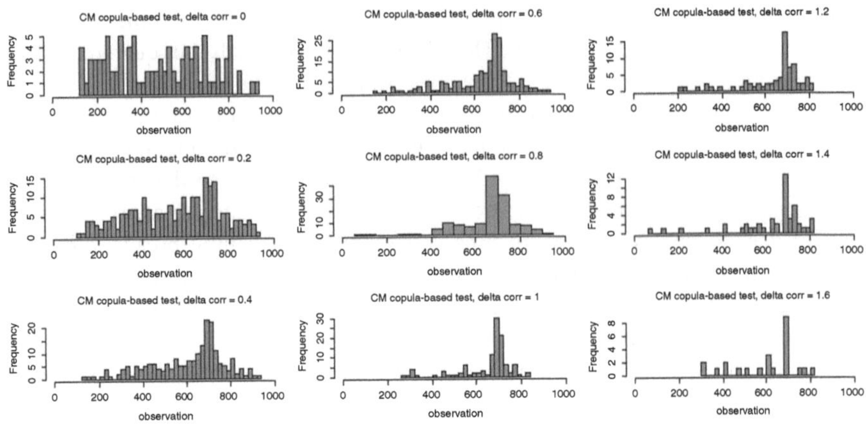

Fig. 3 Cramér–von Mises test results

because now the components of the multi-dimensional vectors of observations are not independent.

These values reveal one interesting fact: for almost all levels of rank correlation, change corresponds to a bigger critical value, but for zero this value is large enough.

Thus, the model procedure of structural break detecting is as follows: first, determine the maximal value of the test statistic and corresponding observation. Then estimate the difference between Kendall's tau of current and lagged values before and after the suspected moment of structural. According to the critical value of the nearest rank correlation change from the critical values tables and for given significance level, we can make a decision about existence of structural break. If the critical value is exceeded, the moment of the statistic's maximum is taken as a estimation of structural break point.

5 Examples of Using the Test

It is worth noting that, in general, the result may significantly affect the use of data in levels or differences, since levels and differences often correspond to different levels of rank correlation. In both the examples below, the data is used in absolute increases of what is displayed on the relevant charts. Also, during computation, abnormally high values in the beginning or end of the series are not noticed, allowing us to use the same weight for observation as in Brodsky et al. (2009) or Penikas (2012).

Following Penikas (2012), the test statistic has been applied to determine the structural shift to quarterly observations of U.S. GDP in the first quarter of 1947 to the second quarter of 2012. There are a total of 262 observations and 261 observations for differences. Tests were made for time series in differences (absolute

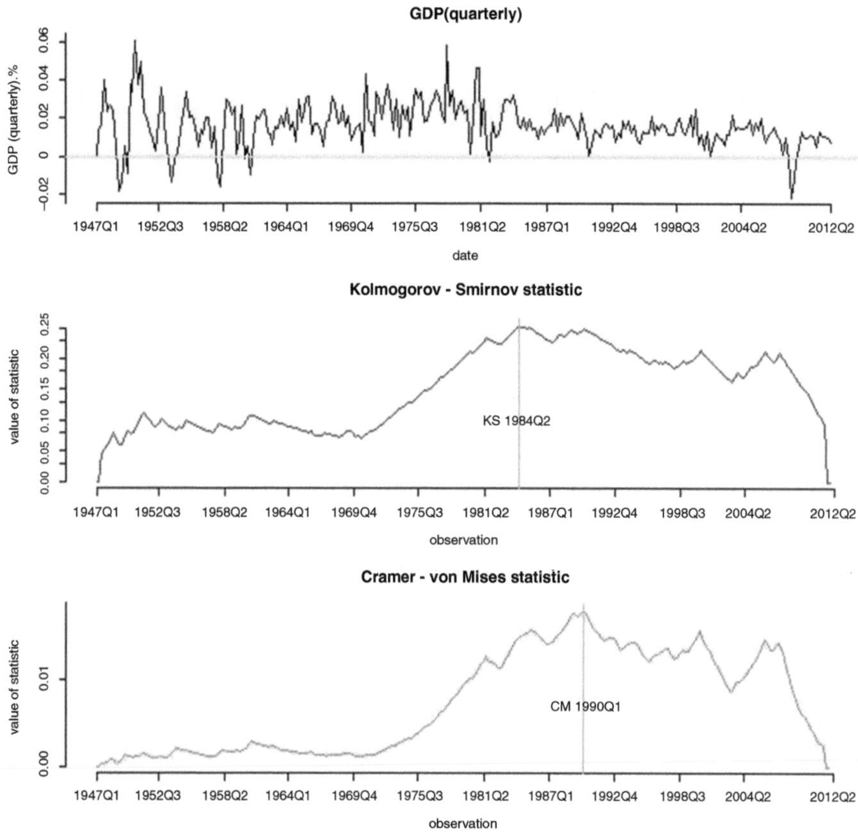

Fig. 4 Copula-Based tests: GDPUSA

increases). In Fig. 4, we present the graphs and Kolmogorov–Smirnov and Cramér–von Mises statistics.

We can see that the two statistics vary in results: Kolmogorov–Smirnov statistic takes the maximum value of 0.2546 in 150th observation, which corresponds to the second quarter of 1982. Kendall's rank correlations between current and lagged observation before and after 150th observation are respectively 0.290 and 0.251; Cramér–von Mises' statistic takes the maximum value of 0.01794 at 173th observation 173, which corresponds to the first quarter of 1990; with corresponding rank correlations 0.278 and 0.243. Two statistics give different results, showing the same behavior. Changes in rank correlation at the supposed structural break points are 0.039 and 0.035 respectively. Therefore, to detect structural break, critical values for length $N = 250$ and zero change in rank correlation should be applied. After comparing obtained values with the table's critical numbers, we could ascertain existence of a structural break with or without change in copula at significance levels

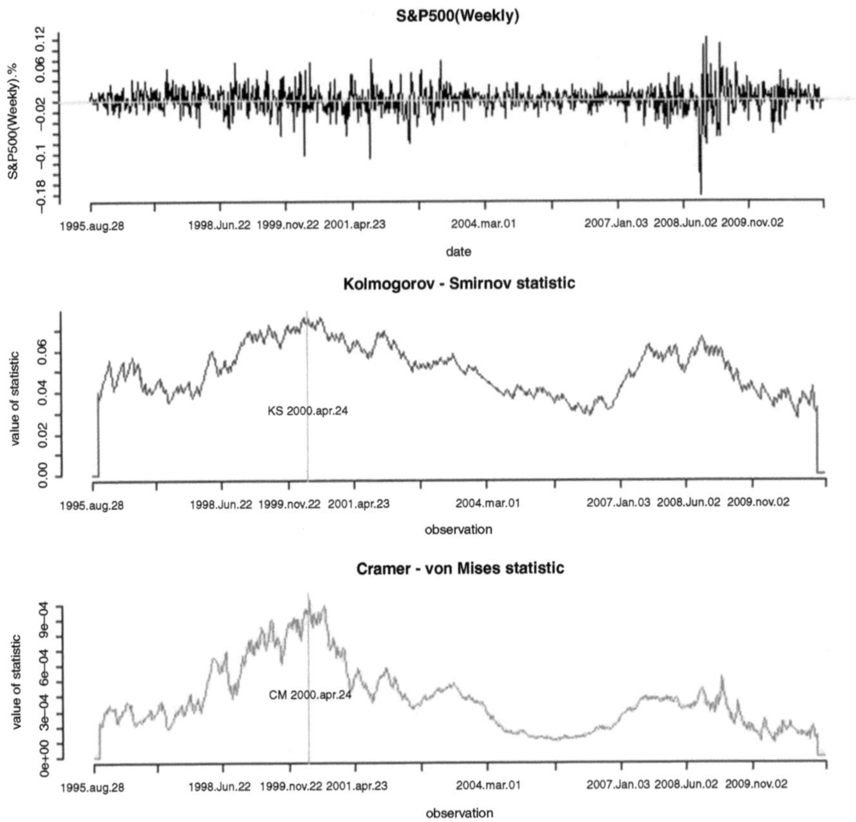

Fig. 5 Copula-Based tests: S&P 500

of 10 and 5 %, but at significance level of 1 %, structural breaks are not revealed (however, estimations of structural break's moments are different for two statistics).

The test procedure was also applied to 823 observations of the S&P 500 stock index, based on market capitalization of the 500 largest public companies traded in USA stock markets.

Data represents weekly closing values of index from the 17th of August, 1995 to 30th of May, 2011 (source: http://finance.yahoo.com/). Results are presented in Fig. 5.

Both statistics attain the maximum at the 244th observation (0.07623 and 0.010328 respectively for Kolmogorov–Smirnov and Cramér–von Mises), which corresponds to the 24th of April, 2000. Rank correlations of current and lagged values before and after the 244th observation are −0.1216 and −0.0317; difference is 0.0899, which corresponds to zero change in rank correlation. Critical values for N = 1,000 were used, and results of the test procedure do not reveal existence of a structural break for this observations of time series for all statistics and all significance levels, whether or not the copula changes. That corresponds with the findings of Patton (2012), whose method also didn't detect a structural break in the same data, thus assuming that the structural break's moment is unknown (we used the test for structural break in rank correlation, based on a parametric bootstrap).

Conclusions
The main contribution of this paper is the formulation and simulation study of the use of a method of structural break detection in the time series, based on nonparametric estimates of a dependence copula of the time series. This study found that in the case of non-linearity of the structure of dependence, our approach gives more reliable results—in contrast to the tests on the structural changes that involve only linear dependence on lagged values. This work made some corrections to the calculation of statistics. It is shown that the accuracy of the method depends on the amount of change of rank correlation with structural changes, as well as the change or non-change of the copula depending upon the structural shift. We formulated a general algorithm for the detection procedure of the structural shift based on the results, and also provided examples of using this algorithm. Further work can be done on a number of extensions, including a generalization of structural change regarding the marginal distribution of the time series, using a broader class of marginal distributions and copula, and using alternative assessment methods copula. This could include other types of non-parametric estimates and the Bayesian approach, using the Monte Carlo Markov chain method.

References

Darsow, W. F., Nguyen, B., & Olsen, E. T. (1992). Copulas and Markov processes. *Illinois Journal of Mathematics, 36*, 600–642.

Ibragimov, R. (2009). Copula-based characterizations for higher order Markov processes. *Econometric Theory, 25*, 819–846.

Penikas, H. (2010). Financial applications of copula-models. *Journal of the New Economic Association, 7*, 24–44.

Brodsky, B., Penikas, H., & Safaryan, I. (2009). Detection of structural breaks in copula models. *Applied Econometrics, 16*, 3–15.

Penikas, H. (2012, preprint) Copula-based univariate time series structural shift identification test.

Azam, K. (2012). Bayesian inference for a semi-parametric copula-based Markov chain. Job Market Paper.

Patton, A. J. (2012). A review of copula models for economic time series. *Journal of Multivariate Analysis Journal of Multivariate Analysis, 110*, 4–18.

Omelka, M., Gijbels, I., & Veraverbeke, N. (2009). Improved kernel estimation of copulas: Weak convergence and goodness-of-fit testing. *Annals of Statistics, 37*, 3023–3058.

Modeling Demand for Mortgage Loans Using Loan-Level Data

Evgeniy Ozhegov

Abstract This paper is concerned with modeling the demand for mortgage loans. The demand for loans can be represented as two functions: probability of borrowing and the loan amount, depending on borrower-specific characteristics, contract terms and set of macrovariables. The decision-making process for borrowing can be described as the sequence of decisions on: (1) choosing the credit program; (2) approving of a borrower; (3) choosing contract terms from a feasible set; (4) and loan performance. The author proposes an econometric approach that deals with endogeneity and self-selection of borrowers when estimating the demand-for-loan equations and specifies the structure of data that is required for implementation.

Keywords Demand for loans • Endogeneity • Sample selection

JEL Classification C31, D12, D14, G21

1 Introduction and Literature Review

Demand for loans in general, and for mortgage loans in particular, is the function of the probability of a credit contract agreement and of credit contract terms based on characteristics of the borrower, the goal of crediting, expected loan performance and some macroeconomic variables.

Econometric estimation of parameters of these functions face inconsistency driven by endogeneity and sample selection. Endogeneity is generated by simultaneity in borrower and credit organization decisions on explanatory variables in demand equations. A sample selection arises when the decision-making process of borrowing is made sequentially and some explanatory variables are partially observed in different stages of crediting.

However, these challenges in estimation process have not been addressed by recent papers that studied the crediting process. Mortgage borrowing as a sequence

E. Ozhegov (✉)
National Research University Higher School of Economics, Research Group for Applied Markets and Enterprises Studies, Perm, Russia
e-mail: tos600@gmail.com

A.K. Bera et al. (eds.), *Financial Econometrics and Empirical Market Microstructure*, DOI 10.1007/978-3-319-09946-0_16

of consumer and bank decisions was introduced by Follain (1990). He defines the borrowing process as a choice of how much to borrow (the Loan-To-Value ratio decision), if and when to refinance or default (the termination decision), and the choice of mortgage instrument itself (the contract decision). Rachlis and Yezer (1993) then suggested a system of four simultaneous equations for mortgage lending analysis: (1) borrower's application, (2) borrower's selection of mortgage terms, (3) lender's endorsement, and (4) borrower's payment according contract or default.

Phillips and Yezer (1996) compared the estimation results of the single-equation approach with those of the bivariate probit model. They showed that discrimination estimation is biased if the lender's rejection decision is decoupled from the borrower's self-selection of loan programs, or if the lender's underwriting decision is decoupled from the borrower's refusal decision.

Ross (2000) studied the link between loan approval and loan default and found that most of the approval equation parameters have the opposite sign, compared with the same from the default equation after correction for the sample selection.

Previous models that tackled sample selection bias in lending analysis are not appropriate to estimate the loan amount or LTV ratio. The probit model of Ross (2000) and bivariate probit model used by Philips et al. (1994) and Philips and Yezer (1996) are suitable for estimating a binary outcome. The following papers studied the dependence of the decision on loan amount as well as different endogenous variables on the exogenous ones.

Ambrose et al. (2004) constructed a simultaneous equation system of LTV and house value, which is used as a proxy for loan amount to account for endogeneity. Bocian et al. (2008) used three-stage Least Squares for the simultaneous decisions on pricing and credit rating and found empirical evidence that non-white borrowers are more likely to receive higher-priced subprime credit than similar white borrowers. Zhang (2010) investigated the sample selection bias and interaction between pricing and underwriting decisions using the standard Heckman model.

Other literature on mortgage choice has focused on the optimal mortgage contract, given uncertainty about future house prices, household income, risk preferences, and, in some papers, mobility risk. Leece (2001) found the choice between ARM and FRM in the UK market dependent on the expected level of rates. Thus, with sustainable low interest rates, a household intends to lock into a FRM. In order to construct consistent and unbiased estimates, he used a linear additive model with time-dependent explanation variables.

Campbell and Cocco (2003) examine household choice between FRM and ARM in an environment with uncertain inflation, borrowing constraints, and income and mobility risk. They demonstrate that an ARM is generally attractive, but less so for a risk-averse household with a large mortgage, risky income, high default cost, or low probability of moving. Coulibaly and Li (2009), using survey data, also found evidence that borrowers who were more risk-averse, with risky income and low probability of future move prefer fixed rate mortgage contracts.

Forthowski et al. (2011) studied the demand for mortgage loans from the point of choosing an ARM versus FRM as a function of expected mobility. They find that, with all else equal, those who choose ARM estimate their probability of moving in the future as relatively high.

Firestone et al. (2007) analyzed the prepayment behavior of low- and moderate-income (LMI) borrowers. Using the data containing the performance of 1.3 million loans originated from 1993 to 1997 they found that lower-income borrowers prepay more slowly than with higher income and this results are stable over time. Courchane (2007) studied differences in pricing for different ethnicities after controlling for other pricing and underwriting parameters. LaCour-Little (2007) also focused on the question of choosing a credit program among LMI borrowers. Using the loan level data from only one financial organization, he founds that LMI borrowers are more likely to choose Federal Housing Administration-insured mortgage programs and special programs that assumed less down payments and higher scores of expected risks due to high levels of current debt or weaker credit history. He also found that nonprime loans were preferred for those borrowers who are time-limited in providing full documentation.

Some recent papers discussed the theoretical framework of optimal mortgage contraction. For instance, Nichols et al. (2005) showed that rejection rates vary directly with interest rates in the mortgage market and inversely in the personal loan market. Theoretically they demonstrated that the discrete levels of mortgage credit supply and the positive relationship between interest and rejection rates arise from a separating equilibrium in the mortgage market. This separation does rely on the simple observation that processing an application through the underwriting process is costly, and is only partially covered by the application fee. When a subprime lender tries to locate too closely (in credit risk space) to prime lenders, the application costs overwhelm credit losses to the point where it is less costly to lower credit standards and accept a higher proportion of applicants. Equilibrium requires that the subprime lender move a substantial distance from prime lenders, thus leading to a discrete and segmented mortgage market of those borrowers who may apply for prime mortgages and for those who apply for subprime mortgages.

Ghent (2011) discussed the dynamic demand for mortgage loans and steady state equilibrium for borrowers with hyperbolic, compared to exponential, discounting, and the preference of such borrowers on the set of traditional fully amortizing mortgages and no-down-payment mortgages. The main findings of this paper was that young households and retirees are more likely to choose NDP mortgages that arise when those households behave hyperbolically and the age of borrower also explains decision-making process.

Piskorski and Tchistyi (2010, 2011) follow DeMarzo and Sannikov (2006) and pose the theoretical model of choosing the optimal mortgage contract that maximizes both lender's and borrower's combined surplus. These papers provide a prediction of higher default rates for adjusted rate mortgages when the interest rate increases but shows that, nevertheless, ARM is an optimal mechanism for mortgage contraction.

Karlan and Zinman (2009) found a different method to solve the endogeneity problem when modeling the loan amount equation. They generated a truly random sample of credit proposals by sending letters to former borrowers. Using a simple

Heckman model, they estimated the elasticities of demand for consumer credits to maturity and interest rates for different risk types of borrowers.

Attanasio et al. (2008) introduced a more progressive approach of managing the sample selection problem when modeling the empirical demand for a loan equation. They studied the existence of credit constraints in different income segments. Using loan-level data of car loans, they found that low-income households have positive elasticity of demand for car loans on the maturity and zero reaction of demand to interest rate change. This means that those households have credit constraints. Attanasio et al. (2008) used a three-stage estimation methodology. At the first stage, they estimated the participance equation. At the second stage, the endogenous variables equations were estimated by semi-parametric regression with correction for self-selection. Then endogenous variables in the demand equation were replaced by fitted values and the parameters were estimated also by semi-parametric regression. The only motivation of using semiparametric regression is that the error terms of the loan amount, endogenous variables error terms and error term from the participation equation are correlated in a non-linear way.

The main contribution of this paper is construction of a structural and econometric model that can provide consistent estimates of the demand-for-loan function, using loan-level individual data.

2 Structural and Econometric Model

Demand-for-mortgage function can be represented by the following equation:

$$ln\ L = \beta_L D + \gamma_L C + \delta_L F + \psi_L P + \mu_L M + e_L \tag{1}$$

where L is usually the loan amount (or LTV ratio), D are socio-demographic characteristics of borrower, C are the contract terms, F are specific variables that describe property, P are contract performance characteristics, and M are macroeconomic and financial variables. All of them can be divided as endogenous and exogenous ones, as described in Table 1.

The borrowing process can be represented by the following sequence of decisions:

1. Application of borrower. Potential borrower realizes the necessity of borrowing, chooses the credit organization and credit program that match her preferences, and fills out an application form with demographic characteristics.
2. Approval of borrower. Considering the application form and recent credit history, the credit organization endorses the application or not, inquires about the form data and set the limit of loan amount when endorsed.
3. Choice of credit terms. The approved borrower makes a choice on contract agreement and, when agreed, on property to buy and credit terms from a feasible

Table 1 Explanatory variables in demand equation

Variables	Endogenous for borrower	Endogenous for credit organization	Exogenous
Contract terms	Down payment; maturity; annual payment; date of contract agreement; program choice (ARM/FRM, prime/nonprime, conventional/special/FHA programs); self-selection for participation in mortgage	Loan limit; program parameters (minimum down payment, maximum maturity)	Program parameters (interest rate, insurance, Government Subsidied Enterprises); cost of application
Socio-demographic characteristics	Number of co-borrowers; aggregated income of co-borrowers; aggregated expenses of co-borrowers; income of borrower; providing of full documentation	Probability of creditworthness (FICO score of riskiness); flag of endorsement	Expenses of borrower; age; number of children; marriage status; level of education; parameters of job; nationality/race; expected mobility; recent credit history
Desired property	Value		Specification the property
Loan performance	Month of first delinquency; date of first delinquency; flag of delinquency; default, refinancing, prepayment		Loss given default
Macro- and financial variables			Yield on treasury notes; refinancing rate; volatility of interest rate; unemployment rate; volume of new construction

set: approved loan amount, down payment, annual payment, rate and maturity determined by the credit program.

4. Loan performance. Borrower chooses the strategy of loan payment: to pay in respect to contract terms or to default, prepay or refinance the loan.

Econometric model repeats the steps of structural one:

1. Using instrumental variables for endogenous demographic characteristics:

$$D^{en} = Z_D \beta_D + e_D, \tag{2}$$

where D^{en} is a vector of endogenous socio-demographic characteristics, Z_D are instrumental variables for demographics.

2. Modeling the probability of application:

$$y_1 = \begin{cases} 1, \ if \ D\beta_D^1 + M\beta_M^1 + e_1 \geq \alpha_1 \\ 0, \ if \ D\beta_D^1 + M\beta_M^1 + e_1 < \alpha_1, \end{cases} \tag{3}$$

where $y_1 = 1$ is an application decision, $D = \left(D^{ex}, \widehat{D}^{en}\right)$ is a vector of exogenous demographics and fitted endogenous demographics, M—macrovariables.

3. Modeling the probability of approval for all applied:

$$\left(y_2 \middle| y_1 = 1\right) = \begin{cases} 1, \ if \ D\beta_D^2 + M\beta_M^2 + e_2 \geq \alpha_2 \\ 0, \ if \ D\beta_D^2 + M\beta_M^2 + e_2 < \alpha_2 \end{cases} \tag{4}$$

where $y_2 = 1$ is an approval decision.

4. Choice of loan amount limit for all endorsed:

$$\left(\overline{L} \middle| y_2 = 1\right) = D\beta_D^{\overline{L}} + M\beta_M^{\overline{L}} + e_{\overline{L}} \tag{5}$$

where $\widehat{L}$ is a decision on loan limit.

5. Modeling the probability of contract agreement:

$$\left(y_3 \middle| y_2 = 1\right) = \begin{cases} 1, \ if \ D\beta_D^3 + M\beta_M^3 + \widehat{L}\beta_{\overline{L}}^3 + e_3 \geq \alpha_3 \\ 0, \ if \ D\beta_D^3 + M\beta_M^3 + \widehat{L}\beta_{\overline{L}}^3 + e_3 < \alpha_3 \end{cases} \tag{6}$$

where $y_3 = 1$ is an agreement decision; $\widehat{L}$ is a fitted value of loan amount limit.

6. Choice of credit terms and property:

$$\begin{cases} \left(C_1 \middle| y_3 = 1, C \in \overline{C}\right) = D\beta_D^{C_1} + M\beta_M^{C_1} + C_{-1}\beta_{C_{-1}}^{C_1} + F\beta_F^{C_1} + e_{C_1} \\ \qquad \cdots \\ \left(C_k \middle| y_3 = 1, C \in \overline{C}\right) = D\beta_D^{C_k} + M\beta_M^{C_k} + C_{-k}\beta_{C_{-k}}^{C_k} + F\beta_F^{C_k} + e_{C_k} \end{cases} \tag{7}$$

where $C = (C_i, C_{-i})$ is a vector of contract terms (LTV, annual payment, maturity, interest rate), $\overline{C}$ is a feasible set of contract terms determined by credit program, F is property characteristics.

7. Modeling the probability of contract events and loss given credit event:

$$\begin{cases} \left(y_4 \middle| y_3 = 1\right) = j, \ if \ \ D\beta_D^4 + M\beta_M^4 + \widehat{C}\beta_C^4 + U_j\beta_{U_j}^4 + e_4 \in \mathcal{L}_j \\ \qquad\qquad\qquad \left(U_j \middle| y_4 = 1\right) = M\beta_M^4 + \widehat{C}\beta_C^4 + e_{U_j} \end{cases} \tag{8}$$

where $y4 = j$ is a fact of j-th credit event, $\widehat{C}$ are fitted values of credit terms, U_j is a loss given j-th event.

Conclusion and Discussion

The proposed modelcan take care of endogeneity problem caused by simultaneity by instrumenting and fitting endogenous explanatory variables using a multistage estimation procedure.

Inconsistency of estimates due the sample selection will be released by introduction and estimation of the bias terms in outcome equations. Effectiveness of this correction depends on accuracy of assumptions about distribution of error terms in selection equations. Thus, it is appropriate to use inverse Mills ratio in outcome equations when selection equation terms are normally distributed. More general assumptions about the error term distributions can be achieved through the use of semi-parametric methods for correction for sample selection bias. But these estimates will be less effective in terms of standard errors.

Questions could be raised about the rationality of borrower and credit organizations' decisions. Sequential estimation procedures like the multivariate probit or multistage Heckman procedure, make no assumptions about rationality of agents. We use partially observed data in selection equations to consider lack of borrower's ability to predict decisions made by her and the credit organization. Full rationality of agents assumes that a borrower in every stage of the decision-making process can predict outcomes of next stages, and this prediction affects her present choice. A model of the fully rational borrowing process should contain fitted predictions on future outcomes as explanatory variables in all Eqs. (2)–(8) which should be estimated as a system of simultaneous equations. This strategy is very complex for estimation purposes because of the discrete and continuous variable equations compounded by the sample selection problems.

Acknowledgements This study (research grant no. 14-01-0104) was supported by The National Research University–Higher School of Economics' Academic Fund Program in 2014–2015.

References

Ambrose, B., LaCour-Little, M., & Sanders, A. (2004). The effect of conforming loan status on mortgage yield spreads: A loan level analysis. *Real Estate Economics, 32*, 541–569.

Attanasio, O. P., Goldberg, P. K., & Kyriazidou, E. (2008). Credit constraints in the market for donsumer durables: Evidence from micro data on car loans. *International Economic Review, 49*, 401–436.

Bocian, D., Ernst, K., & Li, W. (2008). Race, ethnicity and subprime home loan pricing. *Journal of Economics and Business, 60*, 110–124.

Campbell, J. Y., & Cocco, J. (2003). Household risk management and optimal mortgage choice. *The Quarterly Journal of Economics, 118*, 1449–1494.

Coulibaly, B., & Li, G. (2009). Choice of mortgage contracts: Evidence from the survey of consumer finance. *Real Estate Economics, 37*, 659–673.

Courchane, M. (2007). The pricing of home mortgage loans to minority borrowers: How much of the APR differential can we explain? *Journal of Real Estate Research, 29*, 399–439.

DeMarzo, P. M., & Sannikov, Y. (2006). A continuous-time agency model of optimal contracting and capital structure. *Journal of Finance, 61*, 2681–2724.

Firestone, S., Van Order, R., & Zorn, P. (2007). The performance of low-income and minority mortgages. *Real Estate Economics, 35*, 479–504.

Follain, J. R. (1990). Mortgage choice. *AREUEA Journal, 18*(2), 125–144.

Forthowski, E., LaCour-Little, M., Rosenblatt, E., & Yao, V. (2011). Housing tenure and mortgage choice. *Journal of Real Estate Finance and Economics, 42*, 162–180.

Ghent, A. (2011). *Subprime mortgages, mortgage choice, and hyperbolic discounting*. Working paper. Zicklin School of Business, Baruch College.

Karlan, D., & Zinman, J. (2009). Observing unobservables: Identifying information asymetries with a consumer credit field experiment. *Econometrica, 77*, 1993–2008.

LaCour-Little, M. (2007). The home purchase mortgage preferences of low- and moderate-income households. *Real Estate Economics, 35*, 265–290.

Leece, D. (2001). Regressive interest rate expectations and mortgage instrument choice in the United Kingdom housing market. *Real Estate Economics, 29*, 589–613.

Nichols, J., Pennington-Cross, A., & Yezer, A. (2005). Borrower self-selection, underwriting costs, and a subprime mortgage credit supply. *Journal of Real Estate Finance and Economics, 30*, 197–202.

Piskorski, T., & Tchistyi, A. (2010). Optimal mortgage design. *Review of Financial Studies, 23*, 3098–3140.

Piskorski, T., & Tchistyi, A. (2011). Stochastic house appreciation and optimal subprime lending. *Review of Financial Studies, 24*, 1407–1446.

Philips, R., & Yezer, A. (1996). Self-selection and tests for bias and risk in mortgage lending: Can you price the mortgage if you don't know the process? *Journal of Real Estate Research, 11*, 87–102.

Philips, R., Trost, R., & Yezer, A. (1994). Bias in estimates of discrimination and default in mortgage lending: The effects of simultaneity and self-selection. *Journal of Real Estate Finance and Economics, 9*, 197–215.

Rachlis, M., & Yezer, A. (1993). Serious flaws in statistical tests for discrimination in mortgage markets. *Journal of Housing Research, 4*, 315–336.

Ross, S. L. (2000). Mortgage lending, sample selection and default. *Real Estate Economics, 28*, 581–621.

Zhang, Y. (2010). *Fair lending analysis of mortgage pricing: Does underwriting matter*? Working paper. Office of the Comptroller of the Currency (OCC), Economics, Working paper 2010-1.

Sample Selection Bias in Mortgage Market Credit Risk Modeling

Agatha Lozinskaia

Abstract The mortgage crisis that started in the U.S. in 2007 and lasted until 2009 was characterized by an unusually large number of defaults on the subprime mortgage market. As a result, it developed into a global economic recession and placed the stability of the world banking system in jeopardy. Therefore, the issues of credit risk modeling showed the shortcomings of the current credit risk practice. Truncation, or partial observability, and simultaneous equations bias causes sample selection bias. As a result, parameter estimates are biased and inconsistent. Firstly, we provide an overview of current approaches in the mortgage literature to control for the sample selection bias correction, such as the Heckman model and bivariate probit model with selection. Secondly, a review of the most significant mortgage studies discussing this problem is introduced. Specifically, different structural models, specific datasets and empirical results are regarded. In addition, we discuss such key credit risk determinants as borrower characteristics, terms of the mortgage contract, mortgage characteristics, and macroeconomic conditions. Finally, we conclude the discussion with possible research questions.

Keywords Credit risk • Default • Mortgage • Sample selection bias

JEL Classification C10, C34, G21

1 Introduction

Different concepts are used to measure credit risk, such as probability of default (PD); loss given default (LGD); exposure at default (EAD); maturity (M) and correlated defaults. Default is arguably more relevant to the recent subprime mortgage market collapse and related spillover effects. During the financial crisis, almost one out of ten mortgages was delinquent.

Default imposes enormous costs on all market participants. First, there are credit organizations and the Institute of the Mortgage Insurance Development (Russian

A. Lozinskaia (✉)
National Research University Higher School of Economics, Perm, Russia
e-mail: AMPoroshina@gmail.com

A.K. Bera et al. (eds.), *Financial Econometrics and Empirical Market Microstructure*, DOI 10.1007/978-3-319-09946-0_17

stock life insurance company "AIGK"). The latter company insures a borrower's liability and financial risks of creditors. Second, a defaulted borrower at a minimum meets the cost of moving and damages the borrower's credit score, making it difficult to buy another house and forcing a period of rental occupancy. In addition, a lower credit score seriously restricts access to credit approval in the near future. Finally, default is associated with additional psychic costs (Guiso et al. 2009).

Therefore, default modeling is an essential element of a risk management system in any credit organization. However, the notion of mortgage default has not yet been incorporated in the Russian legislation.

Usually, estimation results of default are obtained from a single-equation model, which allows for an important inference about the credit risk and key determinants of it. Moreover, test discrimination in the credit underwriting process plays a significant role in the mortgage supply decision. However, such estimates could be biased and inconsistent due to a sample selection bias. It leads to misunderstanding and misinterpretation of the obtained results.

In the first section, we analyze some widely used econometric models for credit underwriting and default processes, and focus on the sample selection bias problem. The second part reviews mortgage literature that discusses the problems of credit risk modeling and the sample selection bias. Then we discuss key credit risk determinants and conclude with main research questions and suggestions for further empirical work.

2 Econometric Models for Credit Underwriting and Default

Traditional credit risk models on the mortgage market employ a parametric approach to estimate regression of the default probability. These are classical binary choice models (probit and logit).

The idea is that we have a regression model:

$$y_i^* = x_i^{/} \beta + \varepsilon_i \tag{1}$$

where

y_i^*—a latent variable, which is not observed,
$x_i^{/}$—a vector of independent variables,
β—a vector of constant coefficients,
ε_i—error term.

We observe a dichotomous variable y_i defined by

$$y_i = \begin{cases} 1, \ if \ \ y_i^* > 0, \\ 0, \ otherwise. \end{cases} \tag{2}$$

In other words, y_i is the PD, taking the value 1 or zero. In the process of credit underwriting y_i^* would be defined as a propensity to receive approval from a credit organization.

$$y_i = \begin{cases} 1, \textit{ if the borrower is defaulted}, \\ 0, \qquad \textit{otherwise}. \end{cases} \tag{3}$$

The probit and logit models differ in the specification of distributional form of the error term ε in (2).

If it is a normal distribution, we have a probit model.

$$F\left(Z_t\right) = \int_{-\infty}^{Z_t/\sigma} \frac{1}{\sqrt{2\pi}} \exp\left(-\frac{t^2}{2}\right) dt \tag{4}$$

If errors follow logistic distribution, we have a logit model.

$$F\left(Z_t\right) = \frac{\exp\left(Z_t\right)}{1 + \exp\left(Z_t\right)} \tag{5}$$

The problem of disproportionate sampling occurs in the credit risk modeling. The number of defaulted borrowers would be much smaller than the number of non-defaulted ones. However, if we use the logit or the probit model, or even the liner probability model, the estimated coefficients are not affected by the unequal sampling rates for two groups. It is only the constant term that is affected (Maddala 1992).

Sometimes, the sample is limited by censoring or truncation. This occurs when we observe the independent variables for the entire sample, but for some observations we have only limited information about the dependent variable. Assume the censored variable y_i is defined as

$$y_i = \begin{cases} y_i^*, \textit{ if } y_i^* > 0, \\ 0, \textit{ if } y_i^* \leq 0. \end{cases} \tag{6}$$

The error term in (1) is normally distributed $\varepsilon_i \sim N\ (0, \sigma^2)$. This model is known as the tobit model (Tobin's probit) or a censored normal regression model (Tobin 1958). Tobin (1958) studied household expenditures on durable goods. Consumers maximize utility by purchasing durable goods under the constraint that total expenditures do not exceed income (Long 1997). A similar case is expenditures on credit products, like a mortgage.

In the case of the truncated regression model, we have no data either y_i^* or x_i for some observations because no samples are drawn if y_i^* is below or above a certain level τ (Maddala 1992).

However, all above-listed models suffer from a sample selection bias, leading to biased estimation. The first reason is the sample selection bias, owing to a simultaneity bias. This problem arises when default modeling does not take into consideration the underwriting process. The decision of approval or decline of a credit application is based on the latter process. Moreover, the truncation or the partial observability causes this bias. We are faced with this issue when information about denied applicants is absent. Therefore, the magnitude of bias depends on the degree of correlation between two processes—the default process and the credit underwriting process. In addition, data completeness including credit history is included in the dataset of defaulted and no-defaulted borrowers' affects on the bias (Ross 2000).

For a long time, the lack of public available mortgage data obstructed the implementation of studies on credit risk modeling for the mortgage market. For this reason, in earlier works small data from private sources of information were used. Modeling credit risk in the mortgage market with the correction for the sample selection bias was complicated because of lack of information about both groups of applicants—approved and declined. Furthermore, to solve the issue of sample selection bias, the affected factors are required only on the credit underwriting process, but not on the probability of default in the sample.

The pioneer works of Heckman (1976, 1979) propose the self-selection model, which has an endogenous discrete variable. It generalizes the tobit and truncated regression models by explicitly modeling the mechanism that selects observations as being censored or uncensored (Long 1997).

The aim is to estimate the model:

$$y_i = x_i^{/}\beta + \varepsilon_i \tag{7}$$

Assume that y_i^* is observed not when y_i^* exceeds particular threshold τ as in the tobit and truncated regression models, but based on the value of a second latent variable z_i^*

$$z_i^* = w_i^{/}\alpha + u_i \tag{8}$$

where

z_i^*—a second latent variable, which is not observed,
$w_i^{/}$—a vector of independent variables, which can have $x_i^{/}$ variables in common,
α—a vector of constant coefficients,
u_i—error term.

For now, we assume that y is observed only when unobserved latent z_i^* variable exceeds a particular threshold τ. In a simple case τ = 0. Otherwise, y is unobserved.

$$y_i = \begin{cases} x_i^{/}\beta + \varepsilon_i, & if\ z_i^* > 0, \\ unobserved, & if\ z_i^* \le 0. \end{cases} \tag{9}$$

Actually, we do not observe y. All we observe is a dichotomous variable z with the value of 1 if z_i^* variable exceeds a particular threshold (put it $\tau = 0$) and 0 otherwise.

$$z_i = \begin{cases} 1, \mathit{if}\ \ z_i^* > 0, \\ 0, \mathit{otherwise}. \end{cases} \tag{10}$$

As a result, we have a basic selection equation (10) and basic outcome equation (9). In the literature it is called the Heckman model, sample selection model, tobit II, or heckit model. In a special case, when $y_i = z_i$ we have the tobit model. Typically, we also make the following assumption about the distribution of, and relationship between, the error terms in (9) and (10) equations:

$$\varepsilon_i \sim \mathrm{N}\ \left(0,\ \sigma^2\right) \tag{11}$$

$$u_i \sim \mathrm{N}\ (0,\ 1) \tag{12}$$

$$corr\left(\varepsilon_i, u_i\right) = \rho_{\varepsilon u} \tag{13}$$

In other words, Eqs. (11)–(13) show that we assume a bivariate normal distribution with zero means and correlation $\rho_{\varepsilon u}$. The sample selection bias problem arises when estimating β in the Eq. (7) if error terms ε_i and u_i are correlated. The conditional mean equals

$$\begin{aligned} &E\left(y_i \middle| y_i = is\ \ observed\right) = E\left(y_i \middle| z_i^* > 0\right) = E\left(y_i \middle| z_i = 1\right) = \\ &= E\left[x_i^{/}\beta + \varepsilon_i \middle| w_i^{/}\alpha + u_i > 0\right] = x_i^{/}\beta + E\left[\varepsilon_i \middle| w_i^{/}\alpha + u_i > 0\right] = \\ &= x_i^{/}\beta + E\left[\varepsilon_i \middle| u_i > -w_i^{/}\alpha\right] \end{aligned} \tag{14}$$

If the errors terms ε_i and u_i are independent, then the last term in (14) simplifies to $E[\varepsilon_i] = 0$ means ε equals 0 and OLS regression of y on x in (7) will give consistent estimates of β. However, any correlation between the two errors means that we need to obtain $E[\varepsilon_i | u_i > -w_i^{/}\alpha]$ when ε_i and u_i are correlated.

Using derivations similar to the tobit model, Greene (2003) noted that

$$E\left[\varepsilon_i \middle| u_i > -w_i^{/}\alpha\right] = \rho\sigma_\varepsilon\lambda_i\left(\frac{-w_i^{/}\alpha}{\sigma_u}\right) \tag{15}$$

$$\lambda_i\left(\frac{-w_i^{/}\alpha}{\sigma_u}\right) = \frac{\phi\left(\frac{-w_i^{/}\alpha}{\sigma_u}\right)}{1 - \Phi\left(\frac{-w_i^{/}\alpha}{\sigma_u}\right)} = \frac{\phi\left(\frac{w_i^{/}\alpha}{\sigma_u}\right)}{\Phi\left(\frac{w_i^{/}\alpha}{\sigma_u}\right)} \tag{16}$$

where

λ_i—Heckman's λ or the inverse Mill's ratio.

Thus, the conditional mean in the Heckman model is:

$$E\left(y_i \middle| y_i = is\ observed\right) = E\left(y_i \middle| z_i^* > 0\right) = x_i^{/}\beta + E\left[\varepsilon_i \middle| u_i > -w_i^{/}\alpha\right]$$
$$= x_i^{/}\beta + \rho\sigma_\varepsilon\lambda_i\left(\frac{-w_i^{/}\alpha}{\sigma_u}\right) = x_i^{/}\beta + \rho\sigma_\varepsilon\left[\frac{\phi\left(\frac{w_i^{/}\alpha}{\sigma_u}\right)}{\Phi\left(\frac{w_i^{/}\alpha}{\sigma_u}\right)}\right] \tag{17}$$

The widely used way to estimate the Heckman model (7)–(13) is Heckman's two-step procedure. It involves first estimating the probit model in Eq. (10) and computing Heckman's λ

$$\hat{\lambda}_i\left(w_i^{/}\hat{\alpha}\right) = \frac{\phi\left(w_i^{/}\hat{\alpha}\right)}{\Phi\left(w_i^{/}\hat{\alpha}\right)} \tag{18}$$

where

$\phi\left(w_i^{/}\hat{\alpha}\right)$—the standard normal density,

$\Phi\left(w_i^{/}\hat{\alpha}\right)$—the cumulative density function.

and then estimating the regression of y on x and $\hat{\lambda}$. The coefficient on the λ indicates if there is sample selection bias. As a result, estimators are consistent and asymptotically normal. Alternatively, an MLE version is used to estimate a Heckman model. However, this procedure is less robust than the two-step procedure and it is sometimes difficult to get it to converge, but it will be more efficient (Wooldridge 2002).

Technically, the Heckman model is identified when the same independent variables in the selection equation (10) appear in the outcome equation (9). However, to avoid issues with identification (multicollinearity and imprecise estimates), we nearly always want at least one independent variable that appears in the selection equation but does not appear in the outcome equation (i.e., we need a variable that affects selection, but not the outcome).

The example of using a Heckman model could be the case when we model different parameters of the credit contract. However, we observe them only for clients who receive approval. It means that the selection equation is a decision of the underwriting process, and the outcome equation is a parameter of the credit contract—like the loan-to-value ratio or the loan amount, the maturity, or the contract rate.

The Heckman model is useful when the outcome equation involves a continuous dependent variable. However, when we are interested in a case where the outcome equation involves a dichotomous dependent variable, a bivariate probit model with selection (BVP with sample selection) or double probit model is used.

The simplest way is to start a bivariate probit model (BVP). The basic idea is that we have two decisions that are interrelated, such as a borrower's default decision and lender's decision in the credit underwriting process. We have two unobservable variables y_1^* and y_2^* related to two binary dependent variables y_1 and y_2.

$$y_1^* = x_1\beta_1 + \varepsilon_1 \tag{19}$$

$$y_2^* = x_2\beta_2 + \varepsilon_2 \tag{20}$$

$$y_1 = \begin{cases} 1, \text{ } if \text{ } y_1^* > 0, \\ 0, \text{ } if \text{ } y_1^* \leq 0. \end{cases} \tag{21}$$

$$y_2 = \begin{cases} 1, \text{ } if \text{ } y_2^* > 0, \\ 0, \text{ } if \text{ } y_2^* \leq 0. \end{cases} \tag{22}$$

In a bivariate probit model, we have two separate probit models with correlated disturbances ε_1 and ε_2. We typically assume that the errors are independent and identically distributed as a standard bivariate normal with correlation ρ.

$$E\left(\varepsilon_1 \middle| x_1, x_2\right) = E\left(\varepsilon_2 \middle| x_1, x_2\right) = 0 \tag{23}$$

$$Var\left(\varepsilon_1 \middle| x_1, x_2\right) = Var\left(\varepsilon_2 \middle| x_1, x_2\right) = 1 \tag{24}$$

$$corr\left(\varepsilon_1, \varepsilon_2\right) = \rho \tag{25}$$

If the errors between the two probit models are independent of one other i.e. $\rho = 0$, then we can just estimate the two probit models separately. In this case, it is possible to obtain consistent results. However, when $\rho \neq 0$, it is more efficient to estimate equations jointly. To do that, we are interested in joint probability of y_1 and y_2 in (21) and (22).

$$\Pr\left(y_{1i} = 1\right) = \Pr\left(\varepsilon_{1i} > -x_1\beta_1\right) \tag{26}$$

$$\Pr\left(y_{2i} = 1\right) = \Pr\left(\varepsilon_{2i} > -x_2\beta_2\right) \tag{27}$$

If two random variables are independent, then their joint probability is just the product of their marginal probabilities. The problem in our situation is that the two probabilities are not independent; we need to calculate joint probabilities for non-independent events. For this reason, we need to assume some joint distribution of y_1 and y_2. Usually, we use a bivariate normal distribution.

To estimate the bivariate probit model, (19)–(25) MLE is used. It can sometimes difficult to get the model to converge. A likelihood ratio test is used to test the hypothesis that the bivariate model fits data better than the separate probits.

By adding two more conditions

$$y_2 = \begin{cases} x_1\beta_1 + \varepsilon_1, \; if \; y_2^* = 1, \\ unobserved, \; otherwise. \end{cases} \tag{28}$$

$$y_2^* \; is \; observed \, for \; all \; classes \tag{29}$$

we receive the BVP model with sample selection, which is an extended version of the classical Heckman model. Similarly, the Heckman's two-step procedure is used to estimate this model.

In the first step (20), a simple binary probit or logit model of mortgage approval (the participation/selection equation) is estimated by using the full sample of approved and declined applicants. Obtained estimators are used to calculate the consistent parameter estimates of Heckman's λ (18) for every observation in the full sample.

The second step includes estimation of a simple binary probit or logit model for the probability of default (19) (the outcome equation) by using the calculated Heckman's λ. This two-stage approach corrects the sample selection bias and provides consistent and computationally efficient probabilities of default. The magnitude of bias is dependent both on the correlation between random error terms in the systems of equations, and Heckman's λ.

3 Literature Review

Since the mid-1990s, data were made publicly available, e.g. American mortgage datasets from the Federal Housing Authority (FHA) foreclosure, the Boston Fed Study, the Home Mortgage Disclosure Act (HMDA), and several studies applied Heckman's model and BVP with sample selection models.

Among these, Rachlis and Yezer (1993) demonstrates a modified Heckman model. They provide the theoretical model based on the simultaneous system, but do not determine key variables of credit risk in the mortgage lending.

$$A_i^* = K_A + X_{Ai}\beta_A + \varepsilon_{Ai}, \tag{30}$$

$$T_i^* = K_T + X_{Ti}\beta_T + \varepsilon_{Ti}, \tag{31}$$

$$E_i^* = K_E + X_{Ei}\beta_E + \varepsilon_{Ei}, \tag{32}$$

$$F_i^* = K_F + X_{Fi}\beta_F + \varepsilon_{Fi}. \tag{33}$$

where

K_A, K_T, K_E, K_F—constant terms.
X_{Ai}, X_{Ti}, X_{Ei}, X_{Fi}—matrices of observed values of independent variables,
ε_{Ai}, ε_{Ti}, ε_{Ei}, ε_{Fi}—identically and independent distributed random error terms.

It includes four latent variables A_i^*, T_i^*, E_i^*, F_i^* indicating the equation for the probability of applying for credit (30), the equation of choosing the particular terms of the mortgage contract (31) e.g., a loan-to-value ratio and term to maturity, the equation of the probability of the endorsement (32), and the equation of the probability of foreclosure, respectively. Maddala and Trost (1982) modeled the mortgage lending process in this manner.

Later on, the empirical study by Yezer et al. (1994) applied the Monte-Carlo experiment to estimate simultaneous system consisted of Eqs. (31)–(33). They assumed that the loan-to-value ratio requested by the borrower can be expressed as a function of the latent variables reflecting the lender's rejection decision, the borrower's default decision, exogenous applicant characteristics (including credit history related to creditworthiness), and demographic information. A rejection decision depended on the same variables, but an endogenous loan-to-value ratio is used instead of the lender's rejection decision. The ex ante probability of default is written as a vector of variables reflecting characteristics of the applicant, the actual property, housing prices, and lender forbearance. However, the loan-to-value after endorsement does not appear in an ex ante default equation. Yezer et al. (1994) found that isolated modeling processes of the credit underwriting and default leads to the biased parameter estimates.

Munnell et al. (1996) attempted to determine whether race discrimination affects mortgage lending decisions in the U.S. market. Both ordinary least squares and binominal logit techniques are used to estimate the probability of mortgage loan application denial depending on variables related to risk of default, cost of default, loan characteristics and personal characteristics. They showed that minorities are more than twice as likely to be denied a mortgage as whites. They used risk and cost of default, but did not model the credit underwriting process simultaneously with the borrower's default decision. However, they mentioned that there could be omitted variables, which could account for the differential treatment found in the paper's results.

Phillips and Yezer (1996) illustrated that correction for sample selection bias rejection and default models by using a BVP model with selection had substantial effect on obtained results. By using demographic characteristics of the household, income, loan amount and characteristics of mortgage, the authors showed empirically that the biased estimates were obtained as a result of isolated modeling processes of the credit underwriting and default.

Ross (2000) supported the findings of Phillips and Yezer (1996) by estimating BVP with sample selection model for the publicly available U.S. mortgage dataset. Empirical findings show that a BVP with sample selection model has higher predictive power compared with unconditional default models. Future research should attempt to investigate the credit risk regarding regional differences and to find better instruments for identifying the correlation between unobservable determinants of approval and default. Specifically, these variables are needed to explain loan approval and clearly do not affect default. For example, they could be credit market conditions at the loan origination or characteristics of the credit organization. This study concludes that if a defaulted and non-defaulted sample does not contain detailed borrower characteristics, the estimated default and the sample-selection correction model will suffer from substantial bias. The use of more borrower characteristics, including credit history and others risk factors, will directly minimize concerns about sample selection bias.

Bajari et al. (2008) applied a BVP model with partial observability, which was first studied in the pioneer work of Poirier (1980). They build the model for the net equity and the probability of default regarding four possible reasons. These reasons include a fall in house prices on the real estate market, lower expected house prices, increase in contract interest rates compared with market interest rates, and inability to pay off the mortgage due to the cash-out—specifically, pressure of income.

$$U_{1,it} = \alpha_{0i} + \frac{V_{it}}{L_{it}}(\alpha_1 + \alpha_2 Eg_{it} + \alpha_3 Vg_{it}) - (\alpha_4 IR_{it} + \alpha_5 MR_{it}) + \varepsilon_{1,it} \tag{34}$$

$$U_{2,it} = \beta_{oi} + \beta_1 Z_{it} + \beta_2 \frac{P_{it}}{Y_{it}} + \beta_3 Z_{it}\left(\frac{P_{it}}{Y_{it}}\right) + \varepsilon_{2,it} \tag{35}$$

where

$U_{1,it}$—the latent utility associated with non-defaulting,
$U_{2,it}$—the latent variable related to the budget constraint of household i at time t,
V_{it}—the value of borrower i's home at time t,
L_{it}—the outstanding principal on i's mortgage at time t,
g_{it}—the nominal rate of increase in home prices,
Eg_{it}—borrower i's expectation in period t, given her current information, about the future growth rate in home prices,
IR_{it}—interest rates,
MR_{it}—the number of months before the next rate reset for borrower i in period t (for fixed-rate mortgages $MR_{it} = 0$),
Z_{it}—a vector of covariates that serve as predictors of creditworthiness and future income,
P_{it}—monthly payment for the mortgage,
Y_{it}—income,
α_{0i}, β_{0i}—constant terms, which capture time-invariant, unobserved borrower heterogeneity in $U_{1,it}$ and $U_{2,it}$, respectively.
α_1, α_2, α_3, α_4, α_5, β_1, β_2, β_3—a vector of constant coefficients,
$\varepsilon_{1,it}$, $\varepsilon_{2,it}$—independent identically distributed error terms, which are jointly normal with a variance of 1 and a covariance of σ.

The vector Z_{it} included credit score (FICO score), whether the borrower has other mortgage loans on the property, the monthly unemployment rate at the county level, and loan characteristics that act as proxy for credit quality, such as the level of documentation for the loan application and the loan-to-value ratio at origination. g_{it} is defined as the Case–Shiller price index, corresponding to the location (MSA) and tercile of the appraised value of i's house at the date of origination.The outcome is the random variable ND_{it}, which equals 1 if household i does not default in period t and as 0 otherwise. The condition for default is as follows:

$$ND_{it} = I\,(U_{1,it} \geq 0) \times I\,(U_{2,it} \geq 0) = 0 \tag{36}$$

where

I(·)—an indicator function.

The authors used a large dataset containing information about 135,000 American mortgagors over multiple months who have 2.6 million defaults. From the data, the value of ND_{it} is observed. However, when default occurs ($ND_{it} = 0$) we do not observe whether it is because $U_{1,it} < 0$, because $U_{2,it} < 0$, or both.

Empirical findings indicate that borrower characteristics, terms of credit contract, and fundamental characteristics play important roles in explaining the default. Specifically, due to the lack of sociodemographic information at the individual level, this author includes proxy county sociodemographic information and county unemployment rate as proxy variables. Moreover, the main driver of default is the nationwide decrease in home prices. The high geographical correlation defaults the subject to the nationwide decrease in home prices.

Despite the fact that BVP models with selection allows controlling for a sample selection bias, they are strictly parameterized. The main limitation is that there is no prior knowledge about true distribution. As a result, the misspecification problem leads to misestimating and wrong inferences (Creel 2008).

Along with the problem of sample selection bias, another challenge is the problem of endogenity, which leads to inconsistent results. However, previous studies did not pay particular attention to it; they made assumptions about the exogenous nature of explanatory variables.

Moreover, most of the published works focused on the U.S. mortgage market. There are not many published works about drivers of default in the Russian mortgage market that empirically test the sample selection bias.

4 Key Drivers of Default

According to the mortgage literature, the key determinants of default initially include observable socio-demographic characteristics, terms of the mortgage contract, mortgage characteristics, and macroeconomic conditions.

Terms of a credit contract are practically used as proxy variables to estimate the risk of a particular borrower. For example, mortgages with low loan-to value ratio (LTV) are attractive for non-liquid borrowers. The probability that they could face a serious problem of repayment of a loan is much higher. Moreover, borrowers with LTVs higher than 90 % think as holders, because they do not invest a lot of their own capital and are less motivated to overcome obstacles with repayment of a loan. For this reason, mortgages with high LTV are riskier, and lenders offer higher interest rates for these mortgage products. Loans that default tend to be adjustable-rate mortgages, are associated with higher initial LTV, and tend to be issued to borrowers with lower credit scores (Bajari et al. 2008). Campbell and Cocco (2011) empirically supported this idea by using simulated data. Their findings confirm that high LTV (loan-to-value) increases the probability of default.

Typically, mortgages have two types of interest rates—adjustable (ARM) and fixed (FRM). Fixed-rate mortgages are riskier, and the level of their interest rates depends on the stock index. According to Bajari et al. (2008) borrowers tend to have higher interest rates than the market rate.

Socio-demographic characteristics such as income of a particular borrower play a significant role to predict default, because they directly influence the ability to repay a mortgage. However, the debt-to-income ratio has larger effect on borrowers with low credit quality. The level of education could be regarded as a proxy for the level of financial literacy of a particular borrower, which could influence the probability of default as well.

A new stream of mortgage studies originated from the financial turmoil of 2007–2009 that began in the U.S. Several recent empirical findings by Dell'Ariccia et al. (2012), An and Qi (2012), Demyanyk and Van Hemert (2011), Ashcraft et al. (2011), and Mian and Sufi (2009) confirm the highly statistical significance of macroeconomic conditions in explaining mortgage default. Obtained results are consistent with the notion that a relaxation of lending standards, triggered by an increased demand for loans, contributed to the boom and the ensuing crisis, together with other supply-side explanations. Specifically, such supply factors include house price appreciation and mortgage securitization (Keys et al. 2009; Dell'Ariccia et al. 2012).

In addition, Dell'Ariccia et al. (2012) concluded that the development of the mortgage market led to the classical boom-bust scenario. This concept implies fast growth with subsequent relaxation of credit underwriting standards, debt service deterioration, and drop of market premiums. The warning signal of the onset crisis of 2007–2009 was an explosion of house prices in the real estate market during 2003–2005. A fall in house prices contributes also to an increase in defaults (Gerardi et al. 2009).

The contribution of local economic conditions and change of credit underwriting standards to default are also significant (Cutts and Merrill 2008). However, changes in mortgage default rates are most sensitive to changes in the structural component rather than the level of local unemployment rate (Querica et al. 2011). In addition, Querica et al. (2012) find that mortgage default and prepayment are more sensitive to structural unemployment than cyclical unemployment.

Conclusion

The issues of credit risk modeling and analyzing the key default determinants are now at the center of the mortgage literature. Traditional models to predict the PD suffer from a sample selection bias. For this reason, advanced econometric techniques are applied, e.g. BVP model with selection. Empirical mortgage literature supports the idea that modeling the credit underwriting process and the borrower's default decision simultaneously, with control for a sample selection bias, provides consistent results.

The initial research question is to understand key drivers behind the borrowers' decision to default in the mortgage market, based on commonly observed characteristics. For this reason, a structural model on mortgage lending could be developed. It would take into account not only the probability of approval and the probability of default, but also the probability of selecting a particular credit organization and terms of mortgage. The second research question is to assess the existence and the impact of sample bias in an empirical setting such as the Russian mortgage market. Available data sets include recent observations, allowing us to focus on the drivers behind the recent wave of mortgage defaults. The level of detail in the data allows us to control for various loan terms and borrower risk factors, and thus to control for a more comprehensive list of potential drivers of default.

The structural credit risk model could be incorporated into the decision making process of credit experts regarding the mortgage market and to contribute to the development of an effective risk management system. As a result, it leads to effective allocation of capital and will be beneficial for all credit market participants.

Acknowledgements The author would like to thank Anil K. Bera, Andreas A. Woudenberg and Alexander Karminsky for their helpful comments. This study was carried out with support from "The National Research University Higher School of Economics' Academic Fund Program in 2013–2014, Research Grant No. 12-01-0130." This survey was presented at the Perm Winter School-2013. The author is responsible for any errors that remain.

References

Ashcraft, A., Goldsmith-Pinkham, P., Hull, P., & Vickery, J. (2011). Credit ratings and security prices in the subprime MBS market. *The American Economic Review, 101*(3), 115–119.

An, M. Y., & Qi, Z. (2012). Competing risks models using mortgage duration data under the proportional hazards assumption. *The Journal of Real Estate Research, 34*(1), 1–26.

Bajari, P., Chu, C. S., & Park, M. (2008). *An empirical model of subprime mortgage default from 2000 to 2007*. NBER, Working Paper 14625.

Campbell, J.Y., & Cocco, J. (2011). *A model of mortgage default*. NBE, Working Paper No. 17516.

Creel, M. (2008). Some possible pitfalls of parametric inference. *Quantile, 4*, 1–6.

Cutts, A. C., & Merrill, W. A. (2008). *Interventions in mortgage default: Policies and practices to prevent home loss and lower costs*. Working Paper 08-01. Freddie Mac.
Dell'Ariccia, G., Igan, D., & Laeven, L. (2012). Credit booms and lending standards: Evidence from the subprime mortgage market. *Journal of Money, Credit and Banking, 44*(2–3), 367–384.
Demyanyk, Y., & Van Hemert, O. (2011). Understanding the subprime mortgage crisis. *Review of Financial Studies, 24*(6), 1848–1880.
Gerardi, K., Shapiro, A. H., & Willen, P. (2009). *Decomposing the foreclosure crisis: House price depreciation versus bad underwriting*. Working Paper No. 2009-25. Federal Reserve Bank of Atlanta.
Greene, W. H. (2003). *Econometic analysis* (5th ed.). Upper Saddle River, NJ: Prentice Hall.
Guiso, L., Sapienza, P., & Zingales, L. (2009). *Moral and social constraints to strategic default on mortgages*. Working Paper No. 15145. NBER.
Heckman, J. (1976). The common structure of statistical models of truncation, sample selection, and limited dependent variables and a sample estimator for such models. *Annals of Economic and Social Measurement, 5*, 475–492.
Heckman, J. (1979). Sample selection bias as a specification error. *Econometrica: Journal of Econometric Society, 47*(1), 153–161.
Keys, B., Mukherjee, T., Seru, A., & Vig, V. (2009). Did securitization lead to lax screening? Evidence from subprime loans. *Quarterly Journal of Economics, 125*, 307–362.
Long, S. J. (1997). *Regression models for categorical and limited dependent variables*. Los Angeles, CA: Sage.
Maddala, G. S. (1992). *Introduction to econometrics*. Hoboken, NJ: Wiley.
Maddala, G. S., & Trost, P. R. (1982). On measuring discrimination in loan markets. *Housing Finance Review, 1*(3), 245–266.
Mian, A., & Sufi, A. (2009). The consequences of mortgage credit expansion: Evidence from the U.S. mortgage default crisis. *Quarterly Journal of Economics, 124*, 1449–1496.
Munnell, A., Tootell, G., Browne, L., & McEneaney, J. (1996). Mortgage lending in Boston: Interpreting HMDA data. *American Economic Review, 86*, 25–53.
Phillips, R., & Yezer, A. (1996). Self-selection and tests for bias and risk in mortgage lending: Can you price the mortgage if you don't know the process? *Journal of Real Estate Research, 11*, 87–102.
Yezer, M. J., Phillips, R. F., & Trost, R. P. (1994). Bias in estimates of discrimination and default in mortgage lending: The effects of simultaneity and self-selection. *The Journal of Real Estate Finance and Economics, 9*(3), 197–215.
Poirier, D. (1980). Partial observability in bivariate probit models. *Journal of Econometrics, 12*(2), 209–217.
Querica, R. G., Pennington-Cross, A., & Tian, C. Y. (2011). *Mortgage default risk and local unemployment*. Center for Community Capital, the University of North Carolina, Chapel Hill.
Querica, R. G., Pennington-Cross, A., & Tian, C. Y. (2012). *Differential impacts of structural and cyclical unemployment on mortgage default and prepayment*. Center for Community Capital, The University of North Carolina, Chapel Hill.
Rachlis, M. B., & Yezer, A. M. J. (1993). Serious flaws in statistical tests for discrimination in mortgage markets. *Journal of Housing Research, 4*(2), 315–336.
Ross, S. L. (2000). Mortgage lending, sample selection and default. *Real Estate Economics, 28*, 581–621.
Tobin, J. (1958). Estimation of relationships for limited dependent variables. *Econometrica, 26*(1), 24–36.
Wooldridge, J. (2002). *Econometric analysis of cross section and panel data*. Cambridge, MA: MIT.

Global Risk Factor Theory and Risk Scenario Generation Based on the Rogov-Causality Test of Time Series Time-Warped Longest Common Subsequence

Mikhail Rogov

Abstract The paper is concerned with the global risk factor theory and the Rogov-causality test of time-warped longest common subsequence for risk management purposes, including the prospects of hedging and portfolio diversification of operational risks. The author discusses the interaction of all types of risk, the role of human error and the effect of space weather (geomagnetic activity taking into account the interplanetary magnetic field (IMF) polarity). The RogovIndex© family of global risk factor indices is described as part of the risk indicator time series database. The paper discusses the apparatus of time series data, mining including hierarchical clustering based on time-warped longest common subsequence similarity (T-WLCSS). The Rogov-causality test is offered for risk scenario generation. The test involves analyzing the time-lag cumulative distribution function for the longest common subsequence of time series.

Keywords Default rate • Financial risk • Geomagnetic activity • Global risk factor • Human error • Key risk indicator • Longest common subsequence • Operational risk • Risk management • Rogov-causality • Space weather • Time series • Time warp • Volatility

JEL Classification C22, C32, C53, C81, C82, D81, G32

M. Rogov (✉)
Dubna International University, Dubna, Moscow Oblast, Russia
e-mail: rogovm@hotmail.com

A.K. Bera et al. (eds.), *Financial Econometrics and Empirical Market Microstructure*, DOI 10.1007/978-3-319-09946-0_18

1 Introduction

Modern risk management is currently experiencing an ideological crisis, showing the following symptoms:

- Failure to understand the nature of the majority of financial risks
- Eclecticism of methods and concepts, in both technologies and standards of risk management
- Disregard of the interaction between operational risk, credit risk and market risk, lack of continuity in management processes, and lack of common rating scales for the assessment of various risks
- Inadequate tools for operational risk assessment
- Virtual absence of portfolio approach to operational risk management
- Difficulties with forecasting stress and crisis scenarios generation
- Difficulties explaining the nature of chaotic market processes
- Problem of the recently increased relevance of some previously uncommon factors, of which the following ones are thought by the author to be most important:
  - cyber-terrorism and industrial terrorism
  - influence of social networks
  - High Frequency Trading (HFT)
  - threat of antibiotic resistance

The author believes that the next decades will see the development of the following branches of risk management (Rogov 2011):

- human error
- transfer of operational risks including hedging and portfolio diversification
- prediction markets
- new concepts of key risk indicator (KRI)
- risk management of small and medium enterprises (SMEs) and households
- crowdsourcing, including platforms like Ushahidi, Wiki
- new generations of publicly available risk indices
- emergence of new asset classes

The basis for the development of the global risk factor theory

- Jevons (1878), Chizhevsky (1936)
- Advent of modern heliobiology and its findings
- Findings of the sciences of human factors, human errors
- Findings of the sciences of risk management and financial mathematics
- Accumulation of statistical data (statistics of disasters, volatility, defaults other events and indices)

The following postulates can be confirmed or refuted by explaining the causal relationships and by statistical analysis.

Postulate 1. Risks Are Interrelated There are relationships between financial risks of all types (market, credit, operational).

2 The Mechanism of Causal Relationships

Risk interactions play the most important role, because of the existence of close economic, organizational and technological ties between risk owners. The occurrence of some risks (operational, credit, market ones) for some parties implies the emergence of risks for their counterparties. The subsequent chain reaction of credit and market risks propagate through exchange within the economy. In recent decades, these relations have been developing more intensively than ever before, because of market globalization and technological progress.

This causal relationship can be illustrated by a typical example of the domino effect in business environment: discontent of the local population (i.e. political risk, part of operational risks) in Nigeria led to the explosion of a pipeline operated by Royal Dutch Shell on December 21, 2005.[1] As a result, the output was cut by 180,000 barrels per day (operational risk of business interruption); the company declared "force majeure," which meant its failure to perform contract obligations (credit risks for the counterparties), and the oil price went up by 48 cents per barrel (commodity market risk).

The mechanism of a risk factor's influence on the emergence of credit and market risks can be illustrated using the well-known Merton approach (Merton 1974), the basis of the Expected Default Frequency (EDF) methodology: distance to default of a firm (i.e. credit risks of its counterparties) is determined by risks associated with the firm's operations and expressed by the volatility of the market value of the firm's assets exposed to various types of risk: operational, market, credit ones. The assets volatility determines the volatility of the market capitalization (market risks of investors).

[1]Prior to the incident, there had been a change in the storm-time variation index Dst (a space weather indicator) by 43 nT over a period from 4.00 p.m. (local time) on December 19, 2005 to 6.00 a.m. on December 20, 2005. Although this was, of course, only an increment of variation rather than an absolute value, one should keep in mind that a variation level of −50 nT is equivalent to a mild storm rated on the National Oceanic and Atmospheric Administration (NOAA) geomagnetic storm scales as a G1 (such geomagnetic storms sometimes affect the start of animal migration, cause fluctuations in electric power systems, etc.).

3 Statistical Analysis of Relationships

Correlation and co-integration of market and credit risks are well known and can be explained by changes of risk premium; however, relationships of these risks with various operational risks cannot be adequately explained without identifying a common factor.

Let us define the global risk factor as a global-scale correlator of risk factor volatilities.

Postulate 2. Risks Are Anthropic Human error is the global risk factor.

Human error is not the only risk factor, but it has acquired a global nature.

4 Causal Relationships of Risk Generation Mechanism

The principal cause of the global influence of the human factor is that it often and strongly affects the sensitivity of assets performance to the majority of other risk factors, no matter what their own nature. In the past decades, the influence of the human factor has been growing due to the operator's increasing role in business processes and globalization. This is reflected by the increasing correlation of different types of risks.

Investigations of the occurrences of technological operational risk in almost all sectors and regions show that most such events in the last half-century were initially caused by human error rather than technical failure (e.g. Randazzo et al. 2004). And moreover, when caused by technical failure, risk events were mostly the result of accumulated hidden defects due to accumulated maintenance errors caused by organizational errors—again, the human factor. This can be confirmed by many examples, some of which are given below.

The human factor is the main trigger behind the vast majority of transport accidents and disasters. Human errors are responsible for 90 % of all motor vehicle accidents. National statistics of individual countries do not differ much from the world average figures. The human factor accounts for 70–80 % of accidents in air and water transport, and for about 50 % of accidents in railway transport. The human factor is also the dominant cause of industrial accidents and injuries. For instance, about 85 % of lifting crane accidents are associated with violations of labor or technical discipline.

There are about 200 best-known techniques for human factors analysis and assessment, including HAZOP, FTA, ETA, SHERPA, SPEAR, CREAM, THERP, SAPHIRE. For example, the Human Factors Analysis and Classification System (HFACS) (Shappell and Wiegmann 2000) is based on the "Swiss Cheese" model (Reason 2000). The model illustrates errors passing through "holes" (weaknesses) in business processes. According to this theory, there are unsafe acts (errors), preconditions for unsafe acts (including the operator's psychic factors), unsafe supervision and organizational influences.

Postulate 3. Risks Are Heliogeotropic Human errors and failures (the human factor) depend substantially on preconditions such as the effects of heliogeophysical factors (geomagnetic disturbances, etc.).

Geomagnetic activity depends on solar activity. According to the Svalgaard–Mansurov effect (Mansurov 1969; Svalgaard 1968), the variations of the Earth's magnetic field are influenced by the sector structure of the interplanetary magnetic field (IMF). These two major factors can disturb the heart rate (Otsuka et al. 2000) and cause human errors, which in turn, according to postulate 1, trigger chain reactions. These result in the occurrence of all types of financial risks (market, credit, operational ones) all over the world, depending on the assets' sensitivity to the risk factors. ("Geomagnetic Storms" OECD/IFP Futures Project on "Future Global Shocks" CENTRA Technology, Inc., on behalf of Office of Risk Management and Analysis, United States Department of Homeland Security 14.01.2011 IFP/WKP/FGS (2011), Jansen et al. (2000).). Moreover, human intuition and emotions are enhanced during the periods of geomagnetic disturbances, and this enhancement influences market expectations (Krivelyova and Robotti 2003). As related to operational risks caused by risk factors non-correlating with heliogeophysical conditions, their impact depends on the asset sensitivity to these risk factors, while the asset sensitivity itself is heliogeotropic due to the human factor influence.

For a considerable number of risks, the dynamics of risk events can be explained by that of human errors under changing space weather that has a planetary effect. This risk source was termed "the global risk factor" (Rogov 2005, 2006). Astrophysicists have shown the chaotic nature of solar and geomagnetic activity (Spiegel 1993), and this can explain (based on the global risk factor theory) the nature of the observed widely discussed chaotic processes in the markets (Rogov 2003).

5 Global Risk Factor Index

There are many indices of solar and geomagnetic activity, e.g., Wolf numbers, indices aa, am, Kp, Dst, AE etc. The objective was to choose the best indicator for adequate description of the global risk factor or to develop a new one. In the author's opinion, the best global risk factor index should meet the following requirements:

- Fully explain the behavior of market, credit and operational risks
- Allow for possible regularities discovered in heliobiology (the Mansurov effect)
- Be based on uniquely determinable or measurable values (heliogeophysical data)
- Allow real-time updating (Fig. 1)

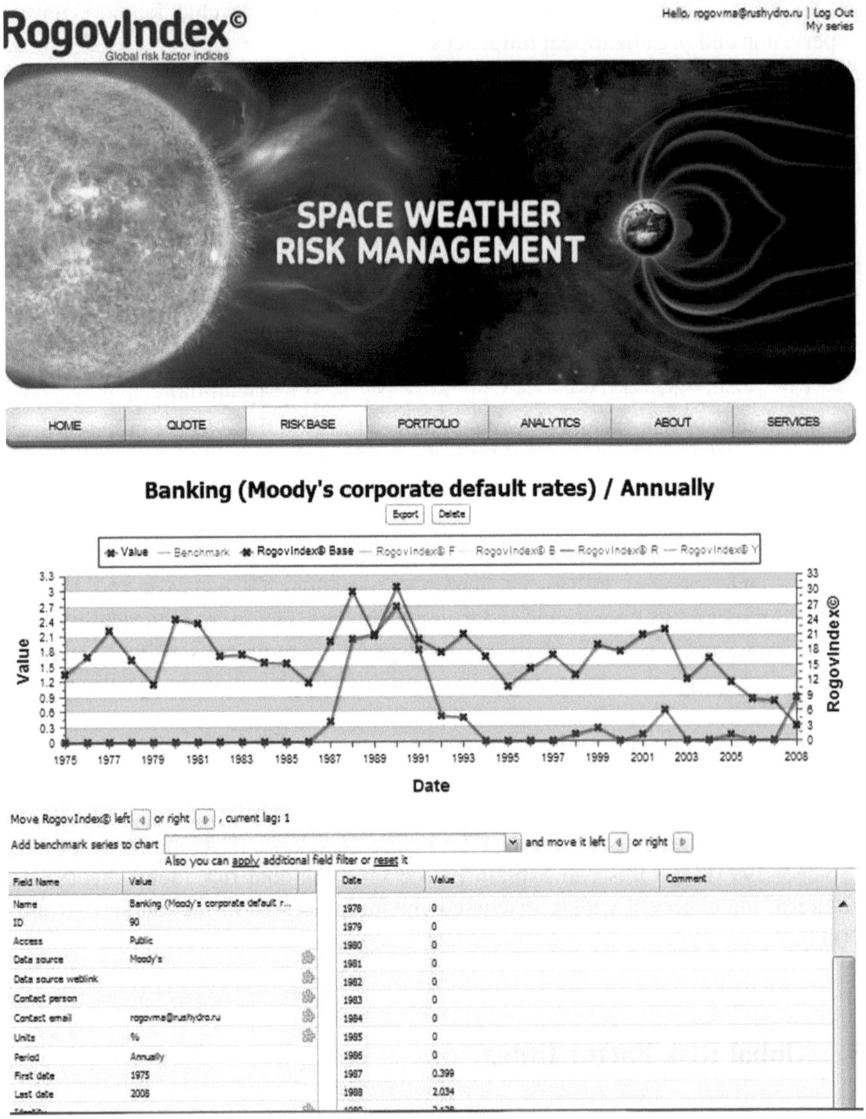

Fig. 1 Global risk factor vs. credit risk: screenshot, www.rogovindex.com

The indices of solar activity are not suitable for describing the global risk factor. This is the very reason for the skepticism of modern science toward the ideas of prominent scholars of the past, particularly Jevons (1878) and Chizhevsky (1936). The failure to find correlations with solar activity (the Wolf number, also known as the sunspot number) has led to the substitution of this idea in modern science with the general idea of accounting for random factors in economics. Economists

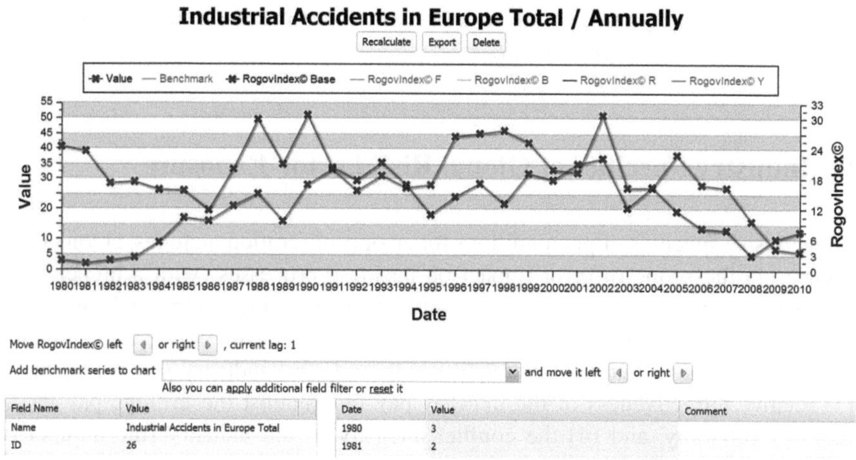

Fig. 2 Global risk factor vs. operational risk: screenshot, www.rogovindex.com

rebranded the term "sunspots" by completely stripping it of the implication of Sun-Earth relationships and using it to denote an external non-fundamental variable that influences human behavior (Cass and Shell 1983) (Fig. 2).

The RogovIndex© family of indices was developed for adequate description of the global risk factor; these indices satisfy the above requirements and are based on the widely accepted index of geomagnetic field variation, averaged over several stations (storm-time variation Dst). The conclusion that the effect of heliogeophysical factors on risk is best described by storm-time variation than by any other of the great variety of indices is consistent with the findings of heliobiological research. Indices from the RogovIndex© family can take into account the Mansurov effect by using properly chosen weights:

$$RogovIndex\,\text{I'}\,[t] = -\frac{\sum_{i=1}^{T} v_i Dst_i}{T} \tag{1}$$

where:

RogovIndex©[t] is a t-period-average value of a global risk factor index;
T is the duration of time period t, hours;
Dst_i is the value of storm-time variation index at an i-th hour, nT;
v_i is a dimensionless weight accounting for the polarity of the interplanetary magnetic field (IMF) in the day to which the i-th hour (Greenwich Mean Time, GMT) belongs:

For RogovIndex©Base all $v_i = 1$

For RogovIndex© B (or RogovIndex©R or RogovIndex©Y) if the day to which the i-th hour belongs is B (or R or Y) in terms of IMF polarity, then $v_i = 1$; else $v_i = 0$

The IMF polarity for each given hour is assessed based on the published IMF polarity data for the respective day.

6 Industry Specifics of Global Risk Factor Exposure

The industry specifics of preconditions for error proliferation includes, among other things: (1) the scope of error impact on business processes (with a higher labor productivity, an error of one operator would affect more performance indicators and, generally, more business processes), (2) the scope of business process regulation (including operator qualification requirements and other industry-specific barriers), (3) relative attractiveness of the industry pay rate against the average pay in the region's economy, and (4) the conflict intensity in the industry (the number of strikes).

Industry specifics result in different global risk factor exposures that should be taken into account by risk managers. For instance, diversified portfolios may be created using the correlation matrix or co-integrating vector approaches that take into account the global risk factor exposures of various assets and consider credit risks in accordance with the industry specifics. A detector of those risks that cannot be explained by the global risk factor behavior allows for planning most topical areas of risk audit for identification of operational risks.

7 Geographical Specifics of Global Risk Factor Exposure

The geographical specifics of global risk factor exposure are related to the distance of the region where the main business process or asset (if appropriate) is located from the magnetic poles, constantly drifting relative to fixed geographic coordinates. For example, the horizontal component of geomagnetic field H, which variation determines the index of geomagnetic activity, is proportional to the sine of magnetic declination. The lines of equal magnetic declination (isogones, agonic lines) are shifting with the secular movement of the Earth's magnetic field.

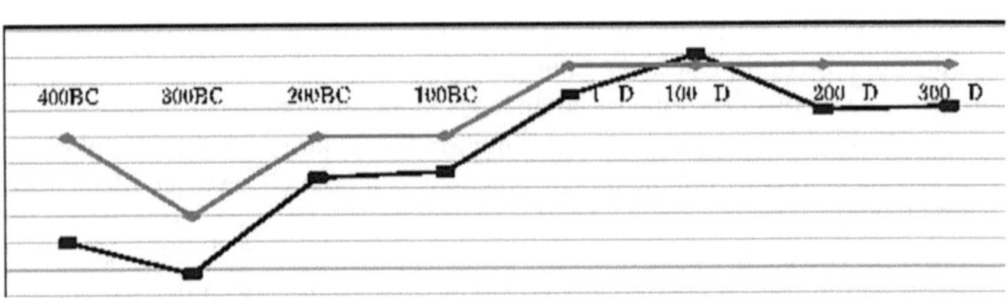

Fig. 3 Roman legal maxima interest rate vs. paleomagnetic declination, Roma (400 BC–300 AD)

Interest rates are known to include the risk premium. Figure 3 shows a similar behavior of secular magnetic declinations in the city of Rome, as found by the author from paleomagnetic maps, and that of maximum interest rate levels established by law in ancient Rome (Roman Legal Maxima Interest Rate) (400 BC–300 AD) (Homer and Sylla 1996).

8 Scenario Generation Based on 22-Year Cycle

Based on the 22 (more exactly, 21.8)-year cycle of geomagnetic activity, one can develop a scenario of the periods of increasing and decreasing risks, e.g., for risks of industrial accidents (Fig. 4).

Under this scenario, we can expect risks to rise sharply in 2013, then decreasing over the subsequent period until 2017–2018, from which point the risks are expected to grow until the mid-2020s, then decrease until 2030, and then rise again. This means that if the scenario proves true, those businesses that will survive the crises of 2013 should plan pm earning capital in the tranquil market over the next 5 years and prepare for future risks in the early twenties.

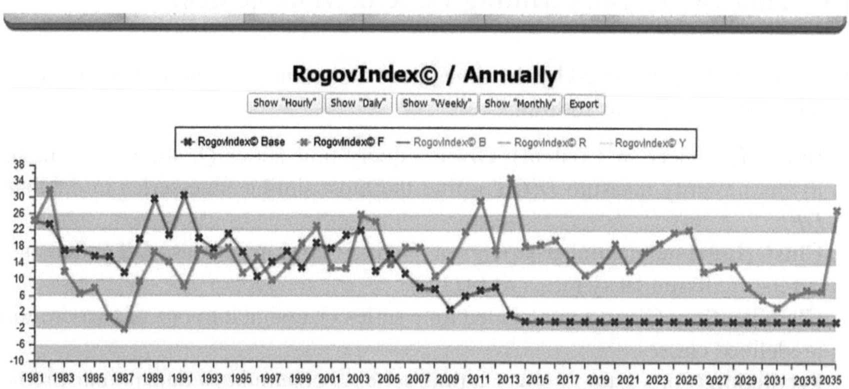

Date	RogovIndex© Base	RogovIndex© F	RogovIndex© B	RogovIndex© R	RogovIndex© Y
1984	17.3584	6.6206	0.9251	4.2919	5.2027
1985	15.7616	8.005	1.7196	4.699	6
1986	15.5847	0.8225	2.7133	4.4963	4.174
1987	11.7578	-2.0873	1.2335	3.9541	2.953
1988	19.916	9.645	2.2653	5.8977	5.8853
1989	29.7201	16.8711	5.5687	5.1008	8.7345
1990	20.9729	14.3453	1.7012	3.4072	8.1725
1991	30.6355	8.4557	3.6401	9.389	6.9597

Fig. 4 RogovIndex© Base vs. RogovIndex© F (scenario generation based on 22-year cycle), screenshot, www.rogovindex.com

9 Benchmarking

The website www.rogovindex.com was created in 2012–2013 to allow for viewing and, if desired, exporting into spreadsheet files the history of hourly, daily, weekly, monthly, or annual (the user may choose to define time zone lag) quotations of global risk factor indices, starting from January 1, 1957 at 1:00 a.m. (GMT).

A market of space weather index derivatives (Hyman (2001), (Rogov 2002) could be developed. For example, the RogovIndex© indices may be used as base assets for new index forwards and options. For this purpose, the website enables the users to build up portfolios of any combination of derivatives and offers analysis tools.

Already, every user of the website can upload to the database his or her own time series of any risk frequency indicators. The users can visually compare their dynamics with that of the RogovIndex© and with other series from this database, updated by the author with various time series of risk indicators (accident rate, failures rate, default rate, volatility, etc.) from publicly available sources. They can compare data of other website users with their own data (the users can set permissions for public access of their data).

10 Time Series Data Mining vs. Risk Management

The major tasks considered by the time series data mining community (Ratanamahatana et al. 2010) are as follows:

- **Indexing** (Query by Content): Given a query time series Q, and some similarity/dissimilarity measure $D(Q;C)$, find the most similar time series in database DB.
- **Clustering**: Find natural groupings of the time series in database DB under some similarity/dissimilarity measure $D(Q;C)$.
- **Classification**: Given an unlabeled time series Q, assign it to one of two or more predefined classes.
- **Prediction** (Forecasting): Given a time series Q containing n data points, predict the value at time $n + 1$.

These tasks can be used to solve problems in risk management (see Table 1).

Let us discuss some tools for time series data mining.

Dynamic time-warping (DTW) (Keogh and Ratanamahatana 2005) is an algorithm for measuring similarity between two sequences that may vary in time or speed. For instance, similarities in walking patterns would be detected, even if in one video the person was walking slowly and if in another he or she were walking more quickly, or even if there were accelerations and decelerations during the course of one observation.

Table 1 Time series data mining vs. risk management tool

Mining time series data task	Risk management tool
Indexing	Benchmarking
Clustering	Risk analysis
Classification	Risk classification
Prediction	Scenario generation
Summarization	Risk map
Anomaly detection	Hidden risk identification
Segmentation	Risk mapping and aggregation into portfolio

Given two time sequences C(m) and Q(n)m, it fills an m by n matrix representing the distances of best possible partial path using a recursive formula:

$$D(i,j) = d(i,j) + min\{D(i.j-1), D(i-1,j), D(i-1,j-1)\},$$
$$1 \leq i \leq n, 1 \leq j \leq m \tag{2}$$

Where D(I,j) represents the distance between Qi and Cj. D(1,1) is initialized to d(1,1). The alignment that results in the minimum distance between the two sequences has value D(m,n).

11 Longest Common Subsequence Similarity (*LCSS*)

The basic idea is to match two sequences by allowing some elements to be unmatched or left out. (Sankoff and Kruskal 1983). Given a sequence C(m), and a sequence Q(n), find a sequence Z, such that Z is the longest sequence that is both a subsequence of C, and a subsequence of Q, The subsequence is defined as a sequence Z(k)m where there exists a strictly increasing sequence i = 1, . . . k of indices of C such for all j = 1 . . . k; Cij = Zj.

$$c_{ij} = \begin{cases} 0, \_if\_i = 0\_or\_j = 0 \\ c_{i-1j-1} + 1, \_if\_i, j > 0, Q_i = C_j \\ \max\{c_{i-1,j}, c_{i,j-1}\}, \_if\_i, j > 0, Q_i \neq C_j \end{cases} \tag{3}$$

Dissimilarity between C and Q

$$LCSS(C,Q) = \frac{m+n-2l}{m+n} \tag{4}$$

Where L is the length of the longest common subsequence.

12 Time-Warped Longest Common Subsequence (*T-WLCS*)

The basic idea is to unite both DTW and LCSS approaches (Guo and Siegelmann 2004)

$$c_{ij} = \begin{cases} 0,\_if\_i = 0\_or\_j = 0 \\ \max\left\{c_{i-1,j}, c_{i,j-1}, c_{i-1j-1} + 1\right\}, \_if\_i, j > 0, Q_i = C_j \\ \max\left\{c_{i-1,j}, c_{i,j-1}\right\}, \_if\_i, j > 0, Q_i \neq C_j \end{cases} \quad (5)$$

Example 1. C = 41516171, Q = 4567, LCS(C,Q) = 4, T-WLCS(C,Q) = 4
Example 2. C = 44556677, Q = 4567, LCS(C,Q) = 4, T-WLCS(C,Q) = 8
Example 3. C = 4455661111177, Q = 4567, LCS(C,Q) = 4, T-WLCS(C,Q) = 8

13 Granger-Causality

A time series X is said to Granger-cause Y if it can be shown, usually through a series of t-tests and F-tests on lagged values of X (and with lagged values of Y also included), that those X-values provide statistically significant information about future values of Y (Granger 1969).

14 Time Series Temporality: Rogov-Causality Test

By an analogy with the Granger-causality test, the author has developed the following temporality test (Rogov-causality):

Assume that

$$Z = LCS\,(X, Y) : Z_i = Y_{t_i} = X_{t_i + lag_i}, i = 1, 2 \ldots l \quad (6)$$

Let us consider the null hypothesis Ho stating that the first time series X from a pair of (X, Y) is not the Rogov-cause of the second time series Y.

This null hypothesis is rejected if there is a long enough (LCSS(X, Y) > 0.5) longest common subsequence of this time series pair, such that the cumulative distribution function (CDF) of time lags lag_i at zero (i.e. the probability of a negative time lag between those values of the time series pair that have fallen within their long enough longest common subsequence) is sufficiently high.

This test is designed for the purposes of risk scenario generation, based on such predictors as lagged time series related in terms of the longest common subsequence similarity (LCSS). The test makes it possible to determine which of the two time series, X or Y, with a long-enough longest common subsequence Z = LCS(X,Y), is most likely not the advanced one (i.e., not the Rogov-cause). It is possible to discover the dynamics of which of the two series can be used (if at all) for scenario generation of the other time series of the pair, based on lagged values.

To avoid misinterpreting the term "causality," one should bear in mind that the presence of Rogov-causality does not mean the existence of a proven cause-effect relationship, but rather characterizes the temporality (the existence of prevailing succession of events in time).

Example: A detective with a limited staff of agents must catch two suspected spies, X and Y, who are visiting different cities around the country, and one of whom is likely to leave reports for the other who follows him. If the detective has an adequate list Z of cities visited sequentially by both spies, he can write out the time delays between the visits of Y and X to the same cities and compose a series of time lags. If the probability of a negative lag (i.e. Y visiting ahead of X) is low, then the visits of Y are unlikely to be the Rogov-cause of the visits of X. If so (i.e. the test has been successfully passed), then, with surveillance of Y alone, one can calculate the confidence interval for the time lag and organize an ambush to catch X in the cities visited by Y an appropriate lag-time ago. Distribution of agents by city should correspond to the lag distribution.

15 Time Series Clustering

For risk benchmarking, the user of the website is provided with tools for hierarchical clustering by various methods on the basis of time series data mining, with the use of either classic metrics (including Pearson correlation) or time series similarity measures based on Longest Common Subsequence Similarity (LCSS), Dynamic Time Warp (DTW). The user can build a dendrogram (phylogenetic tree) of risks and identify potential relationships, similarities and dissimilarities in risk time series dynamics. All these can be used for risk analysis and selection of time series as possible proactive key risk indicators (KRIs) based on crowdsourcing, which makes risk management technologies more easily available to small and medium-size businesses.

Examples of clustering of annual series of US banks' operational risks and default rates by industry and the base index of the global risk factor RogovIndex©Base.

Dendrogram analysis allows for better understanding of the risks in the context, seeing the similarities of close neighbours and dissimilarities of distant ones. Benchmarking sometimes can be used to summarize information and draw conclusions about the common properties of risks with similar dynamics, identify their potential sources, and select proper KRIs for scenario generation.

For example, in Fig. 5 we can see that Internal fraud (ET1) risks could be closely related to Employment practices (ET3) and thus it is possible to prioritize risk

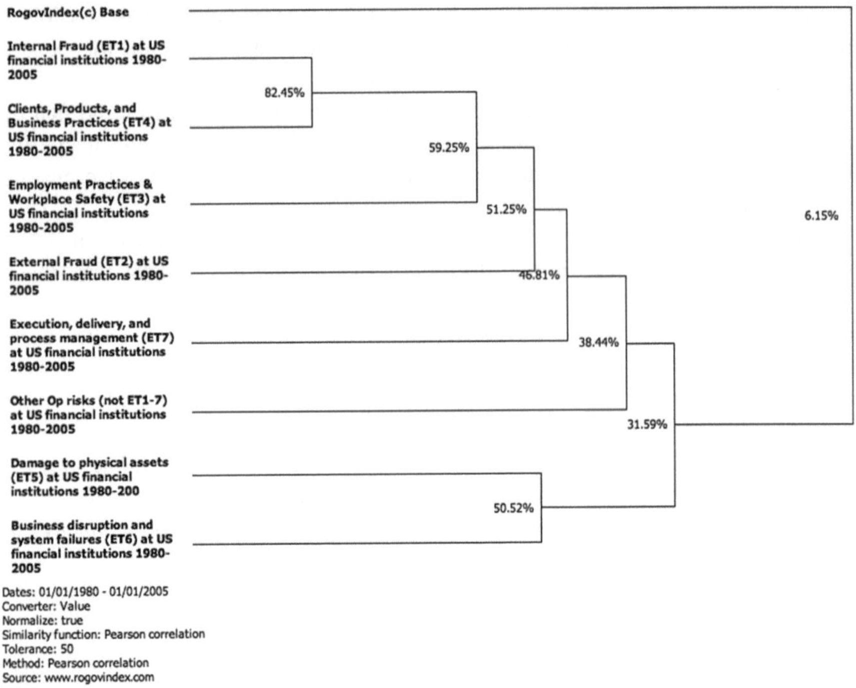

Fig. 5 Operational risks at US Financial Institutions, screenshot, www.rogovindex.com. Clustering using Pearson correlation

treatment measures in areas of employment practice in order to manage internal fraud risks (Fig. 6).

Conclusion

The proposed global risk factor theory describes the frequently observed interaction of different types of risks (market, credit, operational) at different assets and in different business processes. The theory opens new prospects for risk benchmarking, analysis, detection of anomalies and hidden risks, classification of risks, particularly based on hierarchical clustering of time series using the Rogov-causality[2] test. This allows the creation of new proactive risk indicators for monitoring, as well as applying the market mechanisms of operational risk optimization through diversification and hedging with the use of index derivatives.

[2]To avoid misinterpreting the term "Rogov-causality," one should bear in mind that the presence of Rogov-causality does not mean the existence of a proven cause-effect relationship, but rather characterizes the temporality (the existence of prevailing succession of events in time).

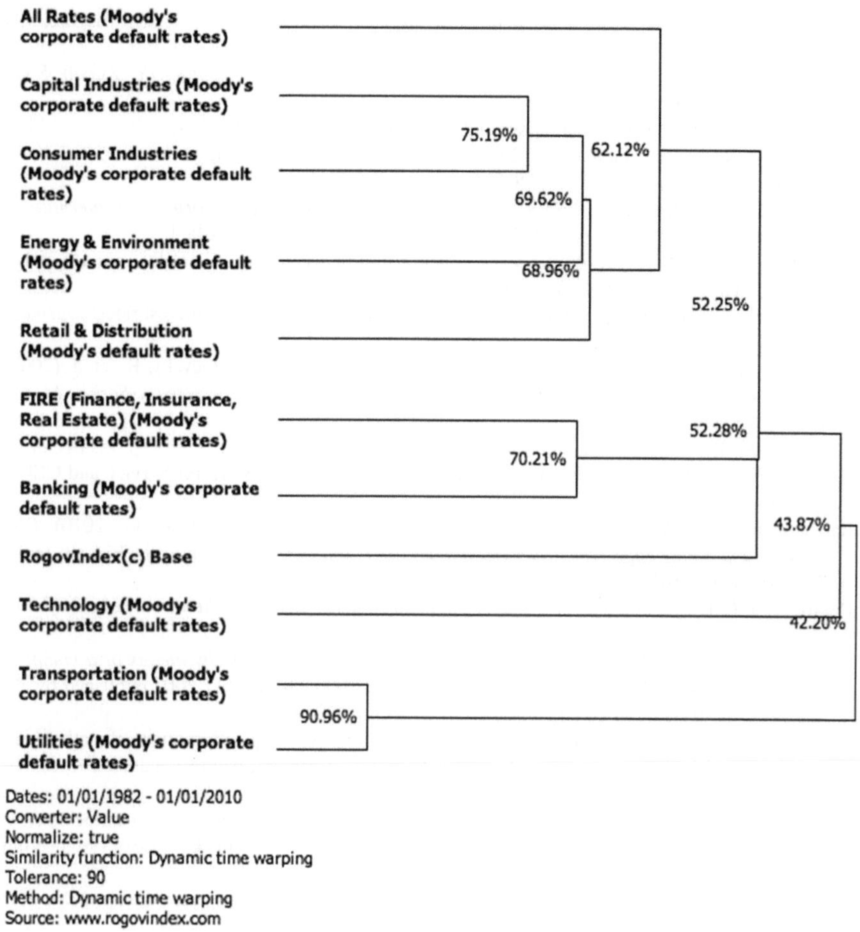

Fig. 6 Moody's corporate default rates, screenshot, rogovindex.com. Clustering using DTW. One of possible dendrograms

References

Cass, D., & Shell, K. (1983). Do sunspots matter? *Journal of Political Economy, 91*(21), 193–228.

Chizhevsky, A. (1936). *The terrestrial echo of solar storms*. Moscow: Mysl.

"Geomagnetic Storms" OECD/IFP Futures Project on "Future Global Shocks" CENTRA Technology, Inc., on behalf of Office of Risk Management and Analysis, United States Department of Homeland Security 14.01.2011 IFP/WKP/FGS (2011).

Granger, C. W. J. (1969). Investigating causal relations by econometric models and cross-spectral methods. *Econometrica, 37*(3), 424–438.

Guo, A., & Siegelmann, H. T. (2004). Time-warped longest common subsequence algorithm for music retrieval. *At proceedings of ISMIR*.

Homer, S., & Sylla, R. (1996). *A history of interest rates*. Rutgers, NJ: Rutgers University Press.

Hyman, A. (2001). *The case for solar weather derivatives: A special to the desk.* Crownsville, MD: Scudder.

Jansen, F., Pirjola, R., & Favre, R. (2000). *Space weather: Hazard to the earth?* Zurich: Swiss Reinsurance Company.

Jevons, W. S. (1878). Commercial crises and sun-spots. *Nature, 19*(14), 33–37.

Keogh, E., & Ratanamahatana, C. A. (2005). Exact indexing of dynamic time warping. *Knowledge and Information Systems, 7*(3), 358–386.

Krivelyova, A., & Robotti, C. (2003). *Playing the field: Geomagnetic storms and international stock markets.* Working Paper 2003–2005. Federal Reserve Bank of Atlanta.

Mansurov, S. M. (1969). New evidence of a relationship between magnetic fields in space and on earth. *Geomagnetic Aeronautics, 9*, 622–623.

Merton, R. C. (1974). The pricing of corporate debt: The risk structure of interest rates. *Journal of Finance, 29*(2), 449–470.

Otsuka, K., Yamanaka, T., Cornelissen, G., Breus, T., Chibisov, S. M., Baevsky, R., et al. (2000). Altered chronome of heart rate variability during span of high magnetic activity. *Scripta Medica (Brno), 73*, 111–116.

Randazzo, M. R., Cappelli, D., Keeney, M., Moore, A. P., & Kowalski, E. (2004). *Insider threat study: Illicit cyber activity in the banking and finance sector.* U.S. Secret Service and CERT® Coordination Center.

Ratanamahatana, C.A., Lin J., Gunopulos D., Keogh E., Vlachos M., and Das G. (2010): *Data Mining and Knowledge Discovery Handbook 2010.* 2nd edn. O. Maimon, L. Rokach (eds.). Springer. Pages 1049–1077, (2010)

Reason, J. T. (2000). Human error: Models and management. *British Medical Journal, 320*(7237), 768–770.

Rogov, M. A. (2002). *Method for forming risk management contracts by means of a computer system.* International patent application PCT/RU2002/000509 of 26.11.2002 published by The International Bureau on 10.06.2004 under No. WO 2004049228.

Rogov, M. A. (2003). Chaos, fractals, the neurofinancial theory and quantum financial mathematics in the new risk management paradigm. In *RelStat conference.* Lecture conducted from TSI, Riga.

Rogov, M. A. (2005). Global risk factors. In *The international symposium on stochastic models in reliability, safety, security and logistics.* Lecture from Negev Academic College of Engineering, Beer Sheva, Israel.

Rogov, M. (2006). Global risk factors. *Journal of Business Economics and Management, 8*(1), 25–28.

Rogov, M. (2011). Financial risk management (FRM) and enterprise risk-management (ERM) convergence (manifesto). *Issues of Risk Analysis, Vol. 8,* 2011, No. 3, page 88 (2011)

Sankoff, D., & Kruskal, J. (1983). *Time warps, string edits, and macromolecules: The theory and practice of sequence comparison.* Boston, MA: Addison Wesley.

Shappell, S. A., & Wiegmann D. A. (2000) *The human factors analysis and classification system.* HFACS Final Report. U.S. Department of Transportation.

Spiegel, E. A. (1993). *Chaotic dynamics of the solar cycle.* Annual Report. Air Force Office of Scientific Research.

Svalgaard, L. (1968). *Sector structure of the interplanetary magnetic field and daily variation of the geomagnetic field at high latitudes.* Geophysical papers R-6, Danish Meteorological Institute, Copenhagen.

Stress-Testing Model for Corporate Borrower Portfolios

Vladimir Seleznev, Denis Surzhko, and Nikolay Khovanskiy

Abstract Despite the significant attention to the stress-testing issues in finances world-wide, the ways of quantitative assessment of the stress impact on the portfolios of non-public (in the absence of equity or debt market quotes) corporate borrowers are currently not sufficiently developed or standardized. The aim of this article is to propose high-level universal requirements to the quantitative models of stress-testing of non-public corporate borrower portfolios, and to describe the model, developed by the authors, which meets such requirements. Details of the model's calibration, implementation (using Monte-Carlo simulations) and some practical issues are covered in the article.

Keywords Credit risk • Quantitative risk assessment • Stress-testing

1 Introduction

Stress-testing has become one of the most important risk-management instruments worldwide. Despite the increasing interest in this subject, currently the problem of constructing stress-testing models for credit portfolios of non-public companies (further—stress-testing models) is covered by research and regulatory papers only fragmentarily and usually at a very high-level. Therefore, the main goals of this article are to formulate clear overall requirements for quantitative stress-testing models, and to propose one of the possible practical implementations of those requirements based on the modification of the Vasicek model—the model that underpins current international capital requirements (IRB approaches of Basel II–III).

According to our view, a quantitative stress-testing model for a portfolio of non-public corporate borrowers should fulfill the following requirements:

1. The approach should not be based only on default event modeling, but the model should also produce estimates of the changes in the portfolio rating structure.

V. Seleznev • D. Surzhko (✉) • N. Khovanskiy
OJSC VTB Bank, Moscow, Russia
e-mail: SurzhkoDA@msk.vtb.ru

A.K. Bera et al. (eds.), *Financial Econometrics and Empirical Market Microstructure*, DOI 10.1007/978-3-319-09946-0_19

This will allow us to estimate potential losses (due to defaults) and RWA-changes (rating migrations) simultaneously and consistently.
2. Historical experience shows that concentration of credit risk in asset portfolios has been one of the major causes of bank distress; therefore the model should take into account concentration risks and correlation between default events.
3. The model should be based on the functional dependence between the defaults and dynamics of macro-variables. This property will allow us to model both potential losses based on real historical experience and losses based on hypothetical but plausible scenarios (produced by macro forecasters). Moreover, this property extends the scope of possible validation procedures, because the model could estimate losses during stress as well as expansion scenarios of economic development.
4. The model should allow us to estimate the marginal contribution of a single borrower to the stress-test results. Therefore, we could determine particular borrowers that are the main source of losses in a stress environment (potentially, it could be taken into account during risk-based pricing).
5. The approach should be universal; for example, it should allow us to make a consistent and transparent transformation of the stress-testing model into a portfolio model. This property will allow us to make a consistent comparison between stress-testing results and economic capital estimates. Moreover, it significantly reduces model development team efforts and increases the scope of possible validation procedures.

2 Methodology

2.1 Modification of the Vasicek Model

We propose one of possible implementations of stress-testing for the credit portfolios of corporate borrowers (further—the Model), which is based on Monte-Carlo simulations and the modified Vasicek model (Vasicek 1987). As will be shown, the Model meets all of the criteria described above.

The single systemic factor Vasicek model is based on the assumption that assets of the companies have two drivers—idiosyncratic (determining the individual properties of each company) and systemic (the overall macroeconomic environment). The change in assets of company A_i, according to the Vasicek model, is equal to the sum of two normally distributed random variables: idiosyncratic – ε_i and systematic Z; the level of the dependence of the borrower from the systematic factor is captured by the correlation coefficient ρ_i:

$$A_i = \varepsilon_i \cdot \sqrt{1-\rho} + Z \cdot \sqrt{\rho} \quad (1)$$

If the value of company assets becomes less than some threshold level default occurs. Usually, the default threshold is defined as a company's debt burden. The default threshold could be calibrated based on the assumption of the normal distribution of asset return values A_i and a given borrower's default probability:

$$P(A_i < Ths_i) = PD_i \Rightarrow N(Ths_i) = PD_i \Rightarrow Ths_i = N^{-1}(PD_i)$$

We propose the following modification of the Vasicek model for stress-testing purposes: The default threshold should be decomposed on the sum of the components, each component consisting of the macro-variable M_j multiplied by coefficient β_{ij}, which defines the degree of dependence between default frequency and macro-variable j.

$$Ths_i = \alpha_i + \sum_{j=1}^{m} \beta_{ij} \cdot M_j \tag{2}$$

The proposed threshold decomposition will allow us to capture historical dependence between defaults and macro-variables, which also serves as a default correlation transmitter due to asset value dependence on the same factors. At the same time, the model contains explicit default correlation parameter ρ_i, by which we take into account the default correlation, which is not detectable through dependence on macro factors.

It is impossible to statistically identify dependence between macro-variables and individual borrowers; therefore companies should be grouped into subsets with similar risk characteristics—rating classes.

It is very important to choose macro-factors for model calibration correctly. As general recommendations, we propose the following selection criteria:

1. Each macro-variable should have significant individual predictive power regarding historical default frequencies (R^2).
2. The correlation between selected macro-variables should be relatively low (for purposes of model stability).

A high correlation between all predictive macro-variables is common for emerging economies (for example, the price of oil in OPEC countries or Russia is the main economic driving force); therefore in order to fulfill the second requirement it is recommended to replace the original dynamics of the macro-variables M_i by the principle components of macro-variables $\tilde{M}_i$ with zero correlation between them.

According to the proposed modifications, the density function for default frequency could be written as:

$$f(x)=\prod_{q=1}^{Q}\binom{n_q}{x_q}\int_0^1\prod_{i=1}^{R}\left(N\left(\frac{\alpha_i+\sum_{j=1}^{m}\beta_{ij}\cdot M_{qj}+\sqrt{\rho_i}\cdot Z}{\sqrt{1-\rho_i}}\right)\right)^{x_q}$$
$$\left(1-N\left(\frac{\alpha_i+\sum_{j=1}^{m}\beta_{ij}\cdot M_{qj}+\sqrt{\rho_i}\cdot Z}{\sqrt{1-\rho_i}}\right)^{n_q-x_q}\right)dN(Z), \tag{3}$$

where

Q—index of the time period (quarter or year).
n_q—number of borrowers in the portfolio during the period q.
x_q—number of defaults during the period q.
N—normal distribution function.
M_{qj}—historical value of j macro-variable during the period q.
R—number of rating classes.

Given the default density function, information of the historical default frequencies by rating classes and historical values of macro-variables, parameters α_i, β_{ij}, ρ_i could be found using the maximum likelihood approach. As a result, we could produce conditional PDs for rating classes given the macro-forecast.

2.2 Monte-Carlo Simulation Schema

One of the key requirements for the stress-testing model is the ability to estimate changes in the rating structure of the portfolio over time. The most obvious approach for this task is to incorporate migration matrixes into the model. Due to the dependence of the rating migration dynamics on the economic cycle, it is recommended to use different migration matrixes for stress and expansion scenarios.

We propose the following Monte-Carlo simulation schema, which takes into account the proposed density function (3) and migration matrixes:

1. For the given macro-variable dynamics (from the macro-forecast) for the stress-testing period, conditional PDs are calculated [using (2)] for each rating class—$\mathrm{Ths_i}$.
2. The normal random variable Z is generated (systemic factor).
3. The normal random variable ε_i is generated for each borrower in the portfolio (idiosyncratic factor).
4. If $\varepsilon_i\cdot\sqrt{1-\rho}+Z\cdot\sqrt{\rho}\ \leq \mathrm{Ths}_i$, a default event is fixed for a borrower during a current period. A defaulted borrower is excluded from the portfolio, and its exposure multiplied by LGD is added to the total portfolio losses within the scenario.

5. If the borrower does not default, its ratings for the next period are changed in accordance with the migration matrix—a uniformly distributed random number is generated $r \in [0; 1]$, and a new rating for the next period is assigned to the borrower according to the probabilistic interval of the migration matrix in which random number r falls.
6. Items 1–5 are repeated until the required forecast horizon is achieved.

The result of MC simulations is an array of losses. This array is a numerical representation of the density function of losses due to borrowers defaulting. On the basis of this distribution, the mean and quantiles of portfolio losses can be estimated.

The marginal contribution of individual borrowers to the stress-test results can be estimated using an approach similar to the Monte-Carlo model, which is described, for example, in (Tasche 2000).

2.3 *Transformation to a Portfolio Model*

The proposed stress-testing model could be easily transformed into a portfolio model (the model dedicated to the estimation of unexpected losses). In the case of a portfolio model, a macro-forecast should be excluded from the model by replacing the forecasted M_{qi} values by the random values $\breve{M}_{qi}$. The distribution function of $\breve{M}_{qi}$ could be calibrated using the historical values of macro-variables.

One of the most flexible approaches that could capture the time evolution of macro-variables is the ARIMA model. The ARIMA model would capture the following aspects of time evolution of macro-variables:

1. Stationary part:
   (a) Long-term trends.
   (b) Auto regression dependence (previous values of macro-variable dynamic influence values for the current period).
   (c) Deviations from trends (prior to the period, error affects the current period's errors).
2. The random component—normally distributed random variables with a zero mean and covariance matrix (estimated on the basis of historical deviations of the real values of the macro factors from the ARIMA model).

In the case of the portfolio model, the Monte Carlo simulation schema should be modified in the following way:

1. Using the ARIMA model, $\breve{M}_{qi}$ values are generated for the estimation period.
2. The MC algorithm for the stress-testing model is started, in which, instead of forecasted macro-variables M_{qi}, random variables $\breve{M}_{qi}$ are used.

Conclusion
The proposed model meets all of the requirements mentioned in this article's introduction. The model could produce estimates both of losses due to borrowers' defaults and changes in the rating structure. The model is based on the functional dependence between dynamics of macro-variables and defaults; therefore it could be calculated for baseline and stress-scenarios. Comparison between the results in different scenarios will give us estimates of the changes of direct losses (defaults) and RWA changes (rating structure) due to stress events. The model could also be easily extended to the credit VAR model; therefore a bank could make consistent comparisons between stress-testing results and unexpected losses.

References

Vasicek, O. A. (1987). *Probability of loss on loan portfolio*. San Francisco, USA: KMV Corporation.

Tasche, D. (2000). *Conditional expectation as quantile derivative*. Working paper. Technische Universitaet München. http://arxiv.org/pdf/math/0104190v1.pdf.

MIX
Papier aus verantwortungsvollen Quellen
Paper from responsible sources
FSC® C105338

If you have any concerns about our products,
you can contact us on
ProductSafety@springernature.com

In case Publisher is established outside the EU,
the EU authorized representative is:
Springer Nature Customer Service Center GmbH
Europaplatz 3, 69115 Heidelberg, Germany

Printed by Libri Plureos GmbH
in Hamburg, Germany